Handbook of Control Science and Engineering

Volume I

Handbook of Control Science and Engineering
Volume I

Edited by **Marques Vang**

Clanrye International
New Jersey

Published by Clanrye International,
55 Van Reypen Street,
Jersey City, NJ 07306, USA
www.clanryeinternational.com

Handbook of Control Science and Engineering: Volume I
Edited by Marques Vang

International Standard Book Number: 978-1-63240-264-6 (Hardback)

Printed in the United States of America.

Contents

Preface

Control science or control systems or control engineering is a field of study and engineering science that integrates automata theory and control theory along with the system designing techniques to build systems with the desired functionality. These systems are designed for self-automation, and the sensors are used to initiate input processing, the measurement converted after the processing of input are further used to initiate output measurement and output controls to be achieved. When a designed system is used without human involvement, they are called automatic control systems. The basic approach to build such systems is mathematical modeling.

Control engineering is playing a large role in designing of control systems, and examples of these control systems can vary from microwave oven to high impact submarines. For such designing, the inputs, outputs, and other important functional components use mathematical modeling to develop controllers for complex systems, and then come the most important task of integrating these controllers with the physical systems using available technological tools. These systems can be of various ranges from mechanical to biological or even financial accounting, they all use control theory in time domains as well as frequency domains, depending on the designing and functional problem.

I especially wish to acknowledge the contributing authors, without whom a work of this magnitude would clearly not be realizable. We thank them for allocating much of their very scarce time to this project. Not only do I appreciate their participation but also their adherence as a group to the time parameters set for this publication. I also thank my publisher for considering me for this project and giving me this incredible opportunity.

Editor

Position Control of a 3-CPU Spherical Parallel Manipulator

Massimo Callegari,[1] Luca Carbonari,[1] Giacomo Palmieri,[2] Matteo-Claudio Palpacelli,[1] and Donatello Tina[1]

[1] *Department of Industrial Engineering & Mathematical Sciences, Polytechnic University of Marche, 60131 Ancona, Italy*
[2] *e-Campus University, Faculty of Engineering, 22060 Novedrate, Italy*

Correspondence should be addressed to Massimo Callegari; m.callegari@univpm.it

Academic Editor: Sabri Cetinkunt

The paper presents the first experimental results on the control of a prototypal robot designed for the orientation of parts or tools. The innovative machine is a spherical parallel manipulator actuated by 3 linear motors; several position control schemes have been tested and compared with the final aim of designing an interaction controller. The relative simplicity of machine kinematics allowed to test algorithms requiring the closed-loop evaluation of both inverse and direct kinematics; the compensation of gravitational terms has been experimented as well.

1. Introduction

Parallel kinematics machines, PKMs, are known to be characterized by many advantages like a lightweight construction and a high stiffness but also present some drawbacks, like the limited workspace, the great number of joints of the mechanical structure, and the complex kinematics, especially for 6-dof machines [1]. Therefore the A.'s proposed to decompose full-mobility operations into elemental subtasks, to be performed by separate reduced mobility machines, similarly to what is already done in conventional machining operations. They envisaged the architecture of a mechatronic system where two parallel robots cooperate in order to perform complex assembly tasks. The kinematics of both machines is based upon the same 3-CPU topology but the joints are differently assembled so as to obtain a translating parallel machine (TPM) with one mechanism and a spherical parallel machine (SPM) with the other.

This solution, at the cost of a more sophisticated controller, would lead to the design of simpler machines that could be used also stand-alone for 3-dof tasks and would increase the modularity and reconfigurability of the robotized industrial process. The two robots are now available at the prototypal stage, and the present paper reports the first experiments on the motion control of the orienting device (SPM).

2. Robot's Architecture and Kinematics

2.1. Mechanical Architecture. Since the detailed description of machine's kinematics and prototype design has been provided already in Callegari et al. [2], hereby only the most relevant aspects are recalled.

The spherical parallel machine under study is made of three identical serial chains connecting the moving platform to the fixed base, as shown in Figure 1; each leg is composed by two links: the first one is connected to the frame by a cylindrical joint (C), while the second link is connected to the first one by a prismatic joint (P) and to the end-effector by a universal joint (U); for this reason its mechanical architecture is commonly called 3-CPU. A few *manufacturing conditions*, already investigated for a general pure rotational tripod by Karouia and Hervè [3], must be fulfilled in order to constraint the end-effector to a spherical motion:

(i) the axes of the cylindrical joints ($\mathbf{a}_i$, $i = 1, 2, 3$) are aligned along the x, y, z axes of the base frame and intersect at the center O of the spherical motion;

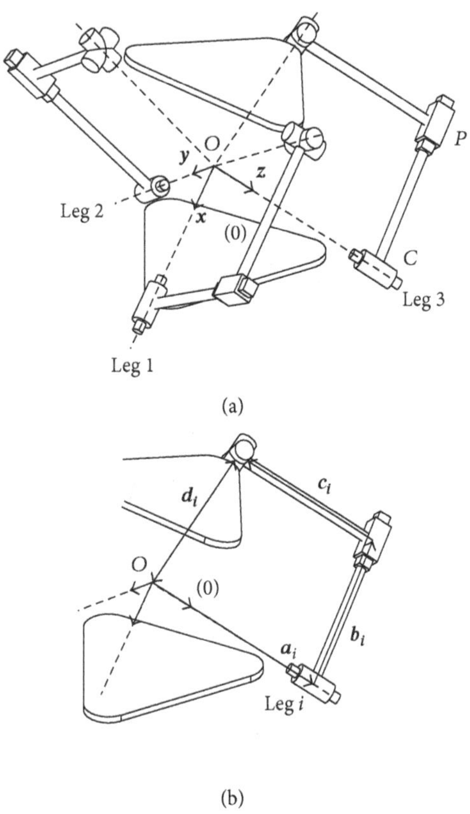

FIGURE 1: Kinematic schemes of the 3-CPU robot (a) and geometry of the legs (b).

(ii) the axis $\mathbf{b}_i$ of each prismatic pair is perpendicular to the axis of the respective cylindrical joint $\mathbf{a}_i$;

(iii) the first axis of each universal joint is perpendicular to the plane of the corresponding leg (plane identified by the axes $\mathbf{a}_i$ and $\mathbf{b}_i$);

(iv) the second axis of the 3 universal joints (resp., for the leg 1, 2, and 3) are aligned along the y_1, z_1, x_1 axes of a local frame centered in P (coincident with O) and attached to the mobile platform.

For a successful operation of the mechanism, one *mounting condition* must be satisfied too; assembly should be operated in such a way that the two frames $O(\mathbf{x}_0, \mathbf{y}_0, \mathbf{z}_0)$ and $P(\mathbf{x}_1, \mathbf{y}_1, \mathbf{z}_1)$ come to coincide when the robot is in its homing position. Such configuration is obtained when the three displacements a_i are equal to the length of the second link c and the displacements of the prismatic joints b_i are equal to the constant distance d. If the mounting conditions are verified, the points P and O remain fixed and coincident while the moving platform performs a spherical motion around them.

2.2. Kinematic Relations. The platform is actuated by driving the strokes of the 3 cylindrical joints; therefore joint space displacements are gathered into the following vector $\mathbf{q}$:

$$\mathbf{q} = \begin{bmatrix} a_1 \\ a_2 \\ a_3 \end{bmatrix}. \qquad (1)$$

The position kinematics of the robot expresses the relation between the orientation of the mobile platform and the displacements of the actuators; the attitude of the machine in space is fully provided by the rotation matrix ${}^{O}_{P}\mathbf{R}$, that can also be conveniently expressed as a composition of elemental rotations. In the development of robot's kinematics, the following Cardan angles set is used:

$$\begin{aligned} {}^{O}_{P}\mathbf{R}(\alpha, \beta, \gamma) &= \mathbf{R}_x(\alpha)\,\mathbf{R}_y(\beta)\,\mathbf{R}_z(\gamma) \\ &= \begin{bmatrix} c\beta c\gamma & -c\beta s\gamma & s\beta \\ s\alpha s\beta c\gamma + c\alpha s\gamma & -s\alpha s\beta s\gamma + c\alpha c\gamma & -s\alpha c\beta \\ -c\alpha s\beta c\gamma + s\alpha s\gamma & c\alpha s\beta s\gamma + s\alpha c\gamma & c\alpha c\beta \end{bmatrix}. \end{aligned} \qquad (2)$$

The position kinematics of the robot is simply expressed by

$$\begin{aligned} r_{12} &= -c\beta s\gamma = \frac{c - a_1}{d}, \\ r_{23} &= -s\alpha c\beta = \frac{c - a_2}{d}, \\ r_{31} &= -c\alpha s\beta c\gamma + s\alpha s\gamma = \frac{c - a_3}{d}, \end{aligned} \qquad (3)$$

where r_{ij} is the element at the ith row and jth column of rotation matrix ${}^{O}_{P}\mathbf{R}$. The solution of the *direct position kinematics* (DPK) problem requires the computation of the rotation matrix ${}^{O}_{P}\mathbf{R}$ as a function of internal coordinates $\mathbf{q}$, which has been solved already by Carbonari et al. [4]. According to Innocenti and Parenti-Castelli [5], a maximum number of 8 different configurations can be worked out; however, a single feasible solution is found when the real workspace of the robot is considered; that is, the actual mobility of the joints is taken into consideration. *Inverse position kinematic* (IPK) problem admits just one solution and it is trivially solved by working out joint displacements $\mathbf{q}$ in (3).

Turning to *differential kinematics*, the expression of the analytic Jacobian $\mathbf{J}_A$ is immediately obtained as a function of the Cardan angles and their rates:

$$\begin{bmatrix} \dot{a}_1 \\ \dot{a}_2 \\ \dot{a}_3 \end{bmatrix} = \mathbf{J}_A \begin{bmatrix} \dot{\alpha} \\ \dot{\beta} \\ \dot{\gamma} \end{bmatrix},$$

$$\mathbf{J}_A = d \begin{bmatrix} 0 & -s\beta s\gamma & c\beta c\gamma \\ c\alpha c\beta & -s\alpha s\beta & 0 \\ -s\alpha s\beta c\gamma - c\alpha s\gamma & c\alpha c\beta c\gamma & -c\alpha s\beta s\gamma - s\alpha c\gamma \end{bmatrix}. \qquad (4)$$

By taking into account the relation between the derivatives of the Cardan angles and the angular velocity $\boldsymbol{\omega}$:

$$\begin{bmatrix} \omega_x \\ \omega_y \\ \omega_z \end{bmatrix} = \begin{bmatrix} 1 & 0 & s\beta \\ 0 & c\alpha & -s\alpha c\beta \\ 0 & s\alpha & c\alpha c\beta \end{bmatrix} \begin{bmatrix} \dot{\alpha} \\ \dot{\beta} \\ \dot{\gamma} \end{bmatrix} = \mathbf{T} \begin{bmatrix} \dot{\alpha} \\ \dot{\beta} \\ \dot{\gamma} \end{bmatrix}, \quad (5)$$

the geometric Jacobian $\mathbf{J}_G$ is easily obtained too:

$$\begin{bmatrix} \dot{a}_1 \\ \dot{a}_2 \\ \dot{a}_3 \end{bmatrix} = \mathbf{J}_A \mathbf{T}^{-1} \begin{bmatrix} \omega_x \\ \omega_y \\ \omega_z \end{bmatrix} = \mathbf{J}_G \begin{bmatrix} \omega_x \\ \omega_y \\ \omega_z \end{bmatrix}, \quad (6)$$

with

$$\mathbf{J}_G = d \begin{bmatrix} 0 & -c\alpha s\beta s\gamma - s\alpha c\gamma & c\alpha c\gamma - s\alpha s\beta s\gamma \\ c\alpha c\beta & 0 & -s\beta \\ -s\alpha s\beta c\gamma - c\alpha s\gamma & c\beta c\gamma & 0 \end{bmatrix}. \quad (7)$$

2.3. User Frames. In order to better define the tasks to be commanded and visualize the obtained results, it is useful to choose a different set of reference frames, as shown in Figure 2. The fixed frame $O^*(\mathbf{x}_0^*, \mathbf{y}_0^*, \mathbf{z}_0^*)$ is defined as follows:

(i) the origin is located at the center of the moving platform when it assumes its initial configuration;

(ii) the $\mathbf{z}_0^*$ axis is aligned to the vector $\mathbf{g}$ of gravity acceleration;

(iii) the $\mathbf{x}_0^*$ axis lies on the upper plane of the platform and points toward the axis $\mathbf{a}_1$ of the cylindrical joint of the first leg;

(iv) the $\mathbf{y}_0^*$ axis is placed according to the right-hand rule.

The mobile frame $P^*(\mathbf{x}_1^*, \mathbf{y}_1^*, \mathbf{z}_1^*)$ is coincident with the fixed frame $O^*(\mathbf{x}_0^*, \mathbf{y}_0^*, \mathbf{z}_0^*)$ when the platform is in its initial configuration. Of course, since the frames are not placed at the center of the spherical motion, the two origins O^* and P^* will be coincident only in the home position.

Once the location of the new frame O^* has been defined by means of the ${}^{O}_{O^*}\mathbf{R}$ rotation matrix, the orientation of the mobile platform can be described in the new frames by

$${}^{O^*}_{P^*}\mathbf{R} = {}^{O}_{O^*}\mathbf{R}^T \, {}^{O}_{P}\mathbf{R} \, {}^{O}_{O^*}\mathbf{R}, \quad (8)$$

where it has been used the identity ${}^{O}_{O^*}\mathbf{R} = {}^{P}_{P^*}\mathbf{R}$. Of course, being the mobile and fixed frames modified, also the Cardan angles $\varphi_x, \varphi_y, \varphi_z$ that yield the rotation matrix ${}^{O^*}_{P^*}\mathbf{R}$ are different from the previously described set (α, β, γ):

$${}^{O^*}_{P^*}\mathbf{R}(\varphi_x, \varphi_y, \varphi_z) = \mathbf{R}_{x^*}(\varphi_x)\,\mathbf{R}_{y^*}(\varphi_y)\,\mathbf{R}_{z^*}(\varphi_z). \quad (9)$$

Henceforth these angles are used to describe the orientation of the manipulator and to assign the tasks of the mobile platform; since they are assumed as external coordinates for the computation of the differential kinematics, the analytic and the geometric Jacobians are worked out again as previously described, providing similar but more complex relations.

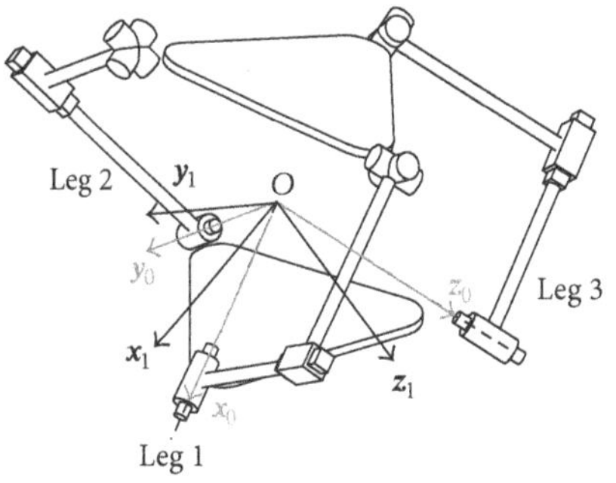

(a)

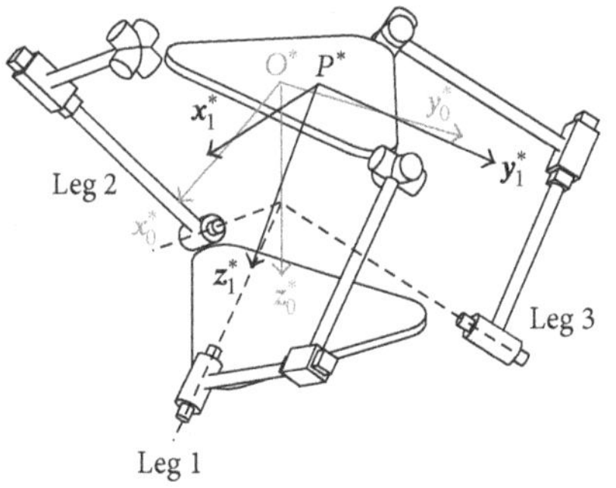

(b)

FIGURE 2: User-defined task frames.

3. Control Algorithms

3.1. Overview. Several kinds of control schemes have been tried on the 3-CPU SPM, with the immediate goal of testing the prototypal robot but aiming at the final design of an efficient co-operative environment for mechanical assembly. In the end, 3 different algorithms have been studied in simulation and then experimentally tested:

(i) a conventional joint resolved PID [6];

(ii) a joint resolved PID with the compensation of gravity forces [7];

(iii) a task-space PID with gravity compensation [8].

In all control schemes, the PID loop has been computed as usual in the following way:

$$\mathbf{u}(t) = \mathbf{K}_P \left(\mathbf{e}(t) + \frac{1}{\mathbf{T}_i} \int_0^t \mathbf{e}(\tau)\, d\tau + \mathbf{T}_D \frac{d}{dt}\mathbf{e}(t) \right), \quad (10)$$

with $\mathbf{u}(t)$ control action and $\mathbf{e}(t)$ input position error; $\mathbf{K}_P$, $\mathbf{T}_I$, and $\mathbf{T}_D$ are, respectively, the proportional gain, integral time, and derivative time matrices of the PID regulator.

3.2. Joint Resolved PID. First, a conventional joint resolved PID has been considered; see Figure 3. The error signal $\tilde{\mathbf{a}}$ is computed in the joint space as a difference between the desired position of the sliders $\mathbf{a}_D$ and their actual values $\mathbf{a}$:

$$\tilde{\mathbf{a}} = \mathbf{a}_D - \mathbf{a}. \quad (11)$$

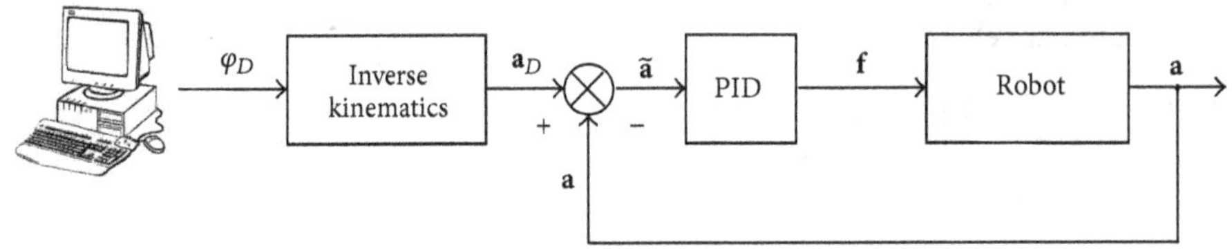

FIGURE 3: Joint space PID controller.

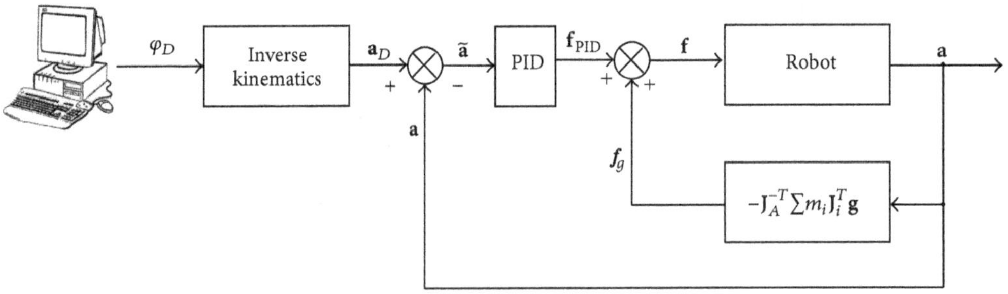

FIGURE 4: Joint space PID controller with gravity compensation.

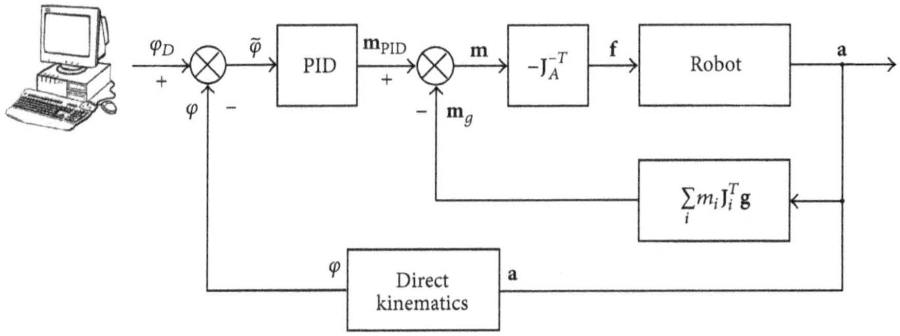

FIGURE 5: Task space PID controller with gravity compensation.

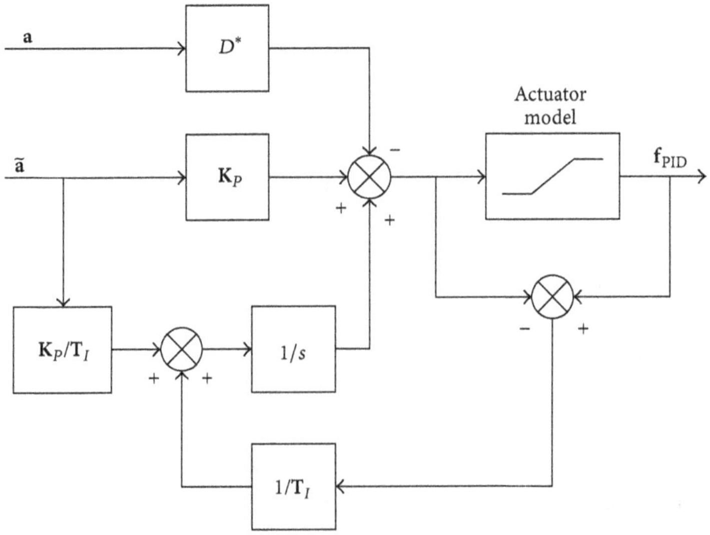

FIGURE 6: Anti-windup modified PID.

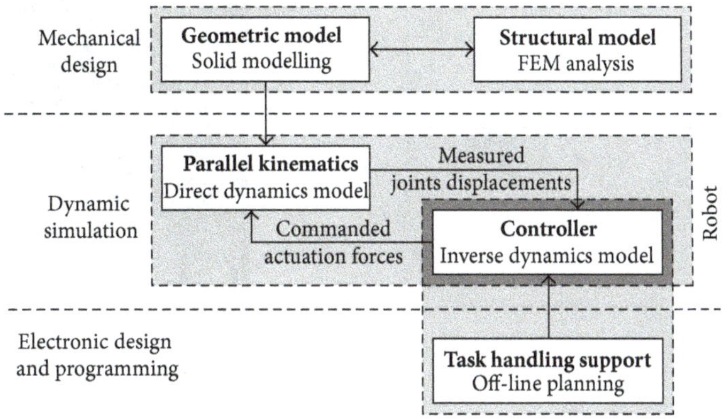

FIGURE 7: Integrated virtual prototyping environment for PKM's analysis and design.

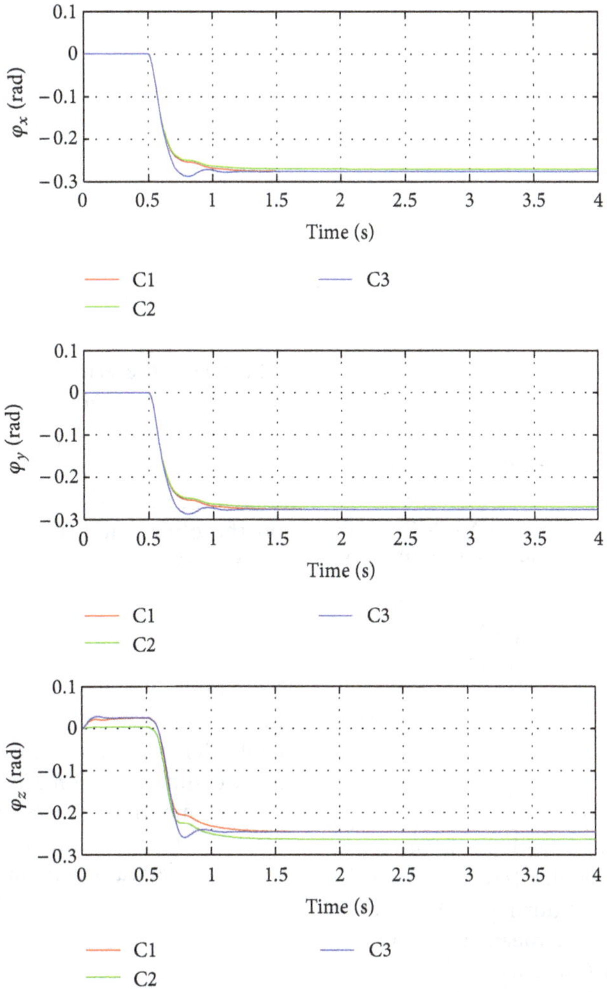

FIGURE 8: Response to step input: task space trajectories.

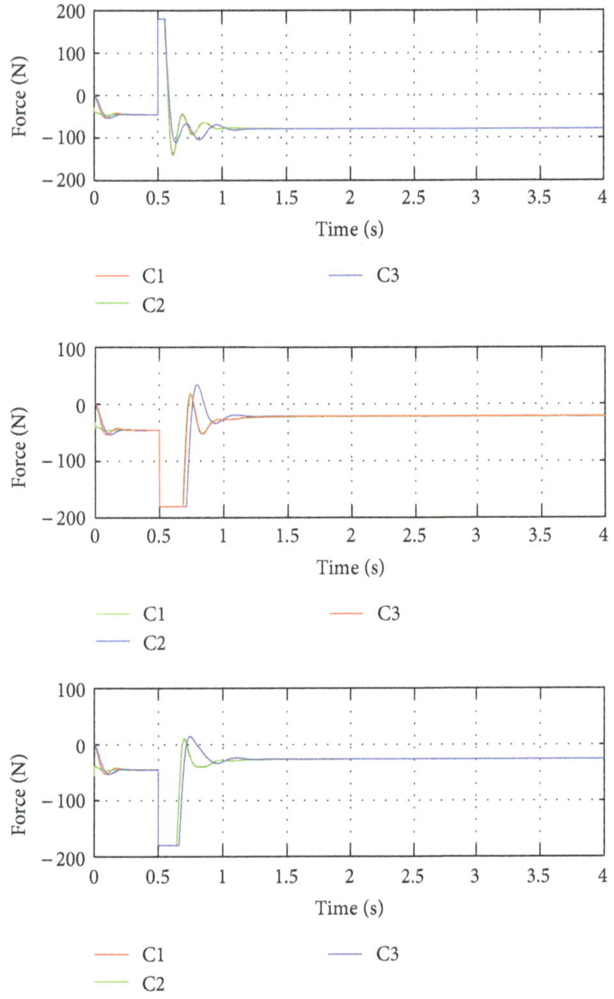

FIGURE 9: Response to step input: control efforts.

Since planning is programmed in the orientation space by assigning the desired configuration of the robot $\boldsymbol{\varphi}_D$, the corresponding position of the actuated joints is computed by means of inverse kinematics relations. The actuation effort of the motors is computed as

$$\mathbf{f} = \mathbf{K}_P \left[\tilde{\mathbf{a}} + \frac{1}{\mathbf{T}_i} \int \tilde{\mathbf{a}}\, dt + \mathbf{T}_D \frac{d\tilde{\mathbf{a}}}{dt} \right], \quad (12)$$

where the diagonal matrices $\mathbf{K}_P$, $\mathbf{T}_I$, and $\mathbf{T}_D$ have been introduced already in the previous section.

3.3. Joint Resolved PID with Gravity Compensation. In robotics the effects of gravitational field are often much more relevant than the other dynamics terms, at least for the small/moderate velocities attained during assembly tasks; such terms can be easily evaluated by means of the virtual work principle, as worked out in Callegari et al. [2]. Thus, a compensation term can be introduced by adding the force vector:

$$\mathbf{f}_g = -\mathbf{J}_A^{-T} \sum_i m_i \mathbf{J}_i^T \mathbf{g}, \quad (13)$$

FIGURE 10: The prototype of the spherical parallel machine.

where $\mathbf{J}_A$ is the analytic Jacobian matrix, m_i is the mass of the ith member, $\mathbf{J}_i$ is the Jacobian that links the velocity of the centre of gravity of the ith member to the vector $\dot{\mathbf{a}}$, and $\mathbf{g}$ is the gravity acceleration. The resulting control scheme is shown in Figure 4.

3.4. Task Space PID with Gravity Compensation. The third control scheme that has been taken into consideration is a task space PID, with the compensation of the gravitational terms; see Figure 5:

$$\mathbf{f} = \mathbf{J}_A^{-T} \left(\mathbf{K}_P' \left[\tilde{\boldsymbol{\varphi}} + \frac{1}{\mathbf{T}_I'} \int \tilde{\boldsymbol{\varphi}}\, dt + \mathbf{T}_D' \frac{d\tilde{\boldsymbol{\varphi}}}{dt} \right] - \sum_i m_i \mathbf{J}_i^T \mathbf{g} \right), \quad (14)$$

where $\tilde{\boldsymbol{\varphi}}$ is the error signal in the task space and the PID gains $\mathbf{K}_P'$, $\mathbf{T}_I'$, and $\mathbf{T}_D'$ are diagonal matrices once again. This algorithm is computationally more expensive than the previous one, since it requires the evaluation of direct kinematics that for PKMs is more complex than inverse kinematics; on the other hand, it could prove useful, for example, in vision assisted assembly tasks with position-based controls, as already experimented on the 3-CPU translating parallel machine by Palmieri et al. [9].

3.5. Implementation in Real-Time Controller. During the implementation of algorithms (12)–(14) on the real-time controller, it was took into consideration the sensitiveness to noise of differentiation. Considering the Laplace transform of (10), the mentioned problem has been numerically mitigated by substituting the classic derivative term $K_P T_D s$ with the following derivative operator:

$$D^*(s) = \frac{K_P T_D s}{1 + s T_D / N}, \quad (15)$$

where N has been chosen equal to 10. Another problem in the implementation of PID controllers over a real-time system is the windup effect. This phenomenon is due to the integral action which saturates the actuators output.

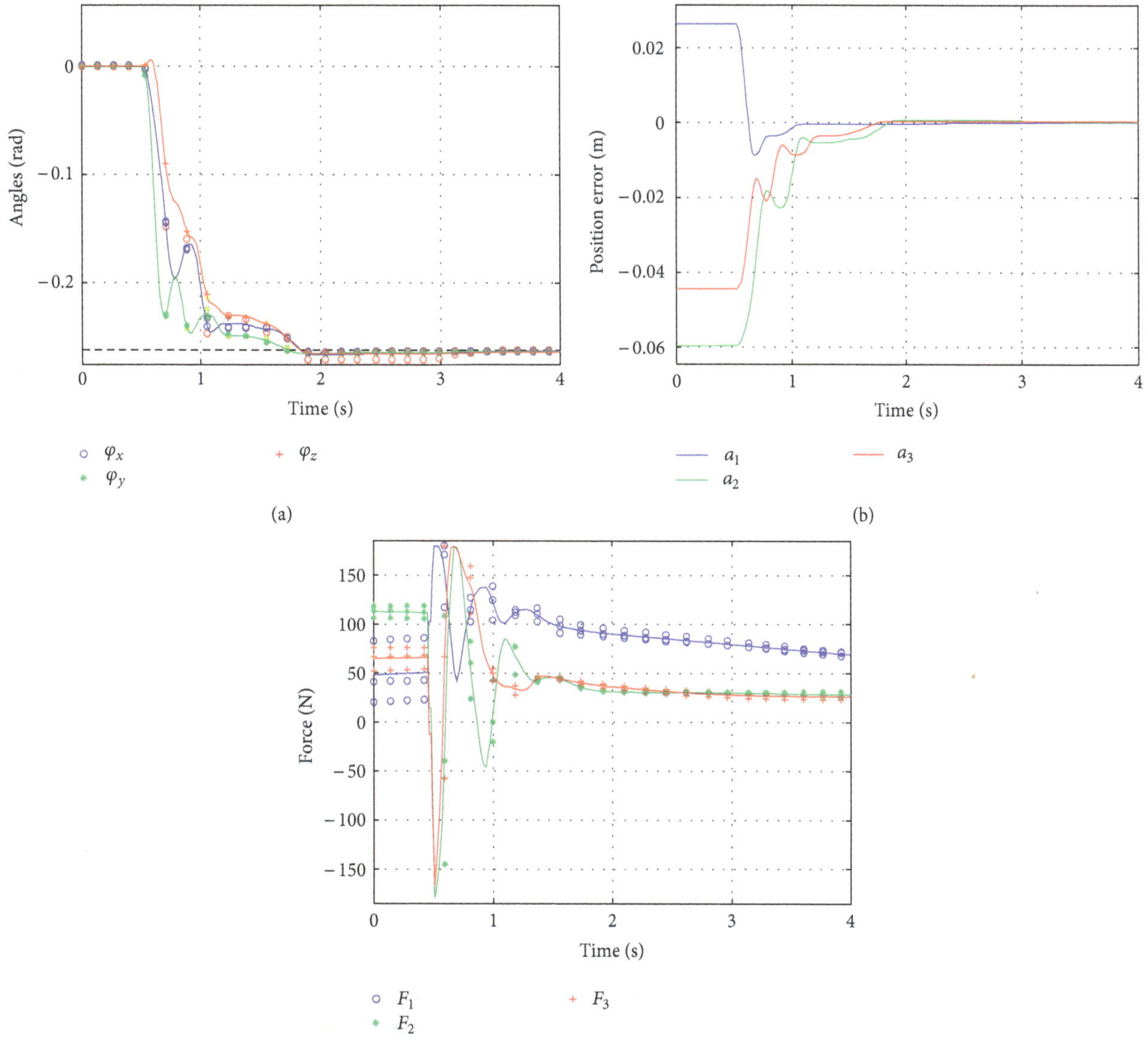

FIGURE 11: Joint-space PID controller: platform's trajectory in task-space (a), joint space errors (b), and motors thrusts (c).

Figure 6 shows a modified scheme of the PID control which implements a typical anti-windup strategy; the model of the actuator, mentioned in the scheme, was easily obtained after identification of the motor mechanical and electrical parameters summarized in Table 2.

4. Simulation Results

4.1. Simulation Environment. Figure 7 shows the virtual prototyping environment used at the Robotics Laboratory of the Polytechnic University of Marche for the design of automated and robotized systems, in particular for the design and virtual testing of parallel kinematic manipulators. The mechanical design is developed through conventional CAD tools, which allow to easily define even the most complex geometries and also to perform, for example, by means of FEM modules, the needed structural analyses; the interface with a multibody code allows to perform closed-loop dynamic analyses, with different levels of difficulty according to the associativity of the used programs. In this case, the LMS Virtual. Lab Motion package has been used, which is able to handle conveniently also complex situations like, for instance, the occurrence of an impact. The multibody package receives in input from the controller the actuation torques and integrates the equation of direct dynamics, providing in output the state variables assumed to be measured. The control system, which is implemented in the Matlab/Simulink environment, computes the control actions by taking into account the commanded task and sometimes, just like the present case, by also exploiting the complete or partial knowledge of robot's dynamics (inverse dynamics model). If the task is constrained by the contact with the environment, like is usually the case for assembly, the contact forces can be evaluated too, to set up more efficient force control schemes. It is noted that, by

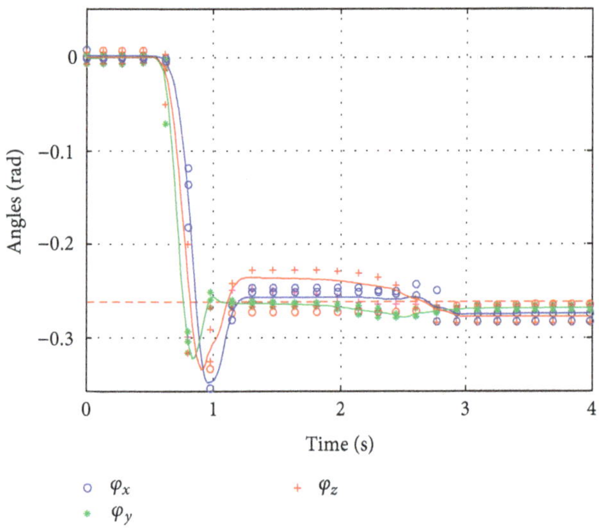

(a)

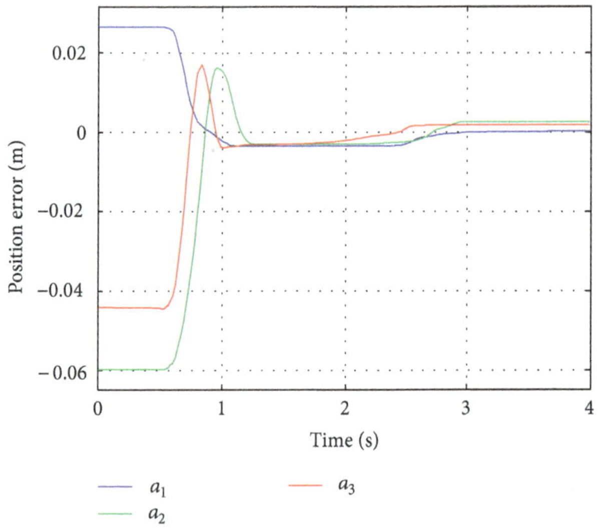

(b)

FIGURE 12: Joint-space PID with gravity compensation: platform's trajectory in task-space (a) and joint space errors (b).

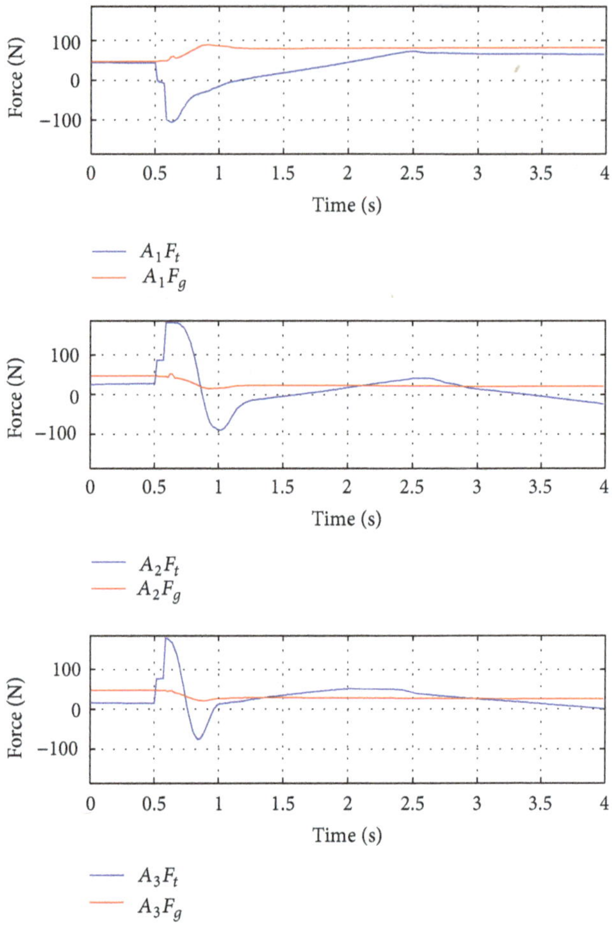

FIGURE 13: Joint-space PID with gravity compensation: F_t total force provided by the motors, F_g gravity component.

TABLE 1: PIDs gains.

	Joint space	Joint space with gravity compensation	Task space with gravity compensation
K_P [N/m]	25000	15000	1000
T_I [s]	200	100	100
T_D [s]	1	0.2	0.25

TABLE 2: Parameters of the linear drives.

Motors properties			
M_s	2.95	kg	Stator mass
K_t	58	N/A	Torque constant
I_n	3	A	Nominal supply current
T_n	184	N	Nominal thrust
v_n	6	m/s	Nominal speed

using the Real-Time Workshop package of the Matlab suite, the same code used during the simulations in the virtual prototyping environment has been directly ported to the real-time control hardware afterwards.

In this way, by means of the mentioned prototyping software, a model of the spherical robot has been made available for the design of the control system and for the tuning of the PID's. Table 1 collects some control gains at the end of the tuning procedure, based on both simulation runs and experimental tests.

4.2. Simulation Analysis. A few test cases have been set up in simulation to evaluate the performances of the 3 PID controllers described in Section 3. The figures show the response of the system when the robot started at rest from

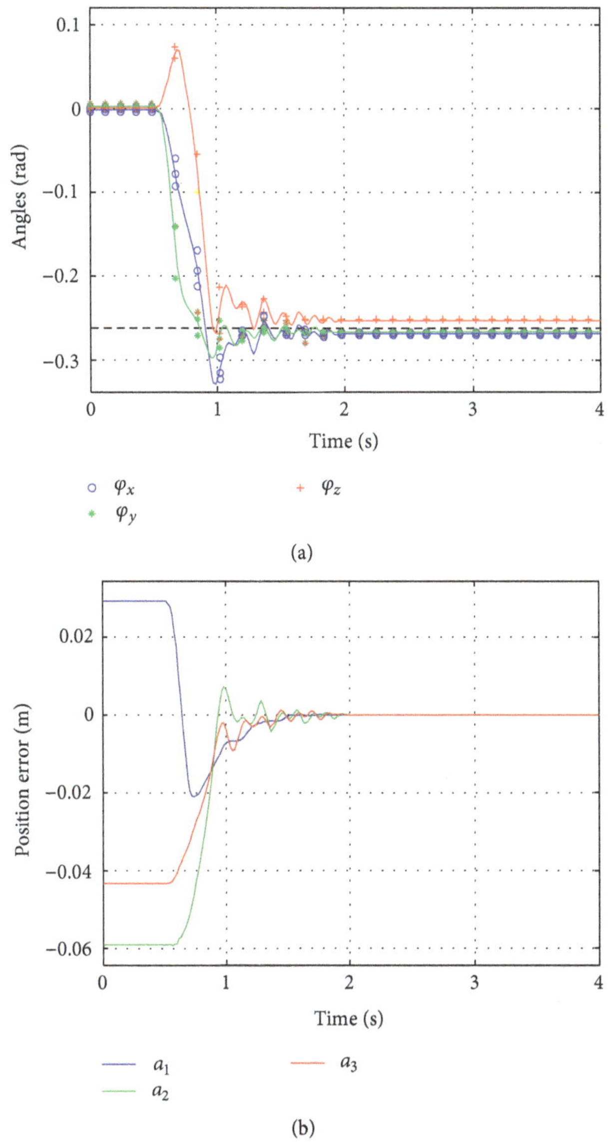

FIGURE 14: Task-space PID with gravity compensation: platform's trajectory in workspace (a) and joint space errors (b).

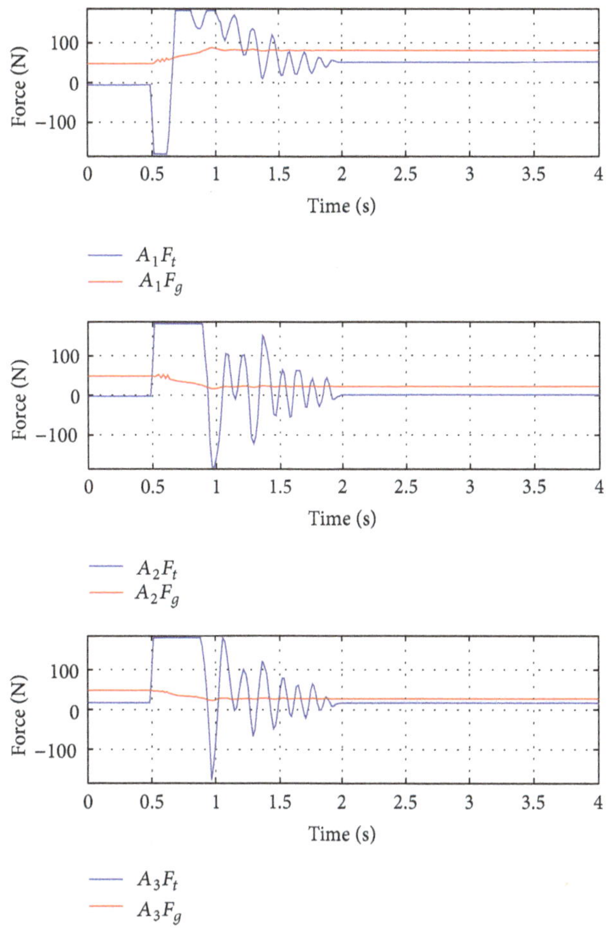

FIGURE 15: Task-space PID with gravity compensation: F_t total force provided by the motors, F_g gravity component.

the home configuration ($\varphi_x = \varphi_y = \varphi_z = 0$) and was required to attain the set point:

$$\boldsymbol{\varphi}_D = \begin{bmatrix} \varphi_{x,D} \\ \varphi_{y,D} \\ \varphi_{z,D} \end{bmatrix} = \begin{bmatrix} -15^\circ \\ -15^\circ \\ -15^\circ \end{bmatrix} = \begin{bmatrix} -0.262 \\ -0.262 \\ -0.262 \end{bmatrix} \text{ rad}, \qquad \dot{\boldsymbol{\varphi}}_D = 0. \tag{16}$$

Such task is very challenging for machine's controller because the set point lies close to a singular configuration of the robot and algorithms (13) and (14) require the inversion of the Jacobian matrix. Figure 8 shows the different performances, in simulation, between the three different controllers: C1, C2, C3 are, respectively, joint resolved PID, joint resolved PID with gravity compensation, and task space PID with gravity compensation. It is noted that the robot is not kept at its home position by means of the brakes but the motors are used to this aim instead, then the set point has been applied in all trials at the time instant $t = 0.5$ s. The orientation trajectories in the task space show the better behaviour of the closed-loop system when it is equipped with the conventional PID algorithm, due to the mentioned presence of singularities.

The simulations also return useful information for what regards the control effort forces, which are plotted in Figure 9: in all cases the application of the set point causes a peak in the required forces, which saturates the actuators.

In the end, it is noted that the task space PID with gravity compensation is more sensitive to parameter variations. This is due to the intrinsic characteristics of robot prototype, which has no external sensor and many singular configurations. In this way, all the information about the task space is obtained through the direct kinematics and the robot Jacobian. Small errors in the computation may affect heavily control system's performance.

5. Experimental Results

5.1. Experimental Setup. The prototype robot is shown in Figure 10; it is actuated by three brushless linear motors by *Phase* and controlled by a *National Instrument* board based

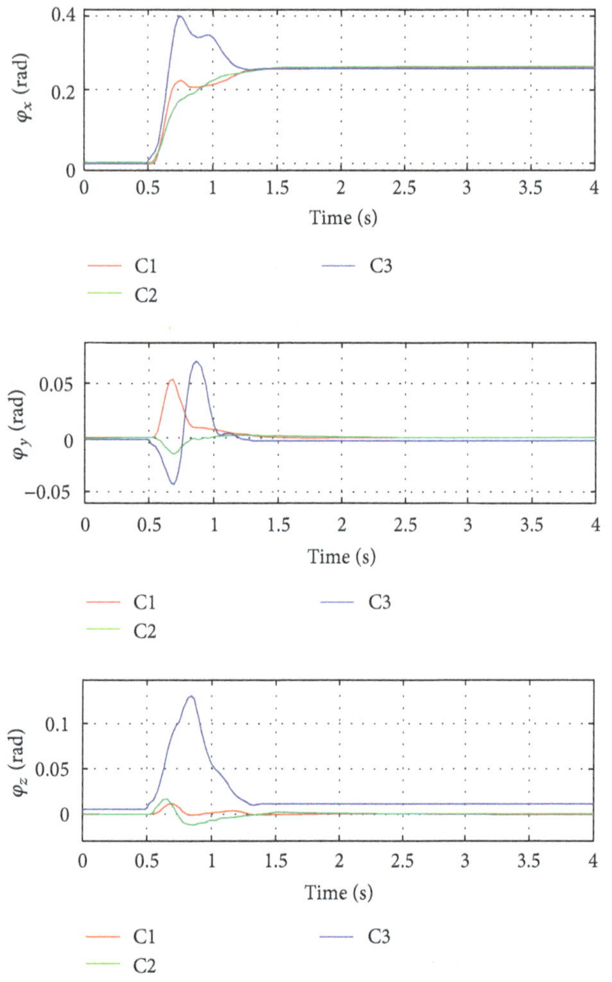

FIGURE 16: Comparative behavior of the three controllers: task-space trajectories.

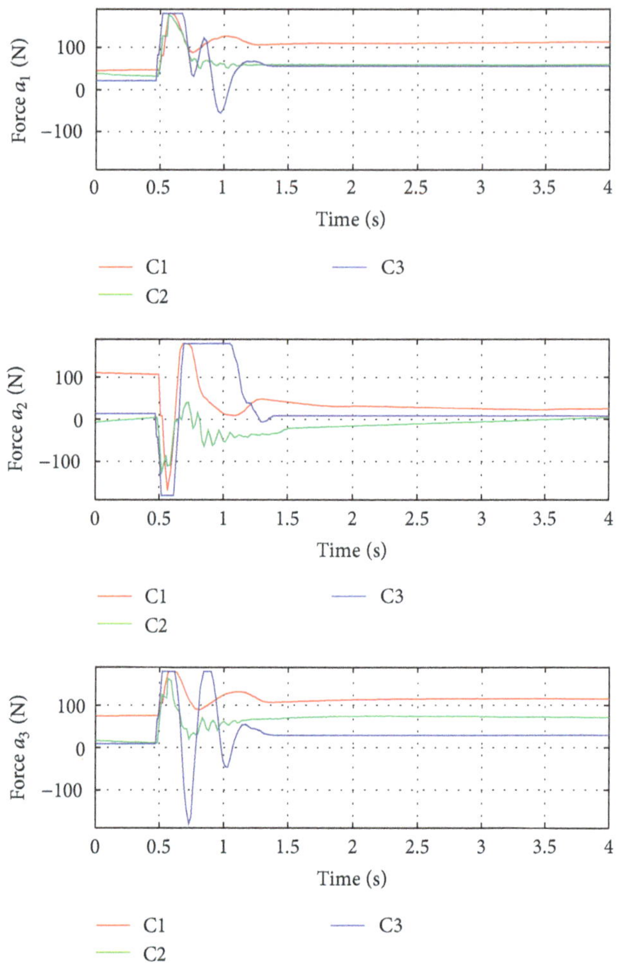

FIGURE 17: Comparative behavior of the three controllers: actuation forces.

on the *PXI/FlexMotion* hardware. The force developed by the sliders is obtained by directly setting the current loop of the drivers, according to the usual relation between the current i and the thrust F:

$$F = K_t i, \tag{17}$$

where the torque constant K_t characterizes the performances of the motor; see Table 2.

With reference to the symbols introduced in Figure 1, the main design data of the prototype are collected in Table 3.

A series of experimental tests have been carried out in order to validate the numerical model described in the previous sections and to experimentally assess the performances of control laws (12)–(14). Results are here reported.

5.2. Case Study A. The first case study was already investigated in simulation, so that numerical and experimental results can now be compared. The platform at the home position has been requested to attain once again the task space set point (16), which corresponds to the following motor strokes:

$$\mathbf{a}_D = \begin{bmatrix} a_{1,D} \\ a_{2,D} \\ a_{3,D} \end{bmatrix} = \begin{bmatrix} 516.4 \\ 430.6 \\ 445.6 \end{bmatrix} [\text{mm}], \qquad \dot{\mathbf{a}}_D = 0. \tag{18}$$

Many tests have been performed for each one of the three control laws (12)–(14) and in the figures some experimental results are presented; values of one of the experimental trials are represented by circle markers while the averaged quantities are represented by solid lines.

Figure 11 presents some results obtained with the conventional PID loop. It is seen that steady state is achieved in less than one second without significant oscillations, due to the pretty high mechanical damping of the system. The corresponding actuation forces F_i are rather large in the first instants, approaching motors' saturation thrusts, then they settle along the static value required for gravity compensation.

By observing Figures 11, 12, 13, and 14, it results that the introduction of a gravity compensation term into the joint loop brings in system's dynamics overshoots that increase

TABLE 3: Mechanical data of the robot prototype.

	Geometrical data	
c	210	mm
d	490	mm
h	280	mm
$a_{i\,\min}$	319	mm
$a_{i\,\max}$	661	mm
$b_{i\,\min}$	130	mm
$b_{i\,\max}$	210	mm
	Mass data	
Slider	7.15	kg
Link 1	1.90	kg
Link 2	2.21	kg
Platform	11.73	kg

the settle time. The task-space controller, on the other hand, requires much more actuation efforts in the first instants of the trials; see Figure 15.

5.3. Case Study B. The second case study was aimed at comparing the performances of the 3 controllers and therefore an elemental task was chosen; the wrist started in quiet at the home pose and was requested to reach the configuration:

$$\boldsymbol{\varphi}_D = \begin{bmatrix} \varphi_{x,D} \\ \varphi_{y,D} \\ \varphi_{z,D} \end{bmatrix} = \begin{bmatrix} 15^\circ \\ 0 \\ 0 \end{bmatrix} = \begin{bmatrix} 0.262\ \text{rad} \\ 0 \\ 0 \end{bmatrix}, \qquad \dot{\boldsymbol{\varphi}}_D = 0, \quad (19)$$

which corresponds to the following motor strokes:

$$\mathbf{a}_D = \begin{bmatrix} a_{1,D} \\ a_{2,D} \\ a_{3,D} \end{bmatrix} = \begin{bmatrix} 470.2 \\ 533.2 \\ 470.2 \end{bmatrix} [\text{mm}], \qquad \dot{\mathbf{a}}_D = 0, \qquad (20)$$

Figure 16 shows the trend of the workspace variable vector $\varphi = [\varphi_x\ \varphi_y\ \varphi_z]^T$ in the three different cases, while Figure 17 plots the actuation forces developed by the motors.

6. Conclusions

The paper presented the first experiments in driving a prototypal SPM developed at the Polytechnic University of Marche by means of linear controllers. The use of a virtual simulation environment can be very profitable in the design of robots' controllers and even in the draft tuning of their parameters. In the present case, three controllers have been first designed in simulation and then implemented on an embedded system for real-time application by means of rapid prototyping software. The task-space controller with the compensation of the gravity terms provided poorer performances than conventional joint-space loops, but in A.'s opinion it could be due to calibration errors, that would heavily affect the computation of direct kinematics, which is required with the present sensing equipment; the use of ANN controllers could be profitable to overcome unmodeled disturbances due to static friction or calibration errors [10].

As a matter of fact, task-space control schemes would be very useful for the realization of interaction controllers, see Siciliano and Villani [11], which is the objective of A.'s coming researches, since they aim at performing mechanical assembly by means of cooperating PKMs [12].

The simplicity of direct kinematics of this machine (in comparison with the usual complexity of PKMs) allows an efficient implementation of algorithms with loop closures in the task space; the same can be said for the easy compensation of the static unbalance of robot's links.

These features and the possible use of visual servoing suggest a possible implementation of control schemes based on force control, where machine's dynamics has to be computed in task-space coordinates, which is rather "natural" for parallel kinematics machines.

References

[1] J. P. Merlet, *Parallel Robots*, Springer, Dordrecht, The Netherlands, 2nd edition, 2006.

[2] M. Callegari, L. Carbonari, G. Palmieri, and M. C. Palpacelli, "Parallel wrists for enhancing grasping performances," in *Grasping in Robotics*, G. Carbone, Ed., pp. 189–219, Springer, New York, NY, USA, 2013.

[3] M. Karouia and J. M. Hervè, "A three-dof tripod for generating spherical rotation," in *Advances in Robot Kinematics*, J. L. Lenarcic and M. M. Stanisic, Eds., pp. 395–402, Kluwer Academic, Dordrecht, The Netherlands, 2000.

[4] L. Carbonari, L. Bruzzone, and M. Callegari, "Impedance control of a spherical parallel platform," *International Journal of Intelligent Mechatronics and Robotics*, vol. 1, no. 1, pp. 40–60, 2011.

[5] C. Innocenti and V. Parenti-Castelli, "Echelon form solution of direct kinematics for the general fully-parallel spherical wrist," *Mechanism and Machine Theory*, vol. 28, no. 4, pp. 553–561, 1993.

[6] I. Cervantes and J. Alvarez-Ramirez, "On the PID tracking control of robot manipulators," *Systems and Control Letters*, vol. 42, no. 1, pp. 37–46, 2001.

[7] C. Yang, Q. Huang, H. Jiang, O. Ogbobe Peter, and J. Han, "PD control with gravity compensation for hydraulic 6-DOF parallel manipulator," *Mechanism and Machine Theory*, vol. 45, no. 4, pp. 666–677, 2010.

[8] R. Kelly and J. Moreno, "Manipulator motion control in operational space using joint velocity inner loops," *Automatica*, vol. 41, no. 8, pp. 1423–1432, 2005.

[9] G. Palmieri, M. C. Palpacelli, M. Battistelli, and M. Callegari, "A comparison between position-based and image-based dynamic visual servoings in the control of a translating parallel manipulator," *Journal of Robotics*, vol. 2012, Article ID 103954, 11 pages, 2012.

[10] D. Tina, L. Carbonari, and M. Callegari, "Design and experimentation of a neural network controller for a spherical parallel machine," in *Proceedings of the 9th International Conference on Informatics in Control, Automation and Robotics (ICINCO '12)*, vol. 1, pp. 250–255, Rome, Italy, 2012.

[11] B. Siciliano and L. Villani, *Robot Force Control*, Springer, Dordrecht, The Netherlands, 2000.

[12] L. Bruzzone and M. Callegari, "Application of the rotation matrix natural invariants to impedance control of purely rotational parallel robots," *Advances in Mechanical Engineering*, vol. 2010, Article ID 284976, 9 pages, 2010.

2

Consensus Formation Control for a Class of Networked Multiple Mobile Robot Systems

Long Sheng, Ya-Jun Pan, and Xiang Gong

Department of Mechanical Engineering, Dalhousie University, Halifax, NS, Canada B3J 2X4

Correspondence should be addressed to Ya-Jun Pan, yajun.pan@dal.ca

Academic Editor: Peter X. Liu

A consensus-based formation control for a class of networked multiple mobile robots is investigated with a virtual leader approach. A novel distributed control algorithm is designed based on the Lyapunov method and linear matrix inequality (LMI) technique for time delay systems. A multiple Lyapunov Krasovskii functional candidate is proposed for investigating the sufficient conditions to linear control gain design for the system with constant time delays. Simulation results as well as experimental studies on Pioneer 3 series mobile robots are shown to verify the effectiveness of the proposed approach.

1. Introduction

Embedded computational resources in autonomous robotic vehicles are becoming more abundant and have enabled improved operational effectiveness of cooperative robotic systems in civilian and military applications. Compared to autonomous robotic vehicles that operate single tasks, cooperative teamwork has greater efficiency and operational capability. Multirobotic vehicle systems have many potential applications, such as platooning of vehicles in urban transportation, the operation of the multiple robots, autonomous underwater vehicles, and formation of aircrafts in military affairs [1–3].

The study of group behaviors for multirobot systems is the main objective of the work. Group cooperative behavior signifies that individuals in the group share a common objective and action according to the interest of the whole group. Group cooperation can be efficient if individuals in the group coordinate their actions well. Each individual can coordinate with other individuals in the group to facilitate group cooperative behavior in two ways, named local coordination and global coordination. For the local coordination, individuals react only to other individuals that are close, such as fish engaged in a school. For the global coordination, each individual can directly coordinate its act with every other individual in the group. Due to communication constraints, most researchers are interested primarily in group cooperation problems where the coordination occurs locally [4–6].

Cooperative control of multirobotic vehicle systems brings us significant theoretical [7–9] and practical challenges. For example, the research objective is defined based on a system of some subsystems rather than a single system; the effects caused by the communication constraints should be considered and how to design coordination strategies so that coordination will result in a group cooperation [10, 11].

As a concrete example of cooperative control, the formation control of multiple autonomous vehicles receives significant interest in recent years [12–15]. It requires that autonomous robotic vehicles collectively maintain a prescribed geometric shape during movement. Maintaining an accurate geometric configuration among multiple robotic vehicles moving in formation can result in less expensive and more capable systems that can accomplish objectives impossible for a single vehicle. The advantages of the formation control for multirobotic systems can be summarized as follows: good feasibility, accuracy, robustness, flexibility, lower cost, energy efficiency, and probability of success. For example, a group robotic vehicles can be used for large objective transferring, terrain model reconnaissance, unknown area exploration, and path obstruction.

Various strategies and approaches, which can be roughly categorized as leader-follower, behavioral, and virtual leader

(a)

(b)

Figure 1: Pioneer 3-DX and 3-AT mobile robots.

approaches [16], have been proposed for the formation control. In the leader-follower approach, one of the vehicles is appointed as the leader, and other vehicles in the group are appointed as the followers. When the group cooperative behavior occurs, the followers should track the trajectory of the leader with some prescribed offset. The basic idea about the behavioral approach is to prescribe several desired behaviors for each vehicle and to make the control action to each vehicle according to each behavior. In the virtual leader structure approach, the entire formation of the group is treated as a single structure. The virtual leader's dynamic or trajectory is converted as the desired action of each vehicle in the group. Tracking control based on consensus algorithm is then needed to tackle this problem.

In the proposed work, a novel distributed control algorithm is designed based on the Lyapunov method and linear matrix inequalities (LMIs) technique for multiple robot systems. Note that Lyapunov-based approaches and LMIs technique have been successfully and widely applied in the area of systems and control. For example, in [17], a survey is in analysing the stability of time-delay systems by using of a Lyapunov-based method, LKF (Lyapunov Krasovskii function) theory to derive sufficient stability conditions in the form of LMIs. Furthermore, in [18], a set-value Lyapunov approach is to prove the stability of multiagent systems under a condition of time-dependent communication links. A Lyapunov-based stability theory and LMIs-based control method are described comprehensively in [19, 20], respectively.

In this work, a consensus-based design scheme is applied to the formation control of multiple-wheeled mobile-robot group with a virtual leader. The group communication configuration is assumed to be a fully coupled system which means decisions made by each robot in the group affect the cost and outcomes of all other members of the group. In this case, what a single robot is going to do is affected by what all other robots in the group are going to do. The distributed formation control architecture is defined to accommodate an arbitrary number of subgroup leaders and arbitrary information flow among the robots. This architecture requires neighbor-to-neighbor information exchange. At the group level, the consensus tracking algorithm is applied to guarantee consensus on the time-varying group reference trajectory in a distributed manner. A consensus-based formation control strategy developed based on the group level consensus tracking algorithm is applied for vehicle level control.

A novel delay-dependent multiple Lyapunov functional candidate related to LKF is constructed to investigate the convergence of the tracking error. The null sums have been added to the new multiple LKF with free weighting matrices introduced to reduce the conservatism in the derivation of the stability conditions. Therefore, the matrices and the sufficient conditions for the stabilization of the proposed control approach can be determined by solving LMIs. The effects caused by the group communication delays are well considered in the proposed approach. In addition to simulation results, the proposed control strategy is experimentally implemented for multiple-wheeled mobile robots under neighbor-to-neighbor information exchange with group communication delay involved.

Notations. The information exchange among robots is usually modelled by graphs. Suppose that a team consists of n mobile robots. $\varsigma_n = (\nu_n, \varepsilon_n)$ is a graph, where $\nu_n = 1, \dots, n$ is a finite nonempty node set, and $\varepsilon_n \subseteq \nu_n \times \nu_n$ is an edge set of ordered pairs of nodes, called edges. The adjacency matrix $A_n = [a_{ij}] \in R^{n\times n}$ of a graph $\varsigma_n = (\nu_n, \varepsilon_n)$ is defined such that a_{ij} is a positive weight if $(j, i) \in \varepsilon_n$ is ture, and $a_{ij} = 0$ if $(j, i) \in \varepsilon_n$ is false.

2. Modelling of Pioneer 3 Mobile Robots

Pioneer mobile robots are durable, differential-drive robots for academic researchers. The most famous advantages of Pioneer 3 robot series are good versatility, reliability, and durability. In this work, two kinds in the series named Pioneer 3-DX (P3-DX) and Pioneer 3-AT (P3-AT) have been used for the experimental implementation.

The P3-DX used in our lab is shown in Figure 1(a). It assembled motors with 500-tick encoders, 19 cm wheels, 8 forward-facing ultrasonic sensors, 8 real-facing sonar, 1, 2, or 3 rechargeable batteries, and a microcontroller which can communicate with a laptop through serial port. P3-DX can reach the maximum speed of 1.6 meters per second and carry a payload of up to 23 kg. It is an all-purpose base and can be

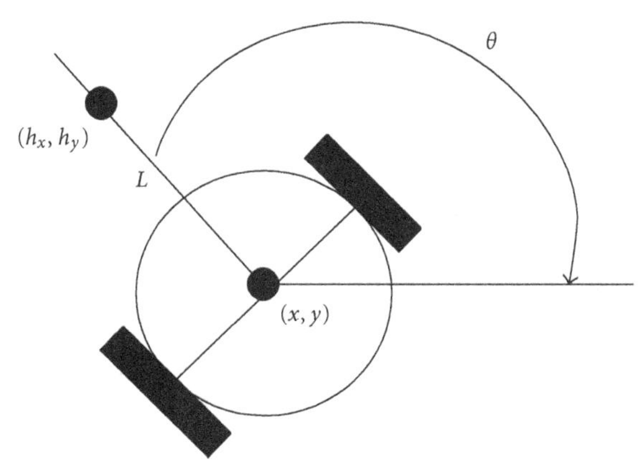

FIGURE 2: Hand position for P3 mobile robot.

used for research and applications involving mapping, teleoperation, localization, monitoring, reconnaissance, vision capture, cooperation, and other behaviors. P3-DX runs best on hard surfaces. It can traverse low sills and household power cords, and it can also climb most wheelchair ramps.

Another robot P3-AT used in this work is shown in Figure 1(b). It is a highly versatile four-wheel drive robotic platform, which is softwarecompatible with other pioneer 3 robots. P3-AT is a popular team performer for outdoor or rough-terrain projects. It has powerful motors and four knobby wheels that can reach the maximum speed of 0.8 meters per second and carry a payload of up to 12 kg. P3-AT uses 100-tick encoders with inertial correction recommended for dead reckoning to compensate for skid steering. Similar with P3-DX, P3-AT also has 8 forward and 8 rear sonar, a microcontroller which can be connected with a laptop through serial port and batteries.

Both of P3-DX and P3-AT have the same kinematic model which can be expressed by the following equation:

$$\begin{aligned} x &= v\cos(\theta), \\ y &= v\cos\theta, \\ \theta &= \omega, \end{aligned} \tag{1}$$

where $[x, y]$ is the inertial position of the P3 mobile robot, θ is the orientation of the robot, and $[v, \omega]$ denote the linear and angular speeds of the robot. Since the P3 mobile robots used in this work have nonholonomic constraints, the coordination problem becomes more complicated. As we know, nonholonomic systems cannot be stabilized with continuous static-state feedback, so the difficulty of the coordination problem for differentially driven mobile robots is that the position and orientation of the center of the robot cannot be simultaneously stabilized with a time-invariant feedback control strategy. Some researchers successfully used discontinuous control laws and time-varying control laws to stabilize the center of a single differentially driven mobile robot; however, the multiple robot case is more complicated for sure.

The most popular way to simplify this complex case is to define a hand position for each robot. As shown in Figure 2, the hand position of the robot usually has been defined at the point $h = [h_x, h_y]^T$ which lies a distance L along the line that is normal to the wheel axis and intersects the wheel axis at the center point $r = [x, y]^T$. The kinematic model of the hand position is holonomic for $L \neq 0$. Instead of considering the coordinating problem at the center of the robot, we consider the problem at the hand position. Another important advantage of defining a hand position for the robot is that the hand position has practical interest. For example, if the task of the robot group is to move an object from one place to another by using the gripper which has been installed at the hand position of each robot, then the control objective for this task is to move the gripper locations in a coordinated fashion.

Now, let us find the kinematic model for the hand position of each robot. First, the hand position can be represented by the following equation:

$$\begin{aligned} h_x &= x + L\cos(\theta), \\ h_y &= y + L\sin(\theta), \end{aligned} \tag{2}$$

where $\mathbf{h} = [h_x, h_y]^T$ is the hand position in x and y plane. Now, let us differentiate (2) with respect to time and substitute (1), then we get

$$\begin{aligned} h_x &= \cos(\theta)v - L\sin(\theta)\omega, \\ h_y &= \sin(\theta)v + L\cos(\theta)\omega. \end{aligned} \tag{3}$$

Define $\mathbf{h} = [h_x, h_y]^T$, $\mathbf{u} = [u_x, u_y]^T$ and let

$$\begin{aligned} v &= \cos(\theta)u_x + \sin(\theta)u_y, \\ \omega &= -\frac{1}{L}\sin(\theta)u_x + \frac{1}{L}\cos(\theta)u_y, \end{aligned} \tag{4}$$

then we have

$$\mathbf{h} = \mathbf{u}, \tag{5}$$

which is the kinematic model of the robot's hand position.

3. Problem Formulation

As described in the above section, the nonlinear kinematic model of the center position of the robot has been simplified and linearized into the form of single-integrator dynamics shown in (5) by defining a hand position for each robot. The control interest has been converted from the center position of the robot to its hand position. So the state of the hand position of the robot is used as the state of the robot itself.

The virtual leader and virtual structure approach is one solution to formation control. Figure 3 shows the example of the virtual leader and virtual structure approach with a formation composed of three vehicles with planar motions. In Figure 3, C_0 is the global coordinate system, $r_i(t) = [x_i(t), y_i(t)]^T$ is the ith vehicle's actual position at time t, and $r_i^d(t) = [x_i^d(t), y_i^d(t)]^T$ is the ith P3 mobile robot's desired positions at time t. Both its actual and desired position at time t are relative to C_0. The group can move with the desired formation shape only if each vehicle can track its desired position accurately.

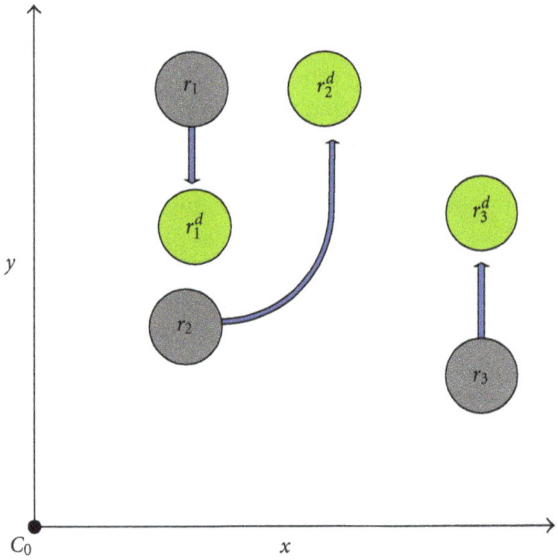

FIGURE 3: The framework for P3 mobile robot team with virtual leaders.

In this paper, we assumed each P3 mobile robot knows the state of its virtual subgroup leader versus time. If each vehicle has inconsistent knowledge of its virtual subgroup leader's states, then the desired formation cannot be maintained.

The linearized model of each robot in the group can be represented by the following equation:

$$\mathbf{r}_i(t) = \mathbf{u}_i(t), \qquad i = 1,\ldots,n, \tag{6}$$

where $\mathbf{r}_i(t) = [x_i(t), y_i(t)]^T \in R^n$ is the state of the ith robot in the group which includes its position and velocity information. i is the index denoting the number of the robot in the group. n is the total number of all robots in the group. $\mathbf{u}_i(t) \in R^m$ is the control input signal.

Now, define the virtual subgroup leader's state, which is also the desired state for each P3 mobile robot in the group, as $\mathbf{r}_i^d(t) = [x_i^d(t), y_i^d(t)]^T$. If $\mathbf{r}_i(t) \to \mathbf{r}_i^d(t), i = 1,\ldots,n$, as $t \to \infty$, then the desired formation shape is maintained, and the group movement follows the desired reference.

4. A Novel Consensus Control Approach

In [21], the controller $\mathbf{u}_i$ is designed for agents in the form of integrators such that all followers track the virtual leader with local interaction in the absence of velocity measurements. Specifically, a distributed consensus tracking algorithm is defined as

$$\mathbf{u}_i(t) = -\alpha\sum_{j=0}^{n}\mathbf{a}_{ij}\left(\mathbf{r}_j - \mathbf{r}_i\right) - \beta \operatorname{sgn}\left[\sum_{j=0}^{n}\mathbf{a}_{ij}\left(\mathbf{r}_i - \mathbf{r}_j\right)\right], \tag{7}$$

where α is a nonnegative constant, β is a positive constant, and $\operatorname{sgn}(\cdot)$ is the signum function, since it considers the tracking problem under fixed and switching network topologies. In the proposed work, a new distributed consensus tracking control algorithm with consideration of some communication delay can be represented as

$$\begin{aligned}\mathbf{u}_i(t) = {} & \mathbf{r}_i^d(t) - \mathbf{k}_i\left(\mathbf{r}_i(t) - \mathbf{r}_i^d(t)\right) \\ & - \sum_{j=1,i\neq j}^{n}\mathbf{a}_{ij}\left(\mathbf{r}_j(t-\tau) - \mathbf{r}_j^d(t-\tau)\right),\end{aligned} \tag{8}$$

where $\mathbf{k}_i$ is the designed control gain. Notice that each robotic vehicle in the group has the same group communication coupling and kinematic model, $\mathbf{k}_i$ can be represented as $\mathbf{k}$ instead, $\mathbf{a}_{ij}$ is the (i, j) entry of adjacency matrix $A_n \in R^{n\times n}$ according to the interaction topology $\varsigma_n = (\nu_n, \varepsilon_n)$ for $\mathbf{r}_i - \mathbf{r}_i^d$, $\mathbf{r}_j - \mathbf{r}_j^d$ is the information from neighbors of the ith P3 mobile robot, which can also be treated as the coupling between the ith P3 mobile robot and its neighbors, and τ is the constant network-induced group communication delay. Submiting (8) into (6), we have

$$\begin{aligned}\mathbf{r}_i(t) = {} & \mathbf{r}_i^d(t) - \mathbf{k}\left(\mathbf{r}_i(t) - \mathbf{r}_i^d(t)\right) \\ & - \sum_{j=1,i\neq j}^{n}\mathbf{a}_{ij}\left(\mathbf{r}_j(t-\tau) - \mathbf{r}_j^d(t-\tau)\right).\end{aligned} \tag{9}$$

If we define

$$\mathbf{e}_i(t) = \mathbf{r}_i(t) - \mathbf{r}_i^d(t), \tag{10}$$

then (9) can be rewritten as

$$\mathbf{e}_i(t) = -\mathbf{k}\mathbf{e}_i(t) - \sum_{j=1,i\neq j}^{n}\mathbf{a}_{ij}\mathbf{e}_j(t-\tau). \tag{11}$$

It is obvious that if $\mathbf{e}_i \to 0, i = 1,\ldots,n$ as $t \to \infty$ which means $\mathbf{r}_i \to \mathbf{r}_i^d, i = 1,\ldots,n$ as $t \to \infty$, then the desired formation shape is maintained, and the group movement follows the desired reference. The next step is to find the sufficient conditions for the design of the control gain $\mathbf{k}$ so that it can stabilize the error dynamics represented by (11).

Lemma 1 (Jensen inequality). *For any constant matrix $E \in \mathcal{R}^{n\times n}$, $E = E^T > 0$, vector function $\omega : [0,\tau] \to \mathcal{R}^n$ such that the integrations concerned are well defined, then*

$$\tau\int_0^{\tau}\omega^T(s)E\,\omega(s)\,ds \geq \left[\int_0^{\tau}\omega(s)ds\right]^T E\left[\int_0^{\tau}\omega(s)ds\right]. \tag{12}$$

Theorem 2. *Consider the error dynamics model represented by (11), for a given time delay τ and the number of members in one group n; if there exist symmetric positive definite matrices $P = \begin{bmatrix} P_{11} & P_{12} \\ P_{12}^T & P_{22}\end{bmatrix} > 0$, $Q = \begin{bmatrix} Q_{11} & 0 \\ 0 & Q_{22}\end{bmatrix} > 0$, $R = \begin{bmatrix} R_{11} & 0 \\ 0 & R_{22}\end{bmatrix} > 0$, matrices M_i, N_i, $i = 1,\ldots,5$ with appropriate dimensions, such that the following inequality holds:*

$$H = \begin{bmatrix} H_{11} & * & * & * & * \\ H_{21} & H_{22} & * & * & * \\ H_{31} & H_{32} & H_{33} & * & * \\ H_{41} & H_{42} & H_{43} & H_{44} & * \\ H_{51} & H_{52} & H_{53} & H_{54} & H_{55}\end{bmatrix} < 0, \tag{13}$$

where

$$
\begin{aligned}
H_{11} &= \hat{P}_{12} + \hat{P}_{12}^T + (n-1)\hat{Q}_{11} + (n-1)\tau\hat{R}_{11} \\
&\quad + N_1 + N_1^T + M_1 K + K^T M_1^T, \\
H_{21} &= -\hat{P}_{12}^T + N_2 - N_1^T + A^T M_1^T + M_2 K, \\
H_{22} &= -(n-1)\hat{Q}_{11} - N_2 - N_2^T + M_2 A + A^T M_2^T, \\
H_{31} &= (n-1)\hat{P}_{11} + N_3 + M_1^T + M_3 K, \\
H_{32} &= -N_3 + M_3 A + M_2^T, \\
H_{33} &= (n-1)\hat{Q}_{22} + (n-1)\tau\hat{R}_{22} + M_3 + M_3^T, \\
H_{41} &= (n-1)\hat{P}_{22} + N_4 + M_4 K, \\
H_{42} &= -(n-1)\hat{P}_{22} - N_4 + M_4 A, \\
H_{43} &= \hat{P}_{12}^T + M_4, \\
H_{52} &= -N_2^T - N_5 + M_5 A, \\
H_{51} &= -N_1^T + N_5 + M_5 K, \\
H_{44} &= -\frac{(n-1)\hat{R}_{11}}{\tau}, \\
H_{53} &= -N_3^T + M_5, \\
H_{54} &= -N_4^T, \\
H_{55} &= -\frac{(n-1)\hat{R}_{22}}{\tau} - N_5 - N_5^T,
\end{aligned} \tag{14}
$$

then system (11) is asymptotically stable, for example, $\mathbf{e}_i(t)$ tends to zero asymptotically which means the ith robot in the group tracks its desired trajectory well.

Proof. For each robot in the group, we take the Lyapunov Krasovskii functional candidate as

$$
\begin{aligned}
V_i &= (n-1)\mathbf{e}_i^T(t)P_{11}\mathbf{e}_i(t) + 2\mathbf{e}_i^T(t)P_{12}\sum_{j=1,j\neq i}^{n}\int_{t-\tau}^{t}\mathbf{e}_j(s)ds \\
&\quad + \sum_{j=1,j\neq i}^{n}\int_{t-\tau}^{t}\mathbf{e}_j^T(s)dsP_{22}\int_{t-\tau}^{t}\mathbf{e}_j(s)ds \\
&\quad + \sum_{j=1,j\neq i}^{n}\int_{t-\tau}^{t}\left[\mathbf{e}_j^T(s),\mathbf{e}_j^T(s)\right]Q\begin{bmatrix}\mathbf{e}_j(s)\\ \mathbf{e}_j(s)\end{bmatrix}ds \\
&\quad + \sum_{j=1,j\neq i}^{n}\int_{-\tau}^{0}\int_{t+\theta}^{t}\left[\mathbf{e}_j^T(s),\mathbf{e}_j^T(s)\right]R\begin{bmatrix}\mathbf{e}_j(s)\\ \mathbf{e}_j(s)\end{bmatrix}dsd\theta,
\end{aligned} \tag{15}
$$

and now we define

$$
\hat{\mathbf{e}}^T(t) = \left[\mathbf{e}_1^T(t),\ldots,\mathbf{e}_n^T(t)\right]_{1\times n}, \tag{16}
$$

and take the multiple Lyapunov Krasovskii functional candidate as

$$
\begin{aligned}
V &= \sum_{i=1}^{n} V_i \\
&= (n-1)\hat{\mathbf{e}}^T(t)\hat{P}_{11}\hat{\mathbf{e}}(t) + 2\hat{\mathbf{e}}^T(t)\hat{P}_{12}\int_{t-\tau}^{t}\hat{\mathbf{e}}(s)ds \\
&\quad + (n-1)\int_{t-\tau}^{t}\hat{\mathbf{e}}^T(s)ds\hat{P}_{22}\int_{t-\tau}^{t}\hat{\mathbf{e}}(s)ds \\
&\quad + (n-1)\int_{t-\tau}^{t}\left[\hat{\mathbf{e}}^T(s),\hat{\mathbf{e}}^T(s)\right]\begin{bmatrix}\hat{Q}_{11} & 0\\ 0 & \hat{Q}_{22}\end{bmatrix}_{2n\times 2n}\begin{bmatrix}\hat{\mathbf{e}}(s)\\ \hat{\mathbf{e}}(s)\end{bmatrix}ds \\
&\quad + (n-1)\int_{-\tau}^{0}\int_{t+\theta}^{t}\left[\hat{\mathbf{e}}^T(s),\hat{\mathbf{e}}^T(s)\right]\begin{bmatrix}\hat{R}_{11} & 0\\ 0 & \hat{R}_{22}\end{bmatrix}_{2n\times 2n} \\
&\qquad \times\begin{bmatrix}\hat{\mathbf{e}}(s)\\ \hat{\mathbf{e}}(s)\end{bmatrix}dsd\theta,
\end{aligned} \tag{17}
$$

where

$$
\begin{aligned}
\hat{P}_{ii} &= \operatorname{diag}\left[P_{ii},\ldots,P_{ii}\right]_{n\times n}, \quad i = 1,2, \\
\hat{P}_{12} &= \begin{bmatrix} 0 & P_{12} & \cdots & P_{12} \\ P_{12} & \ddots & \ddots & \vdots \\ \vdots & \ddots & \ddots & P_{12} \\ P_{12} & \cdots & P_{12} & 0 \end{bmatrix}_{n\times n}, \\
\hat{Q}_{ii} &= \operatorname{diag}\left[Q_{ii},\ldots,Q_{ii}\right]_{n\times n}, \quad i = 1,2, \\
\hat{R}_{ii} &= \operatorname{diag}\left[R_{ii},\ldots,R_{ii}\right]_{n\times n}, \quad i = 1,2.
\end{aligned} \tag{18}
$$

With appropriate dimensions, the following two zero equations hold:

$$
\begin{aligned}
\phi_1 &= 2\mathbf{z}^T N\left[\hat{\mathbf{e}}(t) - \int_{t-\tau}^{t}\hat{\mathbf{e}}(s)ds - \hat{\mathbf{e}}(t-\tau)\right] = 0, \\
\phi_2 &= 2\mathbf{z}^T M\left[\hat{\mathbf{e}}(t) + K\hat{\mathbf{e}}(t) + A\hat{\mathbf{e}}(t-\tau)\right] = 0,
\end{aligned} \tag{19}
$$

where

$$
\begin{aligned}
\mathbf{z} &= \left[\hat{\mathbf{e}}(t),\hat{\mathbf{e}}(t-\tau),\hat{\mathbf{e}}(t),\int_{t-\tau}^{t}\hat{\mathbf{e}}(s)ds,\int_{t-\tau}^{t}\hat{\mathbf{e}}(s)ds\right]^T, \\
N &= \left[N_1,N_2,N_3,N_4,N_5\right]^T, \\
N_i &= \operatorname{diag}\left[n_i,\ldots,n_i\right]_{n\times n}, \quad i = 1,\ldots,5, \\
M &= \left[M_1,M_2,M_3,M_4,M_5\right]^T, \\
M_i &= \operatorname{diag}\left[m_i,\ldots,m_i\right]_{n\times n}, \quad i = 1,\ldots,5, \\
K &= \operatorname{diag}\left[\mathbf{k},\ldots,\mathbf{k}\right]_{n\times n}, \\
A &= \begin{bmatrix} 0 & a_{12} & \cdots & a_{1(n-1)} & a_{1n} \\ a_{21} & 0 & \cdots & a_{2(n-1)} & a_{2n} \\ \vdots & \vdots & \ddots & \vdots & \vdots \\ a_{(n-1)1} & a_{(n-1)2} & \cdots & 0 & a_{(n-1)n} \\ a_{n1} & a_{n2} & \cdots & a_{n(n-1)} & 0 \end{bmatrix}.
\end{aligned} \tag{20}
$$

Then, the derivative of the multiple Lyapunov function candidate is as follows:

$$
\begin{aligned}
V &= V + \phi_1 + \phi_2 \\
&= (n-1)\hat{\mathbf{e}}^T(t)\hat{P}_{11}\hat{\mathbf{e}}(t) + (n-1)\hat{\mathbf{e}}^T(t)\hat{P}_{11}\hat{\mathbf{e}}(t)
\end{aligned}
$$

$$
\begin{aligned}
&+2\hat{\mathbf{e}}^{T}(t)\hat{P}_{12}\int_{t-\tau}^{t}\hat{\mathbf{e}}(s)ds+2\hat{\mathbf{e}}^{T}(t)\hat{P}_{12}\hat{\mathbf{e}}(t)\\
&-2\hat{\mathbf{e}}^{T}(t)\hat{P}_{12}\hat{\mathbf{e}}(t-\tau)+(n-1)\hat{\mathbf{e}}^{T}(t)\hat{P}_{22}\int_{t-\tau}^{t}\hat{\mathbf{e}}(s)ds\\
&-(n-1)\hat{\mathbf{e}}^{T}(t-\tau)\hat{P}_{22}\int_{t-\tau}^{t}\hat{\mathbf{e}}(s)ds\\
&+(n-1)\left(\int_{t-\tau}^{t}\hat{\mathbf{e}}^{T}(s)ds\right)\hat{P}_{22}\hat{\mathbf{e}}(t)\\
&-(n-1)\int_{t-\tau}^{t}\hat{\mathbf{e}}^{T}(s)ds\hat{P}_{22}\hat{\mathbf{e}}(t-\tau)\\
&+(n-1)\hat{\mathbf{e}}^{T}(t)\hat{Q}_{11}\hat{\mathbf{e}}(t)+(n-1)\dot{\hat{\mathbf{e}}}^{T}(t)\hat{Q}_{22}\dot{\hat{\mathbf{e}}}(t)\\
&-(n-1)\hat{\mathbf{e}}^{T}(t-\tau)\hat{Q}_{11}\hat{\mathbf{e}}(t-\tau)\\
&-(n-1)\dot{\hat{\mathbf{e}}}^{T}(t-\tau)\hat{Q}_{22}\dot{\hat{\mathbf{e}}}(t-\tau)\\
&+\tau(n-1)\hat{\mathbf{e}}^{T}(t)\hat{R}_{11}\hat{\mathbf{e}}(t)+\tau(n-1)\dot{\hat{\mathbf{e}}}^{T}(t)\hat{R}_{22}\dot{\hat{\mathbf{e}}}(t)\\
&-(n-1)\int_{t-\tau}^{t}\left[\hat{\mathbf{e}}^{T}(s),\dot{\hat{\mathbf{e}}}^{T}(s)\right]\begin{bmatrix}\hat{R}_{11}&0\\0&\hat{R}_{22}\end{bmatrix}_{2n\times 2n}\begin{bmatrix}\hat{\mathbf{e}}(s)\\\dot{\hat{\mathbf{e}}}(s)\end{bmatrix}ds\\
&+2\mathbf{z}^{T}N\left[\hat{\mathbf{e}}(t)-\int_{t-\tau}^{t}\dot{\hat{\mathbf{e}}}(s)ds-\hat{\mathbf{e}}(t-\tau)\right]\\
&+2\mathbf{z}^{T}M\left[\dot{\hat{\mathbf{e}}}(t)+K\hat{\mathbf{e}}(t)+A\hat{\mathbf{e}}(t-\tau)\right].
\end{aligned}\tag{21}
$$

Using Lemma 1, we have

$$
\begin{aligned}
V\leq{}&(n-1)\dot{\hat{\mathbf{e}}}^{T}(t)\hat{P}_{11}\hat{\mathbf{e}}(t)+(n-1)\hat{\mathbf{e}}^{T}(t)\hat{P}_{11}\dot{\hat{\mathbf{e}}}(t)\\
&+2\dot{\hat{\mathbf{e}}}^{T}(t)\hat{P}_{12}\int_{t-\tau}^{t}\hat{\mathbf{e}}(s)ds+2\hat{\mathbf{e}}^{T}(t)\hat{P}_{12}\hat{\mathbf{e}}(t)\\
&-2\hat{\mathbf{e}}^{T}(t)\hat{P}_{12}\hat{\mathbf{e}}(t-\tau)+(n-1)\hat{\mathbf{e}}^{T}(t)\hat{P}_{22}\int_{t-\tau}^{t}\hat{\mathbf{e}}(s)ds\\
&-(n-1)\hat{\mathbf{e}}^{T}(t-\tau)\hat{P}_{22}\int_{t-\tau}^{t}\hat{\mathbf{e}}(s)ds\\
&+(n-1)\int_{t-\tau}^{t}\hat{\mathbf{e}}^{T}(s)ds\hat{P}_{22}\hat{\mathbf{e}}(t)\\
&-(n-1)\int_{t-\tau}^{t}\hat{\mathbf{e}}^{T}(s)ds\hat{P}_{22}\hat{\mathbf{e}}(t-\tau)\\
&+(n-1)\hat{\mathbf{e}}^{T}(t)\hat{Q}_{11}\hat{\mathbf{e}}(t)+(n-1)\dot{\hat{\mathbf{e}}}^{T}(t)\hat{Q}_{22}\dot{\hat{\mathbf{e}}}(t)\\
&-(n-1)\hat{\mathbf{e}}^{T}(t-\tau)\hat{Q}_{11}\hat{\mathbf{e}}(t-\tau)\\
&+\tau(n-1)\hat{\mathbf{e}}^{T}(t)\hat{R}_{11}\hat{\mathbf{e}}(t)+\tau(n-1)\dot{\hat{\mathbf{e}}}^{T}(t)\hat{R}_{22}\dot{\hat{\mathbf{e}}}(t)\\
&-\frac{n-1}{\tau}\int_{t-\tau}^{t}\hat{\mathbf{e}}^{T}(s)ds\hat{R}_{11}\int_{t-\tau}^{t}\hat{\mathbf{e}}(s)ds\\
&-\frac{n-1}{\tau}\int_{t-\tau}^{t}\dot{\hat{\mathbf{e}}}^{T}(s)ds\hat{R}_{22}\int_{t-\tau}^{t}\dot{\hat{\mathbf{e}}}(s)ds\\
&+2\mathbf{z}^{T}N\left[\hat{\mathbf{e}}(t)-\int_{t-\tau}^{t}\dot{\hat{\mathbf{e}}}(s)ds-\hat{\mathbf{e}}(t-\tau)\right]\\
&+2\mathbf{z}^{T}M\left[\dot{\hat{\mathbf{e}}}(t)+K\hat{\mathbf{e}}(t)+A\hat{\mathbf{e}}(t-\tau)\right]\\
={}&\mathbf{z}^{T}H\mathbf{z},
\end{aligned}\tag{22}
$$

where H is shown in (13). If (13) holds, then from (22), we have $\dot{V}\leq 0$ which means that the system (11) is asymptotically stable, for example, $\mathbf{e}_i(t)$ tends to zero asymptotically which means the ith robot in the group tracks its desired trajectory well. □

Note that the LMI condition in (13) is nonconvex, and hence the following theorem is proposed to be the sufficient condition of (13).

Theorem 3. *Consider the error dynamics model represented by (11), for a given time delay τ, given scalars θ_i, $i=1,\dots,5$ and the number of members in one group n, if there exist matrices $\overline{P}_{11}$, $\overline{P}_{12}$, $\overline{P}_{22}$, $\overline{Q}_{11}$, $\overline{Q}_{22}$, $\overline{R}_{11}$, $\overline{R}_{22}$, X, Y, $\overline{N}_i$, $i=1,\dots,5$ with appropriate dimensions, such that the following inequality holds:*

$$
\begin{bmatrix}
\overline{H}_{11}&*&*&*&*\\
\overline{H}_{21}&\overline{H}_{22}&*&*&*\\
\overline{H}_{31}&\overline{H}_{32}&\overline{H}_{33}&*&*\\
\overline{H}_{41}&\overline{H}_{42}&\overline{H}_{43}&\overline{H}_{44}&*\\
\overline{H}_{51}&\overline{H}_{52}&\overline{H}_{53}&\overline{H}_{54}&\overline{H}_{55}
\end{bmatrix}<0,\tag{23}
$$

where

$$
\begin{aligned}
\overline{H}_{11}&=\overline{P}_{12}+\overline{P}_{12}^{T}+(n-1)\overline{Q}_{11}+(n-1)\tau\overline{R}_{11}\\
&\quad+\overline{N}_{1}+\overline{N}_{1}^{T}+\theta_{1}Y+\theta_{1}Y^{T},\\
\overline{H}_{21}&=-\overline{P}_{12}^{T}+\overline{N}_{2}-\overline{N}_{1}^{T}+\theta_{2}XA^{T}+\theta_{2}Y,\\
\overline{H}_{22}&=-(n-1)\overline{Q}_{11}-\overline{N}_{2}-\overline{N}_{2}^{T}+\theta_{2}AX^{T}+\theta_{2}XA^{T},\\
\overline{H}_{31}&=(n-1)\overline{P}_{11}+\overline{N}_{3}+\theta_{1}^{T}X+\theta_{3}Y,\\
\overline{H}_{32}&=-\overline{N}_{3}+\theta_{3}AX^{T}+\theta_{2}X,\\
\overline{H}_{33}&=(n-1)\overline{Q}_{22}+(n-1)\tau\overline{R}_{22}+\theta_{3}X^{T}+\theta_{3}X,\\
\overline{H}_{41}&=(n-1)\overline{P}_{22}+\overline{N}_{4}+\theta_{4}Y,\\
\overline{H}_{42}&=-(n-1)\overline{P}_{22}-\overline{N}_{4}+\theta_{4}AX^{T},\\
\overline{H}_{43}&=\overline{P}_{12}^{T}+\theta_{4}X^{T},\\
\overline{H}_{52}&=-\overline{N}_{2}^{T}-\overline{N}_{5}+\theta_{5}AX^{T},\\
\overline{H}_{51}&=-\overline{N}_{1}^{T}+\overline{N}_{5}+\theta_{5}Y,\\
\overline{H}_{44}&=-\frac{(n-1)\overline{R}_{11}}{\tau},\\
\overline{H}_{53}&=-\overline{N}_{3}^{T}+\theta_{5}X^{T},\\
\overline{H}_{54}&=-\overline{N}_{4}^{T},\\
\overline{H}_{55}&=-\frac{(n-1)\overline{R}_{22}}{\tau}-\overline{N}_{5}-\overline{N}_{5}^{T},
\end{aligned}\tag{24}
$$

then system (11) is asymptotically stable with control gain $K=YX^{-1}$, for example, $\mathbf{e}_i(t)$ tends to zero asymptotically which means the ith robot in the group tracks its desired trajectory well.

Proof. In order to transform the nonconvex LMI in (13) into a solvable LMI, assume that $M_i=\theta_i M_0$, $i=1,\dots,5$ where θ_i is known and given. Define $X=M_0^{-1}$, $\widehat{W}=\operatorname{diag}(X,X,X,X,X)$, and $Y=KX$. Then by premultiplying the inequality in (13) by $\widehat{W}^T$ and postmultiplying by $\widehat{W}$, we can obtain the inequality (23). □

Remark 4. Equation (23) in Theorem 3 and (13) in Theorem 2 are used to design control gain K. Equation (23) is a sufficient condition of (13) in Theorem 2 due to the simplification of M_i matrices.

5. Simulations

In this subsection, three robots in the group move with a constant delay at 1 second is simulated. In effect, the proposed approach is applicable for any group with multiple robotic systems (greater than three).

The consensus control algorithm in (8) has been applied to adjust the whole group movement performance. The control gain has been designed based on the sufficient condition in Theorem 3 as

$$K_i = YX^{-1} = \begin{bmatrix} 5.6681 & 0 \\ 0 & 5.6681 \end{bmatrix}, \quad (25)$$

where

$$X = \begin{bmatrix} -0.0654 & 0 \\ 0 & -0.0654 \end{bmatrix}, \quad Y = \begin{bmatrix} -0.3705 & 0 \\ 0 & -0.3705 \end{bmatrix}, \quad (26)$$

$i \in [1, 2, 3]$.

The evolution of the group movement has been shown in Figure 4. For robot 1, the desired trajectory is a linear function, $x_{d1} = 60t + 250$; $y_{d1} = t$. For robot 2, it is $x_{d2} = t + 4$; $y_{d2} = t$. And for robot 3, it is $x_{d3} = t$; $y_{d3} = t + 12$. The initial positions for three robots are arranged by every 2 cm at X-axis with 0 mm at Y-axis. Robots 1 and 2 start moving at $(4, 0)$ and $(6, 0)$, respectively, while robot 3 starts at $(8, 0)$ position. As shown in Figure 4 at 1 second after the group starts moving, the initial position information of each robot at $t = 0$ second has not been received by each other. Since the time increment is measured as 0.2 seconds, at the 0.4 seconds, three robots sufficiently started to move toward the desired trajectory from the original positions. In $t = 1$ second, the robots have received the signals between each other, due to the communication delay. Finally, they reached the desired formation after a certain time.

In Figure 5, the tracking error figure shows the tracking error in X- and Y-axis versus time, respectively. Because the tracking error at X-axis corresponded to the Y-axis in term of magnitude, it is accurately sufficient to focus on the tracking error at X-axis only. From this figure, at $t = 0$ second, the tracking errors which are measured at X-axis for robot 3 are highest, approximately 8 cm until the time past $t = 1$ second, which is the time the robots received the signal. Since the control gain was designed well to reduce the effect caused by the delay, the tracking errors gradually converged to zero in the next 2.5 seconds.

To further test the stabilization ability of the proposed control approach, the virtual leaders' trajectories are adopted

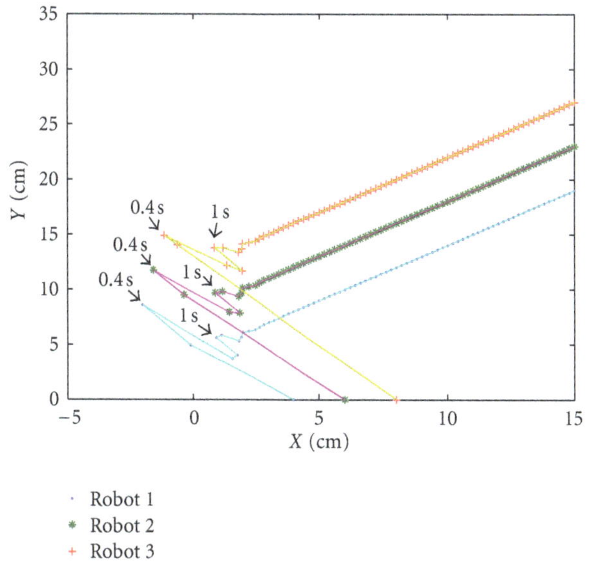

FIGURE 4: Evolution of the robot group movement with long-time communication delay ($\tau = 1$ second).

as nonlinear curve, which is different compared with those in the previous section. For robot 1, the desired trajectory is $x_{d1} = t$; $y_{d1} = \sin(0.5t)$. For robot 2, it is $x_{d2} = t$; $y_{d2} = \sin(0.5t) + 0.5$. And for robot 3, it is $x_{d2} = t$; $y_{d3} = \sin(0.5t) + 1$.

The evolutions of the group movement is as shown in Figure 6. From the figure, it can be seen that finally the group tracks the virtual leaders' trajectory and maintains the desired group formation very well. The tracking errors have been represented in Figure 7 based on both X- and Y-axis. As shown in Figure 7, the tracking errors converge to zero in 2.5 seconds which shows the effectiveness of the proposed consensus control algorithm. It can make the whole networked multirobots reach consensus and maintain the desired group formation well. It also can reduce the effects caused by communication delays.

6. Experimental Results

In this section, the proposed distributed formation control strategy is applied to multiple mobile robots for experimental tests. As shown in the following subsections, there are three main parts: (i) the hardware experimental setup; (ii) the software regarding the control operation system of P3 mobile robots; (iii) discussions on the experimental results.

6.1. Implementation of the Proposed Approach. Experimental tests are implemented in the Advanced Control and Mechatronics Laboratory at Dalhousie University. The multiple mobile robot test platform consists of one Pioneer 3-DX and one Pioneer 3-AT as shown in Figure 8. The two robots are connected with two laptops through serial ports, and the two laptops can communicate with each other through wireless network with TCP/IP protocols. Each robot has encoder for their position and orientation information. All calculations,

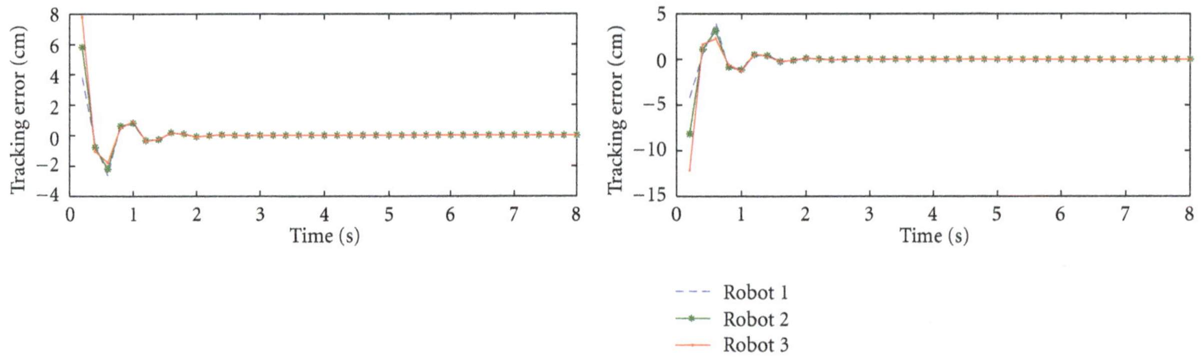

(a) Error in X-axis

(b) Error in Y-axis

FIGURE 5: The tracking error of the robot group movement with time delay ($\tau = 1$ second): (a) X-axis; (b) Y-axis.

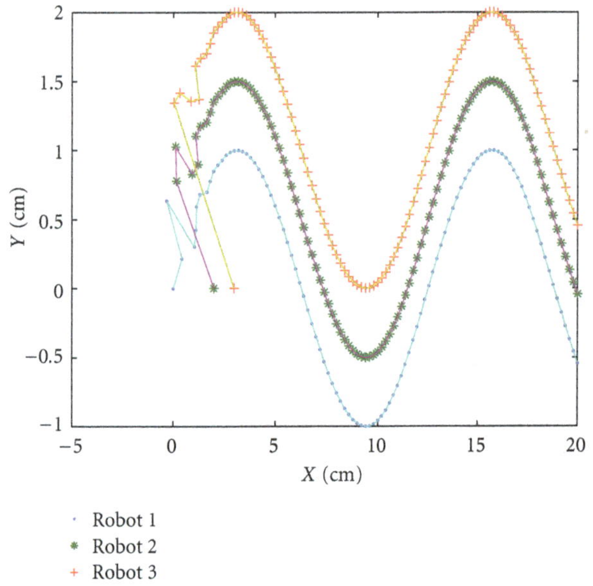

FIGURE 6: Evolution of the group movement with nonlinear formation with time delay ($\tau = 1$ second).

receiving data information from the robots and inputting control signals to the robots, are processed by the laptops.

In the experimental tests, a group of two P3 mobile robots is required to maintain the desired group formation according to its desired trajectory. The kinematic model for each robot is as shown in (1), and this model can be linearized by setting a hand position for each robot, for example, the simplified model represented by (5). After the group starts moving, laptops attached on the robots will keep performing the following tasks once per second: (1) reading position and orientation data from robots; (2) exchanging those data with each other through the wireless network with 1 second communication delay; (3) calculating the control input values according to the distributed consensus formation control algorithm (8) and then calculating the desired linear and orientation speed for the robots according to (4); (4) converting the desired linear and orientation speed of the robots into the desired rotating speed of the left and right wheel of the robots, respectively, based on the following equation:

$$\begin{aligned} \widetilde{R} &= \frac{v}{\omega}, \\ V_L &= \omega\left(\widetilde{R} - 0.5l\right), \\ V_R &= \omega\left(\widetilde{R} + 0.5l\right), \end{aligned} \tag{27}$$

where V_L and V_R are the left and right wheels rotating speed of the robot, $l = 33$ cm is the length between left and right wheels of Pioneer 3-DX, and $l = 38$ cm is that of Pioneer 3-AT, (5) sending signals to robots to set the rotating speeds as the desired values.

6.2. Introduction on Software. In the experimental tests, software programming is another important issue. All commands and calculations processed by the controller (laptop) attached on each robot are programmed in C++ language. For simplifying the programming task, advanced robot interface for applications (ARIAs) has been used in the work.

ARIA is a programming library for C++ programmers who want to access their mobile robot platform and accessories at either a high or low level. Written in the C++ language, ARIA is client-side software for easy, high-performance access to and management of the robot, as well as to the many accessory robot sensors and effectors. ARIA includes many useful utilities for general robot programming and cross-platform (Linux and Windows) programming as well.

ARIA can dynamically control the robot's velocity, heading, relative heading, and other motion parameters either through simple low-level commands or through its high-level "actions" infrastructure. In this work, ARIA also receives odometric position estimates, sonar readings, and all other current operating data sent by the robot platform. Functions defined in ARIA have been invoked to help read position data and set robot's velocity and orientations, which as a result saves lots of time in programming.

Since the two laptops are required to communicate with each other via the wireless network with TCP/IP topology, a library named ArSocket in ARIA is then used for this issue.

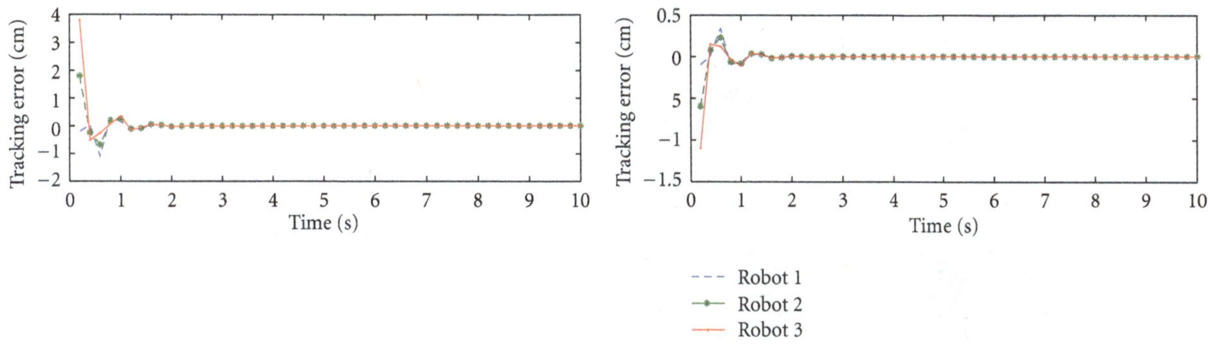

(a) Error in X-axis

(b) Error in Y-axis

Figure 7: Tracking errors of the group with time delay ($\tau = 1$ second): (a) X-axis; (b) Y-axis.

Figure 8: Pioneer 3 mobile robot test platform framework.

With the help of ArSocket, the user can directly create a server with one laptop and create a client with another. At the beginning of the group movement, the server opens a server port for the client and waits for the client to connect. The client is then connected to the server by following the server's IP address. When the server receives the client's call, it will send a call back to the client and start action. Then the client will start action as soon as it receives the call from the server. The server and client can interact synchronously through this way.

6.3. Experimental Results. The two robots in the team start moving at initial position of $(0, 0)$ for robot 1 (P3-AT) and $(1050, 1050)$ for robot 2 (P3-DX) with a unit in mm. As described above, the hand position for each robot has a distance L to the center point of the robot. In the experiment, $L = 30$ cm for both robots, so the initial hand positions of robot 1 and robot 2 are $(300, 300)$ and $(1350, 1350)$, respectively, with a unit in mm. For the first test, the "hands" of the robot group are firstly required to track a straight line as

$$x_d(t) = 60t + 500, \qquad y_d(t) = 60t + 500, \tag{28}$$

where t is the time starting from 0 second, the unit of x_d and y_d is mm, and (28) is according to the coordinate system C_i of the ith robot itself. The group movement trajectory is as shown in Figure 9.

In order to evaluate the group performance, the group movement trajectory was recorded and analyzed using MATLAB. The group trajectory is drawn by Figure 10. Since the position data has been recorded once per second, the time period between each location point in the figure is 1 second. As shown in the figure, P3-AT and P3-DX start moving from initial positions together at 0 second. The group starts getting close to the virtual leader's trajectory or the desired trajectory in the first 8 seconds of the movement. After 8 seconds, the group gets on the desired trajectory. During this process, the desired group formation has been kept very well. Figure 11 shows that the consensus tracking errors for the virtual center position converge more and more close to 0, and the lowest values of the errors are below 2 cm. This convergence property of the tracking error guarantees the good maintenance of the group formation.

For the second test, the group is required to track a curve as

$$x_d(t) = 20t + 500, \qquad y_d(t) = 2t^2 + 500, \tag{29}$$

where t is the time starting from 0 second, the unit of x_d and y_d is mm, and (29) is according to the coordinate system C_i of the ith robot itself. The team movement with fixed group formation is as shown in Figure 12.

For this case, the data has also been recorded and analyzed. Figure 13 shows the group movement trajectory and virtual leader trajectory. As shown in the Figure, the group trajectory of each robot undulates a little bit in the first 8 seconds transition time compared with the previous test as in Figure 10. After 8 seconds, the group gets on desired trajectory and undulates a little bit more than the previous test (the straight line case). Through drawing the consensus tracking errors for virtual center positions in Figure 14, it shows that the tracking error of each robot converges in the first 20 seconds and the minimum value is below 2 cm. And then after 20 seconds, it increases a little bit between 2 cm and 3 cm. That is because the motor speed of each robot, which is calculated and set by the controller once per second, is held for each 1 second during the movement. So the system is not "ideal time continuous." For the curve tracking case,

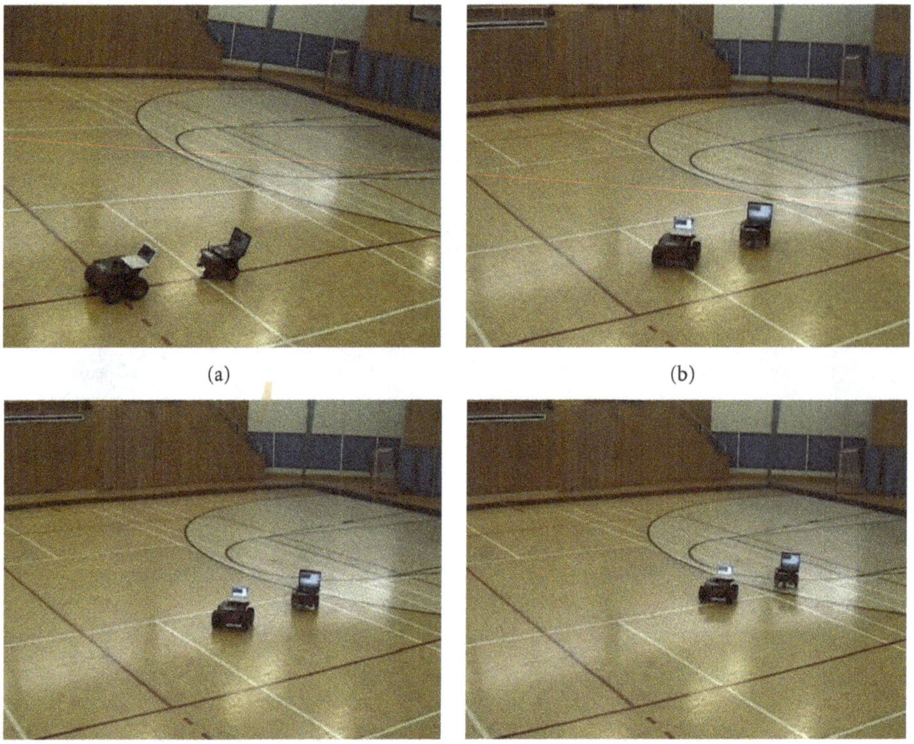

Figure 9: Experimental result of group movement tracking straight line: (a) 0 second; (b) 20 seconds; (c) 40 seconds; (d) 60 seconds.

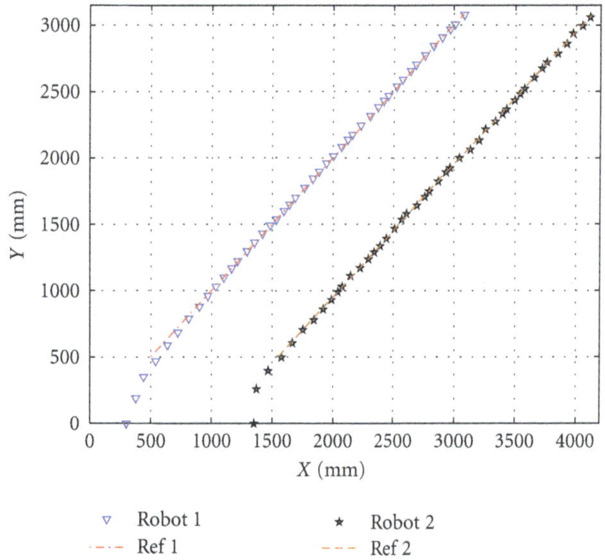

Figure 10: Group movement tracking straight line.

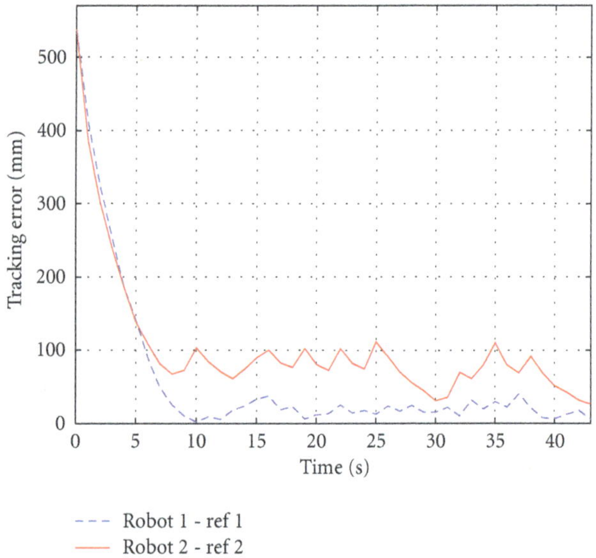

Figure 11: Consensus tracking errors for the virtual position: straight line case.

the virtual leader of each robot moves with an acceleration along Y-axis which is equal to 4 mm/s^2. As the time goes after movement, the virtual leader's velocity increases greatly and becomes bigger and bigger so that it becomes more and more difficult for each robot to react and match the velocity and motion of its virtual leader. This hardware limitation causes the performance of the consensus tracking errors as shown in Figure 14. If the group moves for over 200 seconds, then the speed of the virtual leader along Y-axis becomes more than 0.8 meters per second which is the maximum speed of P3-AT. This is the main difference compared with the previous test (straight line case).

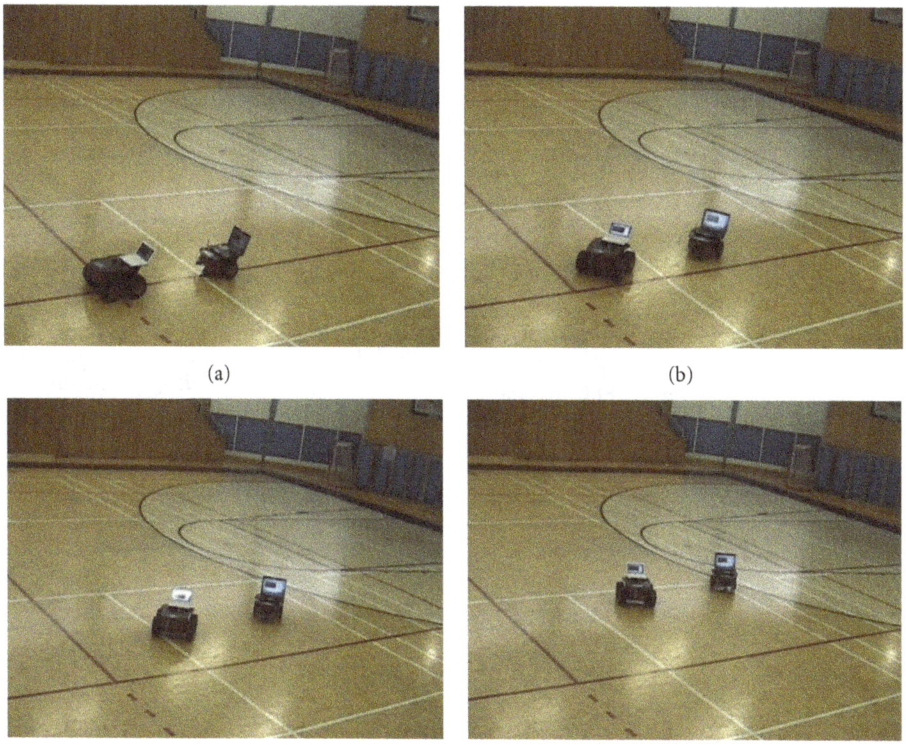

(a) (b)

(c) (d)

FIGURE 12: Experimental result of group movement tracking curve: (a) 0 second; (b) 7 seconds; (c) 20 seconds; (d) 35 seconds.

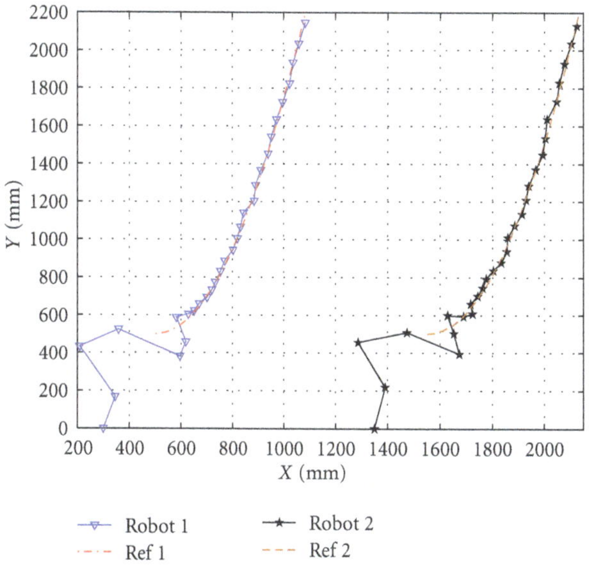

FIGURE 13: Group movement tracking curve.

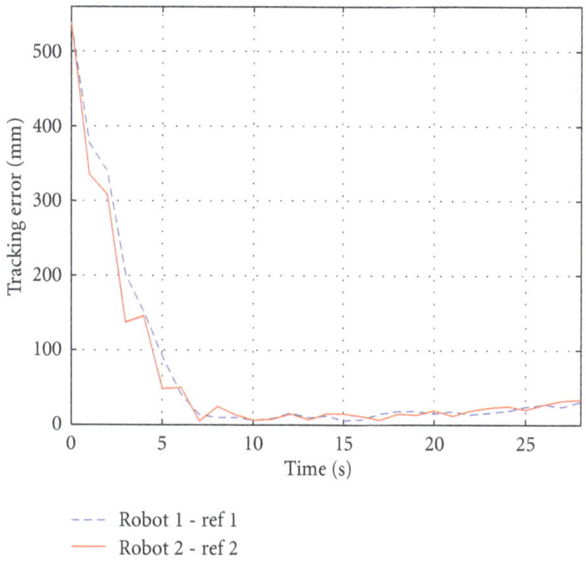

FIGURE 14: Consensus tracking errors for the virtual position: curvilinear case.

From the results, the consensus-based formation control strategies described in Section 4 are applied to the experimental test for P3 robot group. Due to the proper design of the control gain, the mobile robot group can track the virtual leader's trajectory and maintain the desired group formation well.

7. Conclusions

In this work, a consensus-based tracking control strategy was proposed for the virtual leader formation control approach of multiple mobile-robot group with group communication delays. The kinematic model of P3 mobile robot has been introduced. The formation control algorithm has been applied to the linearized dynamic model of P3 mobile robot. A novel multiple Lyapunov functional candidate has been proposed to give the sufficient conditions of the control gain design as shown in theorems. The proposed consensus formation control strategy can be successfully applied to the multiple P3 mobile-robot experimental platform in the host laboratory. The results show a good group performance due to the good feasibility of the proposed approach.

Acknowledgments

The authors would like to thank NSERC and CFI Canada for the funding support.

References

[1] T. R. Smith, H. Hmann, and N. E. Leonard, "Orientation control of multiple underwater vehicles," in *Proceedings of 40th Conference on Decision and Control*, vol. 5, pp. 4598–4603, Orlando, Fla, USA, 2001.

[2] L. E. Buzogany, M. Pachter, and J. J. Azzo, "Automated control of aircraft in formation flight," in *Proceedings of American Institute of Aeronautics and Astronautics Conference Guidance, Navigation and Control (AIAA '01)*, pp. 4598–4603, Monterey, Calif, USA, 2001.

[3] C. A. Kitts and I. Mas, "Cluster space specification and control of mobile multirobot systems," *IEEE/ASME Transactions on Mechatronics*, vol. 14, no. 2, pp. 207–218, 2009.

[4] W. B. Dunbar and R. M. Murry, "Model predictive control of coordinated multi-vehicle formations," in *Proceedings of the Conference on Decision and Control*, pp. 4631–4636, Las Vegas, Nev, USA, 2002.

[5] W. B. Dunbar and R. M. Murray, "Distributed receding horizon control for multi-vehicle formation stabilization," *Automatica*, vol. 42, no. 4, pp. 549–558, 2006.

[6] H. Yamaguchi, "Cooperative hunting behavior by mobile-robot troops," *International Journal of Robotics Research*, vol. 18, no. 9, pp. 931–940, 1999.

[7] B. Mohar, "The laplacian spectrum of graphs," in *Graph Theory, Combinatorics and Applications*, vol. 2, pp. 871–898, 1991.

[8] R. Merris, "Laplacian matrices of graphs: a survey," *Linear Algebra and Its Applications*, vol. 197, pp. 143–176, 1994.

[9] A. Franchi, L. Freda, G. Oriolo, and M. Vendittelli, "The sensor-based random graph method for cooperative robot exploration," *IEEE/ASME Transactions on Mechatronics*, vol. 14, no. 2, pp. 163–175, 2009.

[10] S. Khoo, L. Xie, and Z. Man, "Robust finite-time consensus tracking algorithm for multirobot systems," *IEEE/ASME Transactions on Mechatronics*, vol. 14, no. 2, pp. 219–228, 2009.

[11] C. A. C. Parker and H. Zhang, "Cooperative decision-making in decentralized multiple-robot systems: the best-of-N problem," *IEEE/ASME Transactions on Mechatronics*, vol. 14, no. 2, pp. 240–251, 2009.

[12] A. Fax and R. M. Murray, "Information flow and cooperative control of vehicle formations," *IEEE Transactions on Automatic Control*, vol. 49, no. 9, pp. 1465–1476, 2004.

[13] R. Olfati-Saber and R. M. Murray, "Consensus problems in networks of agents with switching topology and time-delays," *IEEE Transactions on Automatic Control*, vol. 49, no. 9, pp. 1520–1533, 2004.

[14] P. Wang, "Navigation strategies for multiple autonomous mobile robots moving in formation," *Journal of Robotic Systems*, vol. 8, no. 2, pp. 177–195, 1991.

[15] A. Viguria and A. M. Howard, "An integrated approach for achieving multirobot task formations," *IEEE/ASME Transactions on Mechatronics*, vol. 14, no. 2, pp. 176–186, 2009.

[16] M. Egerstedt, X. Hu, and A. Stotsky, "Control of mobile platforms using a virtual vehicle approach," *IEEE Transactions on Automatic Control*, vol. 46, no. 11, pp. 1777–1782, 2001.

[17] S. Xu and J. Lam, "A survey of linear matrix inequality techniques in stability analysis of delay systems," *International Journal of Systems Science*, vol. 39, no. 12, pp. 1095–1113, 2008.

[18] P. Ogren, M. Egerstedt, and X. Hu, "A control Lyapunov function approach to multiagent coordination," *IEEE Transactions on Robotics and Automation*, vol. 18, no. 5, pp. 847–851, 2002.

[19] H. K. Khalil, *Nonlinear Systems*, Prentice Hall, Englewood Cliffs, NJ, USA, 1996.

[20] S. Boyd, L. El Ghaoui, E. Feron, and V. Balakrishnan, *Linear Matrix Inequalities in System and Control Theory*, vol. 15 of *Studies in Applied Mathematics*, 1994.

[21] Y. Cao and W. Ren, "Distributed coordinated tracking with reduced interaction via a variable structure approach," *IEEE Transactions on Automatic Control*, vol. 57, no. 1, pp. 33–48, 2012.

3

Experimental Application of Predictive Controllers

C. H. F. Silva,[1] H. M. Henrique,[2] and L. C. Oliveira-Lopes[2]

[1] *Cemig Geração e Transmissão SA, Avenida Barbacena 1200, 16° Andar Ala B1 (TE/AE), 30190-131 Belo Horizonte, MG, Brazil*
[2] *Faculdade de Engenharia Química, Universidade Federal de Uberlândia, Avenida João Naves de Ávila, 2121, Bloco 1K do Campus Santa Mônica 38408-100 Uberlândia, MG, Brazil*

Correspondence should be addressed to L. C. Oliveira-Lopes, lcol@ufu.br

Academic Editor: Baocang Ding

Model predictive control (MPC) has been used successfully in industry. The basic characteristic of these algorithms is the formulation of an optimization problem in order to compute the sequence of control moves that minimize a performance function on the time horizon with the best information available at each instant, taking into account operation and plant model constraints. The classical algorithms Infinite Horizon Model Predictive Control (IHMPC) and Model Predictive Control with Reference System (RSMPC) were used for the experimental application in the multivariable control of the pilot plant (level and pH). The simulations and experimental results indicate the applicability and limitation of the control technique.

1. Introduction

The process control is related to the application of the automatic control principles to industrial processes. The globalization effect upon industries brought a perception of the importance associated to the product quality over organization profits. Because of that, the process control has been more and more demanded and explored, in order not only to assure that the process is according to acceptable performance levels but also to address legal requirements in terms of safety and quality of the products [1]. In this context, predictive controllers are able to deal with system requirements in a proper way and simple to be implemented. One of the biggest concerns on control theory is related to the stability of closed loop of systems. It is natural that only stable closed-loop response is considered for real implementation. A representative algorithm of this controller class is Infinite Horizon Model Predictive Control (IHMPC) [2].

During the closed-loop project, there are some theoretical tools that can be incorporated into the controllers and aggregate desirable characteristics. This is the case of predictive controllers that incorporate a reference path in their formulation (RSMPC) [3]. In this case, the value of control action is computed from a QDMC (Quadratic Dynamic Matrix Control) problem that has a first-order reference system incorporated directly into the formulation. Consequently, there is an effort for eliminating undesirable situations of too high speed and for balancing the system response.

Most industrial chemical processes have a nonlinear characteristic, although linear controllers are used in many of these systems. The great advantage of this approximation is having an analytical solution to the control problem and also the low computational complexity. This kind of approach is very common in the industry practice.

The pH system is used as a benchmark for applications in process control, mainly because of its strong nonlinear behavior. For the experimental application case addressed in this work, it is a control of multiple inputs and multiple outputs (MIMO), in which inputs are acid and base flows and outputs are reactor level and stream pH output. This system presents not only a nonlinear feature, but also an interaction between inputs and outputs linked to the system directionality. In this kind of system, classical controllers do not work to keep the stability and required performance. In addition, MPC solves the problem of open loop optimal control in real time, considering the current state of the system to be controlled as much as a policy feedback offline. As can be seen in the experimental data, the multivariable system could not be controlled with optimal classical controllers.

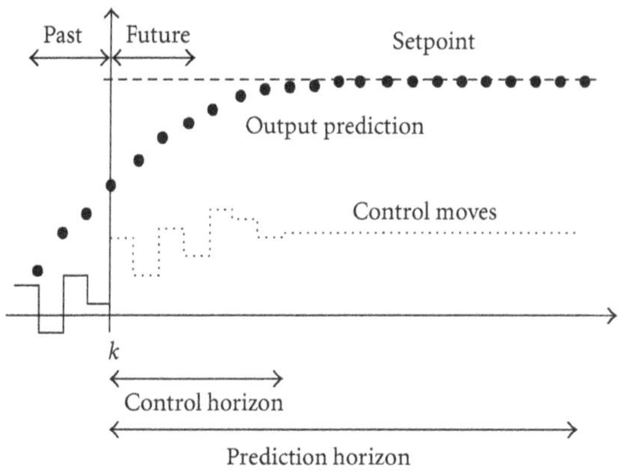

FIGURE 1: MPC and the receding horizon strategy.

That is why it was chosen to implement IHMPC and RSMPC predictive controllers in real time [4].

Section 2 presents the control algorithms used in the implementation described in this work. In Section 3 there is a description of the experimental system. Section 4 presents the simulations with the controllers and experimental results of the closed loop. Section 5 shows the main conclusions of this paper.

2. Model Predictive Control

The model predictive control (MPC) theory was originated in the end of 1970s decade and developed considerably ever since [5–10]. MPC became an important control strategy of industrial applications, mainly due to great success of its implementations in the petrochemical industry. The reason for MPC being so popular might be due to its ability to deal with quite difficult control challenges, although other characteristics such as its ability to control constrained multivariable plants, have been claimed as one of the most important features [6]. The ability to handle input, output, and internal state constraints can be assigned as the most significant contribution to the many successful industrial applications [11–13]. Other benefits to consider are the possibility of embedding safety limits, the possibility of obtaining advantages in using it in highly nonlinear plants as well as in time-varying plants, the consideration of process noise in the formulation, the high level of control performance, reduced maintenance requirements, and improved flexibility and agility [14].

The various predictive control algorithms differ from each other depending on the way the predictive model is used to represent the process, the noise description, and on the cost function to be minimized. There are many control predictive applications well succeeded not only in chemical industry but also in other areas [15]. The use of state-space models, in spite of other formulations, was responsible for a substantial maturing of the predictive control theory during the 1990s decade [16]. The state-space formulation not only allows the application of linear system theorems already known, but also it makes easier to generalize for more complex cases. In this situation, the MPC controller can be understood as a compensator based on a state observer and its stability, performance, and robustness are determined by the observer poles, established directly by parameter adjustment, and by regulator poles, determined by performance horizons and weights [17]. Figure 1 shows the receding horizon strategy, which is the center of the MPC theory.

Figure 2, based on Richalet [18] and Richards and How [19], shows interfacial areas of the MPC control developments.

Considering its application to MIMO processes, MPC deals directly with coordination and balancing interactions between inputs and outputs and also with associated constraints [20]. In spite of being really effective in suitable situations, MPC still has limitations as, for instance, operation difficulties, high maintenance cost (as it demands specialized work) and flexibility loss, which can result in a controller weakness. These limitations are not connected only to the algorithm itself, but also to the necessity of a plant model, for which is necessary maintenance. In practical terms, the limits of MPC applicability and performance should not be connected to the algorithm deficiencies, but to the issues linked to the modeling difficulties, sensor adequacy, and insufficient robustness in the presence of faults [15, 21].

2.1. IHMPC. The IHMPC formulation was introduced by Muske and Rawlings [2]. For simplicity the algorithm will be presented only for stable systems. For unstable systems see the reference section [2]. The discrete dynamical system used is shown in (1) in which $\mathbf{y}$ is the vector output, $\mathbf{u}$ is the vector input, and $\mathbf{x}$ is the states vector:

$$\begin{aligned} \mathbf{x}_{k+1} &= \mathbf{A}\mathbf{x}_{k+1} + \mathbf{B}\mathbf{u}_k, \\ \mathbf{y}_k &= \mathbf{C}\mathbf{x}_k. \end{aligned} \tag{1}$$

The receding horizon regulator is based on the minimization of the infinite horizon open-loop quadratic objective function at time k (2). $\mathbf{Q}$ is a symmetric positive semidefinite penalty matrix on the outputs. $\mathbf{R}$ is a symmetric positive semidefinite penalty matrix on the inputs. $\mathbf{S}$ is a symmetric

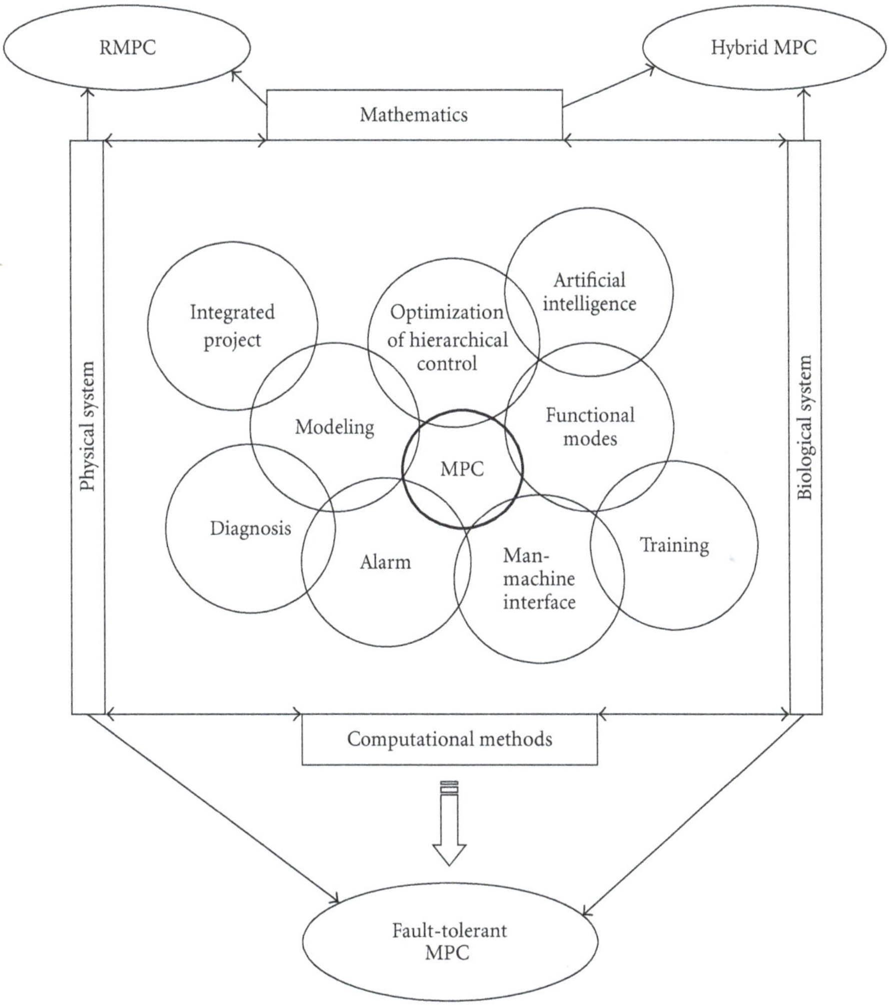

FIGURE 2: MPC and related areas.

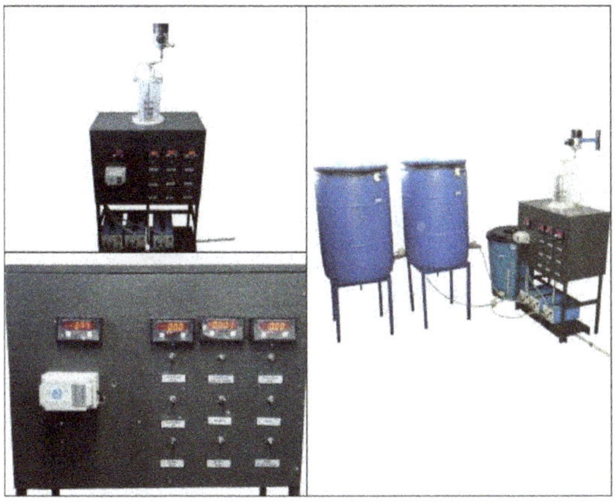

FIGURE 3: Experimental system.

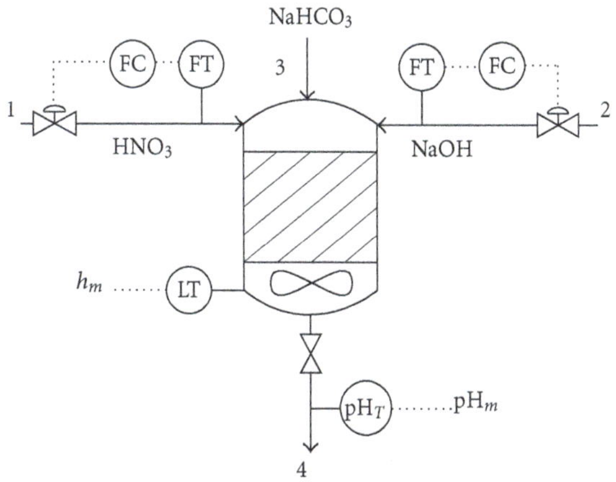

FIGURE 4: Diagram representation of the experimental system.

positive semidefinite penalty matrix on the rate of the input change with $\Delta\mathbf{u} = \mathbf{u}_{k+i} - \mathbf{u}_k$. The vector $\mathbf{u}^N$ contains the N future open-loop control moves (3). The infinite horizon open-loop objective function can be expressed as a finite horizon open-loop objective. For stable system, $\overline{\mathbf{Q}}$ is defined as the infinite sum (4). This infinite sum can be determined from the solution of the discrete Lyapunov equation (5). Using simple algebraic manipulation, the quadratic objective

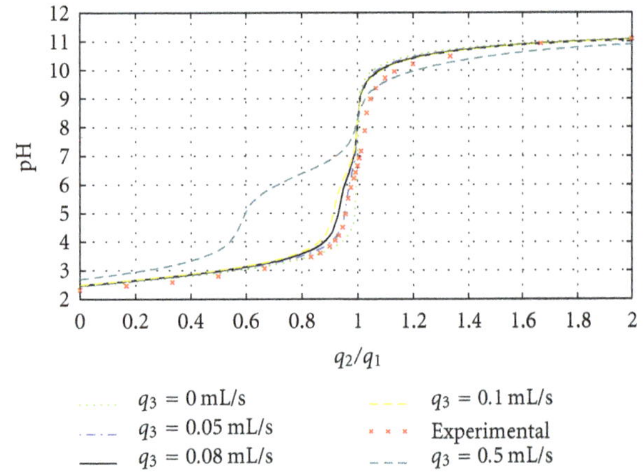

Figure 5: Titration Curves of the neutralization system.

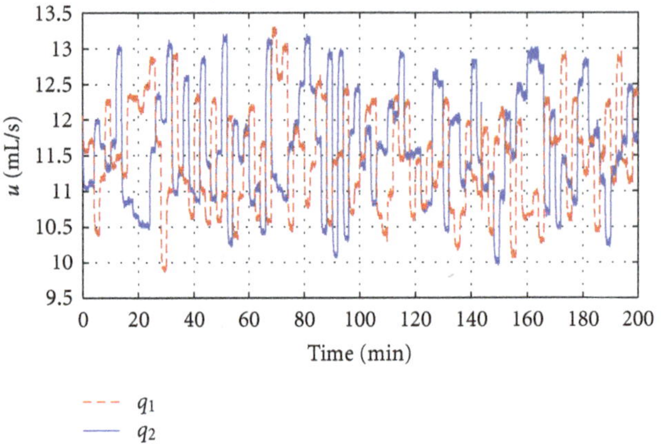

Figure 7: Random inputs for parameter estimation.

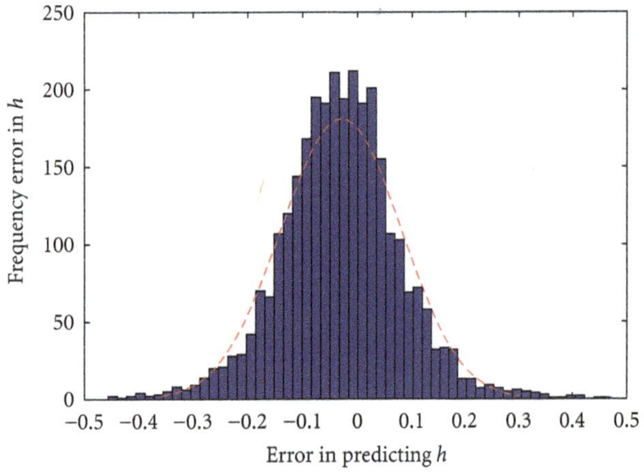

Figure 6: Error in predicting the level of the reactor.

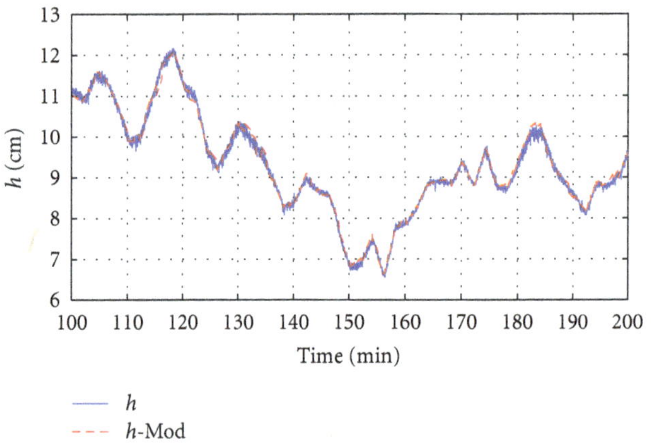

Figure 8: Open-loop run behavior (Mod indicates that it is a result of model simulation).

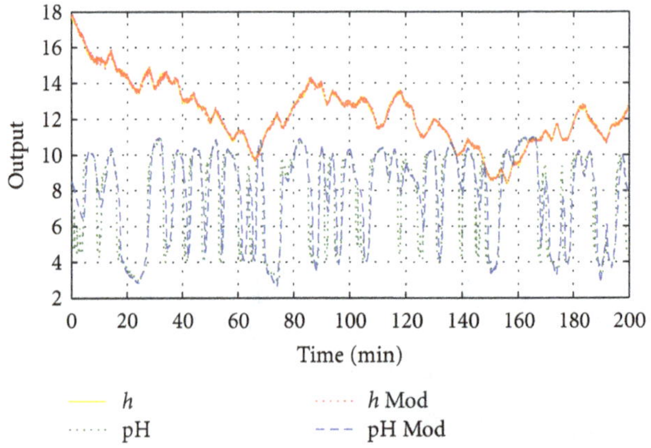

Figure 9: Real data x simulated data (Mod).

is show in (6). The matrices **H**, **G**, and **F** are showed in (7)–(9):

$$J_k = \min_{\mathbf{u}^N} \sum_{i=0}^{\infty} \mathbf{y}_{k+i}^T \mathbf{Q} \mathbf{y}_{k+i} + \mathbf{u}_{k+i}^T \mathbf{R} \mathbf{u}_{k+i} + \Delta \mathbf{u}_{k+i}^T \mathbf{S} \Delta \mathbf{u}_{k+i}, \quad (2)$$

$$\mathbf{u}^N = \begin{bmatrix} \mathbf{u}_k \\ \mathbf{u}_{k+1} \\ \vdots \\ \mathbf{u}_{k+N-1} \end{bmatrix}, \quad (3)$$

$$\overline{\mathbf{Q}} = \sum_{i=0}^{\infty} \mathbf{A}^{T^i} \mathbf{C}^T \mathbf{Q} \mathbf{C} \mathbf{A}^i, \quad (4)$$

$$\overline{\mathbf{Q}} = \mathbf{C}^T \mathbf{Q} \mathbf{C} + \mathbf{A}^T \overline{\mathbf{Q}} \mathbf{A}, \quad (5)$$

$$\min_{\mathbf{u}^N} \Phi_k = \min_{\mathbf{u}^N} \mathbf{u}^{N^T} \mathbf{H} \mathbf{u}^N + 2\mathbf{u}^{N^T} (\mathbf{G}\mathbf{x}_k - \mathbf{F}\mathbf{u}_{k-1}), \quad (6)$$

$$\mathbf{H} = \begin{bmatrix} \mathbf{B}^T \overline{\mathbf{Q}} \mathbf{B} + \mathbf{R} + 2\mathbf{S} & \mathbf{B}^T \mathbf{A}^T \overline{\mathbf{Q}} \mathbf{B} - \mathbf{S} & \cdots & \mathbf{B}^T \mathbf{A}^{T^{N-1}} \overline{\mathbf{Q}} \mathbf{B} \\ \mathbf{B}^T \overline{\mathbf{Q}} \mathbf{A} \mathbf{B} - \mathbf{S} & \mathbf{B}^T \overline{\mathbf{Q}} \mathbf{B} + \mathbf{R} + 2\mathbf{S} & \cdots & \mathbf{B}^T \mathbf{A}^{T^{N-2}} \overline{\mathbf{Q}} \mathbf{B} \\ \vdots & \vdots & \ddots & \vdots \\ \mathbf{B}^T \overline{\mathbf{Q}} \mathbf{A}^{N-1} \mathbf{B} & \mathbf{B}^T \overline{\mathbf{Q}} \mathbf{A}^{N-2} \mathbf{B} & \cdots & \mathbf{B}^T \overline{\mathbf{Q}} \mathbf{B} + \mathbf{R} + 2\mathbf{S} \end{bmatrix}, \quad (7)$$

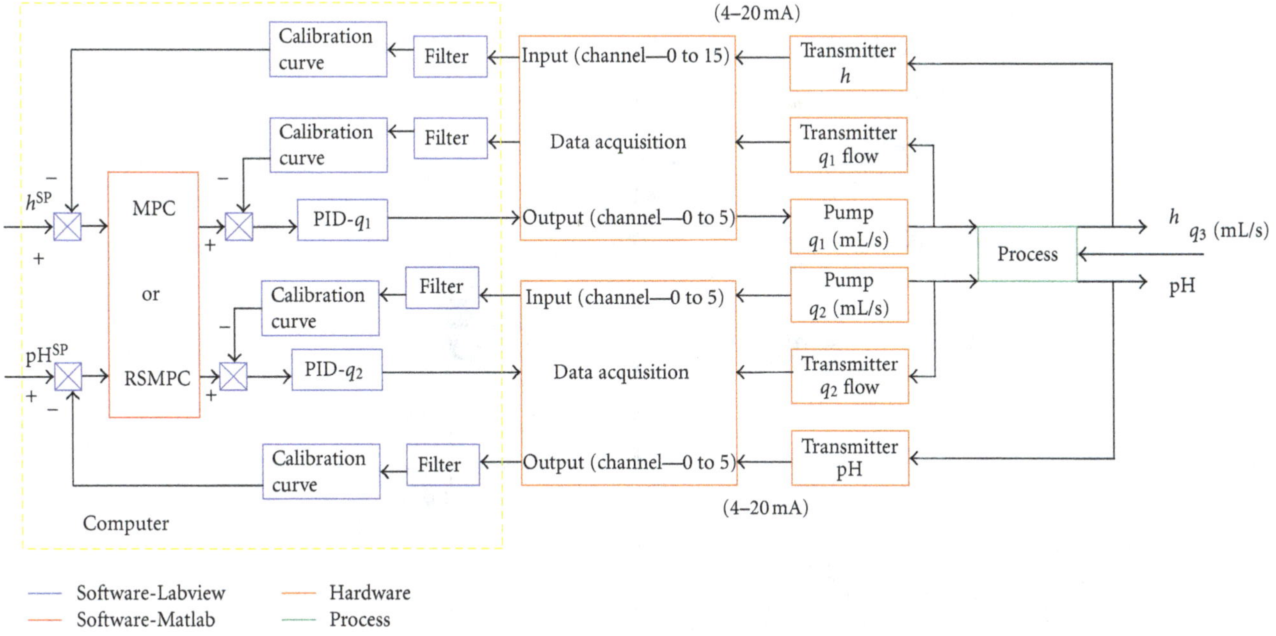

FIGURE 10: Structure of operation—experimental plant.

$$\mathbf{G} = \begin{bmatrix} \mathbf{B}^T\overline{\mathbf{Q}}\mathbf{A} \\ \mathbf{B}^T\overline{\mathbf{Q}}\mathbf{A}^2 \\ \vdots \\ \mathbf{B}^T\overline{\mathbf{Q}}\mathbf{A}^N \end{bmatrix}, \quad (8)$$

$$\mathbf{F} = \begin{bmatrix} \mathbf{S} \\ \mathbf{0} \\ \vdots \\ \mathbf{0} \end{bmatrix}. \quad (9)$$

2.2. RSMPC. The RSMPC strategy is basically a transformation of an optimization problem with constraints into a quadratic programming problem [3]. The RSMPC development is presented next. Consider a general system described by (10). By linearizing (10) with the Taylor's series expansion around the point immediately before the current operational point, one obtains (11). The matrices are defined in (12)–(16). Equation (11) is integrated from t to $t + \Delta t$ and by assuming $\mathbf{u}_k$ constant between samples, one can write (17):

$$\frac{d\mathbf{x}}{dt} = \mathbf{f}(\mathbf{x}, \mathbf{u}), \qquad \mathbf{y} = \mathbf{h}(\mathbf{x}), \quad (10)$$

$$\frac{d\mathbf{x}}{dt} = \mathbf{A}_{k-1}\mathbf{x} + \mathbf{B}_{k-1}\mathbf{u} + \mathbf{f}_{k-1}, \qquad \mathbf{y} = \mathbf{C}_{k-1}\mathbf{x} + \mathbf{h}_{k-1}, \quad (11)$$

$$\mathbf{A}_{k-1} = \left(\frac{\partial \mathbf{f}}{\partial \mathbf{x}}\right)\bigg|_{\mathbf{x}=\mathbf{x}_{k-1},\, \mathbf{u}=\mathbf{u}_{k-1}}, \quad (12)$$

$$\mathbf{B}_{k-1} = \left(\frac{\partial \mathbf{f}}{\partial \mathbf{u}}\right)\bigg|_{\mathbf{x}=\mathbf{x}_{k-1},\, \mathbf{u}=\mathbf{u}_{k-1}}, \quad (13)$$

$$\mathbf{C}_{k-1} = \left(\frac{\partial \mathbf{h}}{\partial \mathbf{x}}\right)\bigg|_{\mathbf{x}=\mathbf{x}_{k-1}}, \quad (14)$$

$$\mathbf{f}_{k-1} = \mathbf{f}(\mathbf{x}, \mathbf{u})\big|_{\mathbf{x}=\mathbf{x}_{k-1},\, \mathbf{u}=\mathbf{u}_{k-1}}, \quad (15)$$

$$\mathbf{h}_{k-1} = \mathbf{h}\big|_{\mathbf{x}=\mathbf{x}_{k-1}}, \quad (16)$$

$$\begin{aligned} \hat{\mathbf{x}}_{k+1} &= \mathbf{\Phi}\mathbf{x}_k + \mathbf{\Psi}\mathbf{B}\mathbf{u}_k + \mathbf{\Psi}\mathbf{f}_{k-1}, \\ \hat{\mathbf{y}}_k &= \mathbf{C}\mathbf{x}_k + \mathbf{h}_{k-1}, \end{aligned} \quad (17)$$

where

$$\mathbf{\Phi} = e^{\mathbf{A}\Delta t}, \quad \mathbf{\Psi} = \mathbf{A}^{-1}\left(e^{\mathbf{A}\Delta t} - \mathbf{I}\right)\mathbf{B}, \quad \mathbf{\Omega} = \mathbf{A}^{-1}\left(e^{\mathbf{A}\Delta t} - \mathbf{I}\right). \quad (18)$$

Since A is not singular, then (17) can be written for each prediction instant from $k = 1$ to $k = P$, where P is the prediction horizon and N is the control horizon, with $P \geq N$ and $\Delta\mathbf{u}_{k+j} = \mathbf{0}$ to $N \leq j \leq P$. The resulting set of equation is showed below, where

$$\hat{\mathbf{y}} = \mathbf{\Gamma}\Delta\mathbf{u} + \boldsymbol{\gamma}, \quad (19)$$

$$\hat{\mathbf{y}} = \left[\hat{\mathbf{y}}_{k+1}\ \hat{\mathbf{y}}_{k+2}\ \cdots\ \hat{\mathbf{y}}_{k+N}\ \cdots\ \hat{\mathbf{y}}_{k+P}\right], \quad (20)$$

$$\Delta\mathbf{u} = \left[\Delta\mathbf{u}_{k+1}\ \Delta\mathbf{u}_{k+2}\ \cdots\ \Delta\mathbf{u}_{k+N}\right]. \quad (21)$$

The controller is designed in order to transform the closed-loop behavior in a fist-order system behavior according to (24). To overcome the problem of infeasible solutions of the MPC optimization problem, a slack variable ($\boldsymbol{\lambda}$) is introduced in order to allow the system to deviate from reference system and to satisfy hard constraints [3] resulting in (25). This equation can be written for each prediction instant (26):

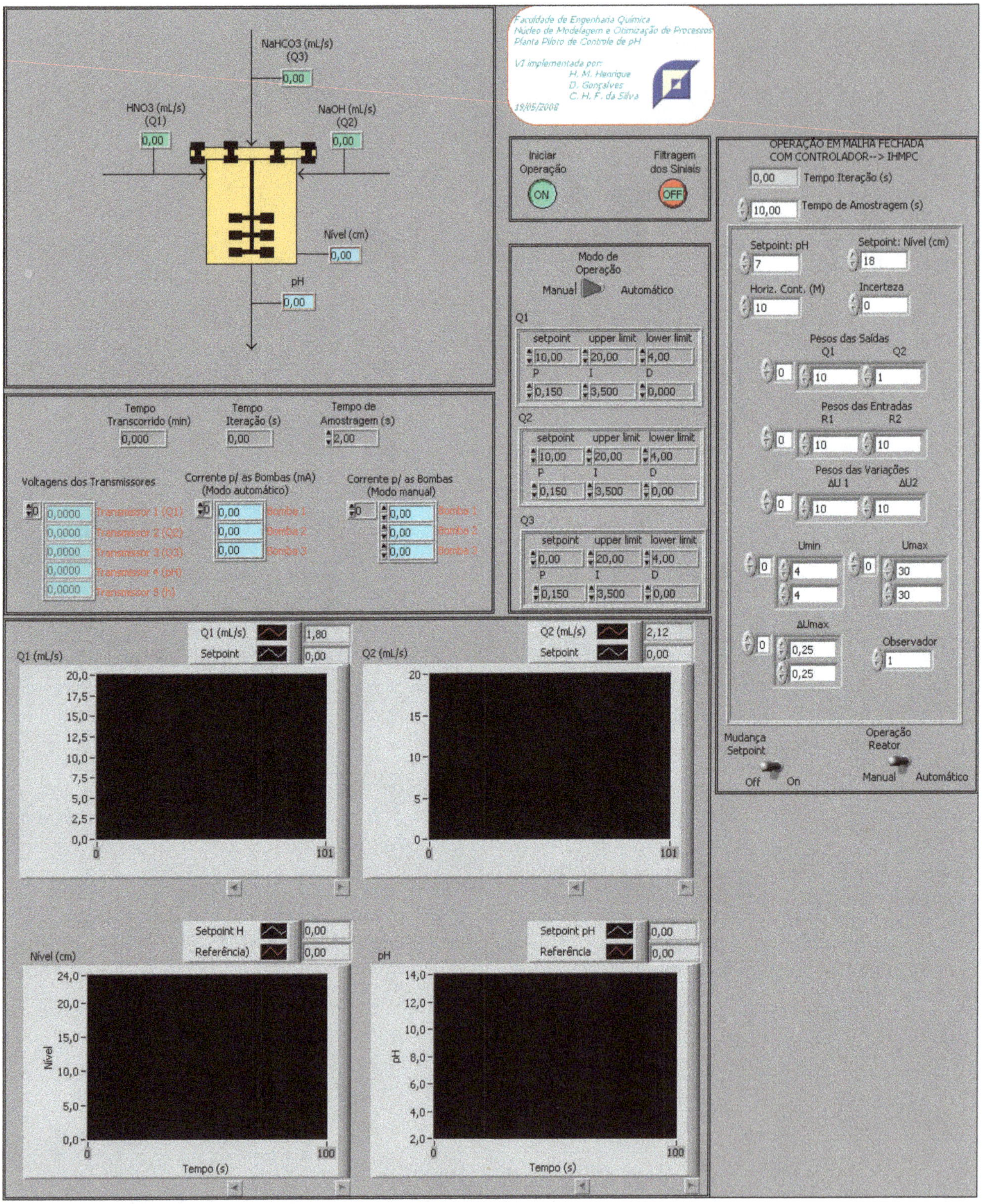

Figure 11: Frontal panel—IHMPC.

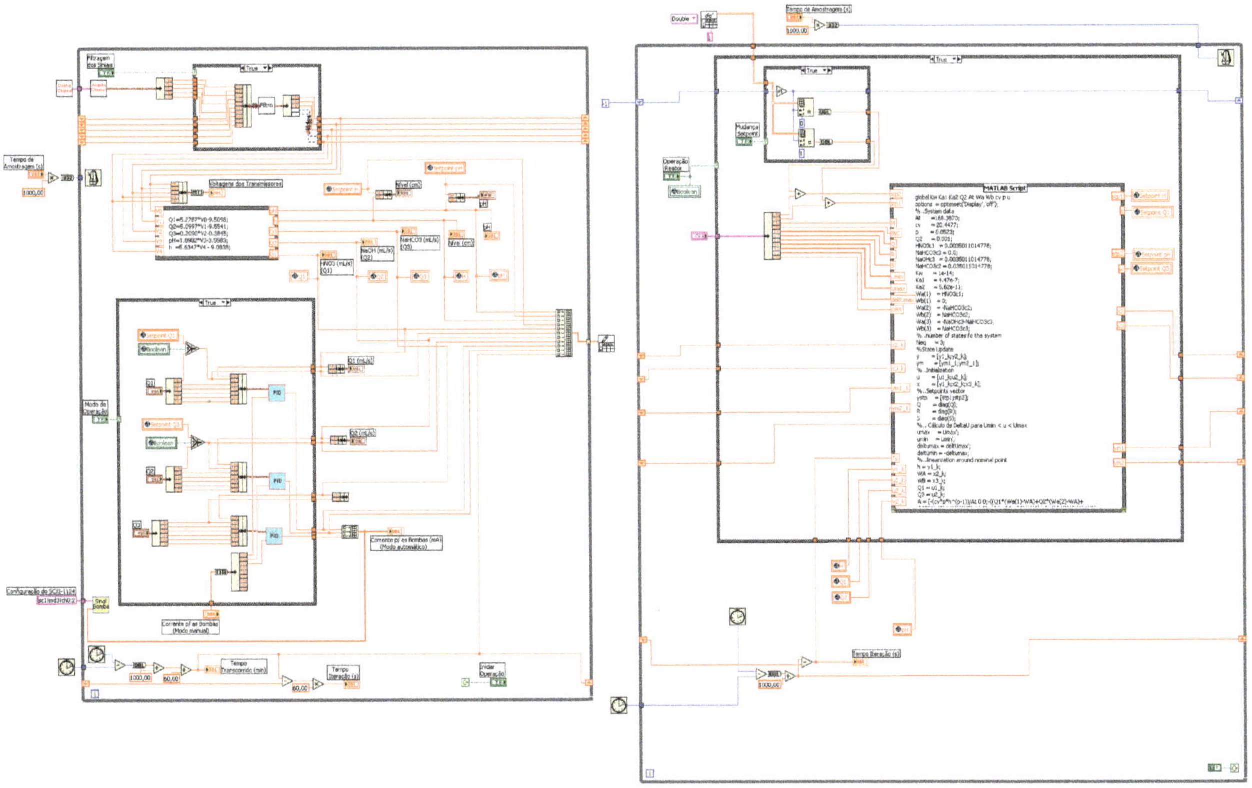

FIGURE 12: Block diagram—IHMPC.

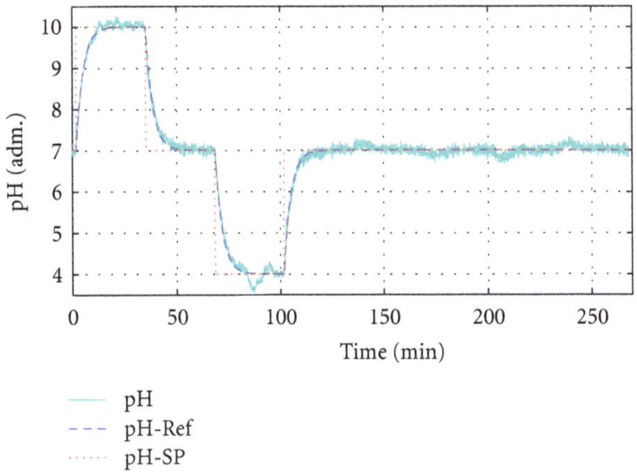

FIGURE 13: Closed-loop simulation: RSMPC—pH (setpoint: SP and reference system: Ref).

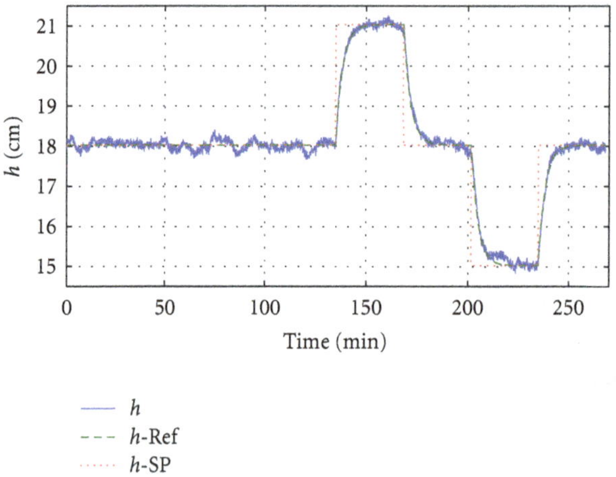

FIGURE 14: Closed-loop simulation: RSMPC—h (setpoint: SP and reference system: Ref).

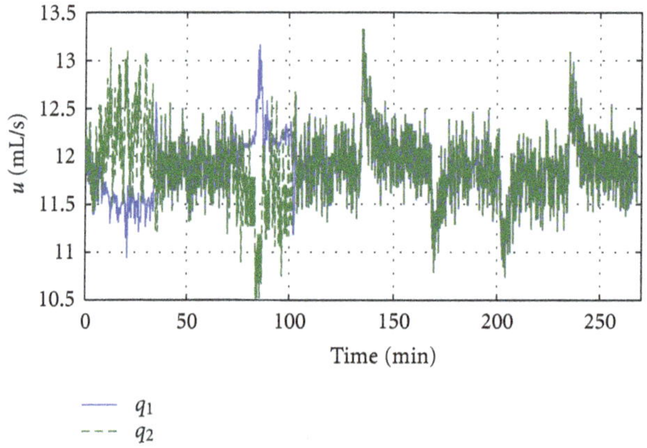

FIGURE 15: Closed-loop simulation: RSMPC—control moves.

$$\boldsymbol{\Gamma} = \begin{bmatrix} \mathbf{C}\boldsymbol{\Psi} & \mathbf{0} & \cdots & \mathbf{0} \\ \vdots & \vdots & & \vdots \\ \mathbf{C}(\boldsymbol{\Phi}+\mathbf{I})\boldsymbol{\Psi} & \mathbf{C}\boldsymbol{\Psi} & \cdots & \mathbf{0} \\ \vdots & \vdots & & \vdots \\ \mathbf{C}\left(\sum_{i=1}^{N}\boldsymbol{\Phi}^{i-1}\right)\boldsymbol{\Psi} & \mathbf{C}\left(\sum_{i=1}^{N-1}\boldsymbol{\Phi}^{i-1}\right)\boldsymbol{\Psi} & \cdots & \mathbf{C}\boldsymbol{\Psi} \\ \vdots & \vdots & & \vdots \\ \mathbf{C}\left(\sum_{i=1}^{P}\boldsymbol{\Phi}^{i-1}\right)\boldsymbol{\Psi} & \mathbf{C}\left(\sum_{i=1}^{P-1}\boldsymbol{\Phi}^{i-1}\right)\boldsymbol{\Psi} & \cdots & \mathbf{C}\left(\sum_{i=1}^{P-N}\boldsymbol{\Phi}^{i-1}\right)\boldsymbol{\Psi} \end{bmatrix}, \tag{22}$$

$$\gamma = \begin{bmatrix} \mathbf{y}_k + \mathbf{C}\boldsymbol{\Phi}\Delta\mathbf{x}_k \\ \mathbf{y}_k + \mathbf{C}\left(\sum_{i=1}^{2}\boldsymbol{\Phi}^{i}\right)\Delta\mathbf{x}_k \\ \vdots \\ \mathbf{y}_k + \mathbf{C}\left(\sum_{i=1}^{N}\boldsymbol{\Phi}^{i}\right)\Delta\mathbf{x}_k \\ \vdots \\ \mathbf{y}_k + \mathbf{C}\left(\sum_{i=1}^{P}\boldsymbol{\Phi}^{i}\right)\Delta\mathbf{x}_k \end{bmatrix}, \tag{23}$$

$$\frac{d\mathbf{y}}{dt} = \mathbf{K}\left(\mathbf{y}^{\mathrm{SP}} - \mathbf{y}\right), \tag{24}$$

$$\mathbf{C}\mathbf{f}_{k-1} + \mathbf{C}\mathbf{A}_{k-1}\mathbf{x}_k + \mathbf{C}\mathbf{B}_{k-1}\mathbf{u}_k + \boldsymbol{\lambda}_k = \mathbf{K}\left(\mathbf{y}^{\mathrm{SP}} - \mathbf{y}\right), \tag{25}$$

$$\mathbf{D}\boldsymbol{v} = \mathbf{b}, \tag{26}$$

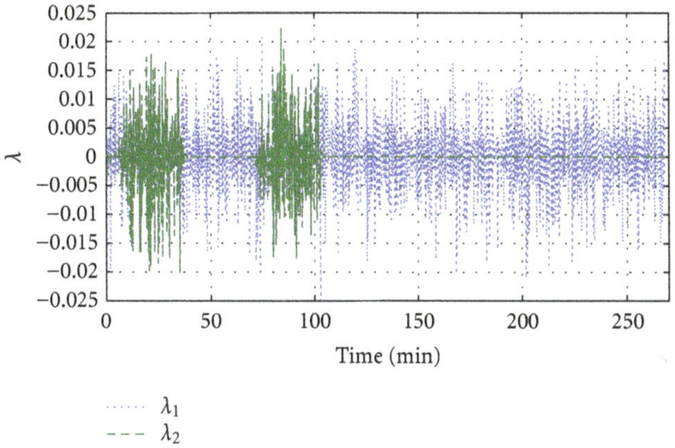

FIGURE 16: Closed-loop simulation: RSMPC—slack variables.

where

$$v = \begin{bmatrix} \Delta\mathbf{u}_k & \cdots & \Delta\mathbf{u}_{k+N} & \boldsymbol{\lambda}_k & \cdots & \boldsymbol{\lambda}_{k+P} \end{bmatrix}, \quad (27)$$

$$\mathbf{b} = \begin{bmatrix} \mathbf{K}\mathbf{y}_k^{SP} - (\mathbf{CA}+\mathbf{KC})\mathbf{x}_k - \mathbf{C}\mathbf{f}_{k-1} \\ \mathbf{K}\mathbf{y}_{k+1}^{SP} - (\mathbf{CA}+\mathbf{KC})\boldsymbol{\Phi}\mathbf{x}_k \\ \mathbf{K}\mathbf{y}_{k+2}^{SP} - (\mathbf{CA}+\mathbf{KC})\boldsymbol{\Phi}^2\mathbf{x}_k \\ \vdots \\ \mathbf{K}\mathbf{y}_{k+P}^{SP} - (\mathbf{CA}+\mathbf{KC})\boldsymbol{\Phi}^P\mathbf{x}_k \end{bmatrix}, \quad (28)$$

$$\mathbf{D} = \begin{bmatrix} \mathbf{CB} & \mathbf{0} & \cdots & \mathbf{0} & \mathbf{I} & \mathbf{0} & \mathbf{0} & \mathbf{0} & \cdots & \mathbf{0} \\ (\mathbf{CA}+\mathbf{KC})\boldsymbol{\Psi} & \mathbf{CB} & \cdots & \mathbf{0} & -\mathbf{I} & \mathbf{I} & \mathbf{0} & \mathbf{0} & \cdots & \mathbf{0} \\ (\mathbf{CA}+\mathbf{KC})\boldsymbol{\Psi}\mathbf{C} & (\mathbf{CA}+\mathbf{KC})\boldsymbol{\Psi} & \cdots & \mathbf{0} & \mathbf{0} & -\mathbf{I} & \mathbf{I} & \mathbf{0} & \cdots & \mathbf{0} \\ \vdots & \vdots & \ddots & \vdots & \vdots & \vdots & \vdots & \vdots & \ddots & \vdots \\ (\mathbf{CA}+\mathbf{KC})\boldsymbol{\Phi}^{P-1}\boldsymbol{\Psi} & (\mathbf{CA}+\mathbf{KC})\boldsymbol{\Phi}^{P-2}\boldsymbol{\Psi} & \cdots & (\mathbf{CA}+\mathbf{KC})\boldsymbol{\Phi}^{P-M-1}\boldsymbol{\Psi} & \mathbf{0} & \mathbf{0} & \mathbf{0} & \mathbf{0} & \cdots & \mathbf{I} \end{bmatrix}. \quad (29)$$

The cost function can be written as (30). This equation can also be reorganized as a quadratic programming problem (31):

$$\min_{\Delta\mathbf{u}(k),\ldots,\Delta\mathbf{u}(k+N_m)\boldsymbol{\lambda}(k),\ldots,\boldsymbol{\lambda}(k+N_p)} J = \frac{1}{2}\left(\Delta\mathbf{u}^T\mathbf{R}\Delta\mathbf{u} + \boldsymbol{\lambda}^T\mathbf{S}\,\boldsymbol{\lambda}\right), \quad (30)$$

$$\begin{aligned} \min_{v(k),\ldots,v(k+N_m+N_p+2)} \quad & J = \frac{1}{2}v^T\boldsymbol{\varepsilon}v \\ \text{subject to:} \quad & \mathbf{D}v = \mathbf{b} \\ & \mathbf{u}_{\min k+j} \le \mathbf{u}_{k+j} \le \mathbf{u}_{\max k+j} \\ & -|\Delta\mathbf{u}_{\min}|_{k+j} \le \Delta\mathbf{u}_{k+j} \le |\Delta\mathbf{u}_{\max}|_{k+j} \\ & j = 0,\ldots,P \end{aligned} \quad (31)$$

in which

$$\boldsymbol{\varepsilon} = \begin{bmatrix} \mathbf{R} & [\mathbf{0}] \\ [\mathbf{0}] & \mathbf{S} \end{bmatrix}. \quad (32)$$

The controller cannot eliminate offset unless (33) is handled [3]:

$$\mathbf{C}\mathbf{f}_{k-1} \approx \frac{\mathbf{y}_k - \mathbf{y}_{k-1}}{\Delta t}. \quad (33)$$

3. The Experimental System

The experimental system built was a neutralization process that occurs in a shaken reactor tank. Figure 3 presents a photo of the experimental system.

Figure 4 shows a diagram indicating how the system works. The numbers indicate acid flow (1), base flow (2), buffer flow (3), and output flow (4). This number is used in the indexes of the modeling of the experimental system. The flows of acid (q_1) and base (q_2) are the manipulated variables ($\mathbf{u}$); the flow of buffer solution (q_3) is the unmeasured or measured disturbance ($\mathbf{d}$), depending on how it is treated in the control formulation. For the case of this study, the buffer solution flow was used only to make simulation tests. The level height of the reactor (h) and the output flow pH (q_4) are the controlled variables or the outputs ($\mathbf{y}$).

The system was specially chosen for this study because of its strong nonlinear dynamic behavior and low operational cost, considering the consumption of inputs as electrical energy and raw material [3, 22]. The system becomes interesting under the control perspective, as it is a multivariable and nonlinear process in which the static gain suffers strong variations inside the flow operation band. Figure 5 presents the system titration curve. It can be observed that the process pH changes from 4 to 9 in the [0.87; 1.22] interval of the ratio q_2/q_1. This change in the process gain is just what makes the neutralization problem difficult to solve under the control perspective, as small modifications on the relation q_2/q_1 can cause big or small pH variations, depending on where the operating point is in the operational region. It has to be emphasized that, in an analytical process of titration, only one drop is enough to change the pH from 4 to 9. Thus, it can be inferred that the sensitivity of the dynamic process associated to this offers a challenge problem for control in real time. It can be noticed that, once a buffer solution is added, the nonlinear behavior of the system smoothes up, so that the controller action gets

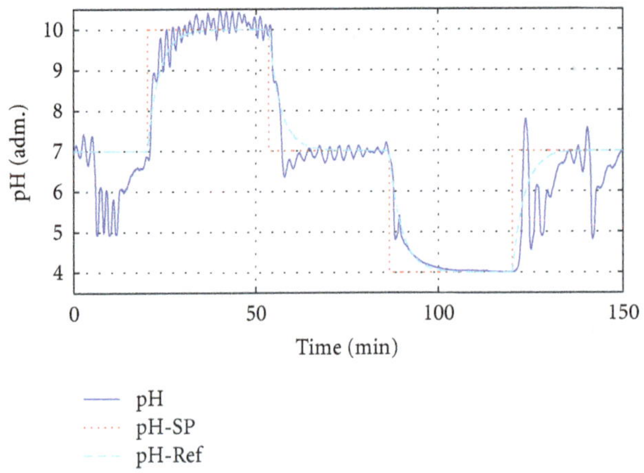

Figure 17: Experimental response RSMPC—pH (setpoint: SP and reference system: Ref).

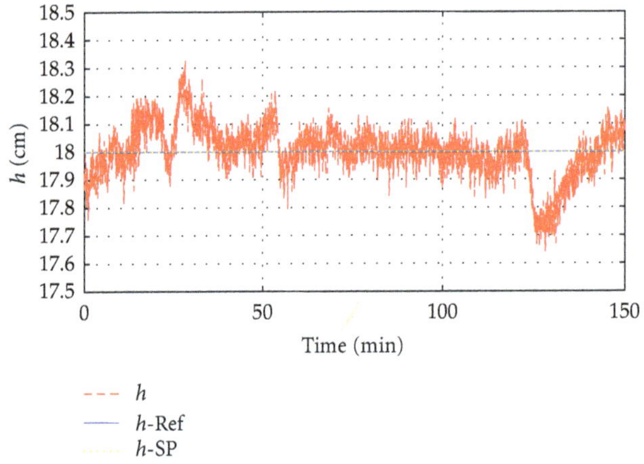

Figure 18: Experimental response: RSMPC—h (setpoint: SP and reference system: Ref).

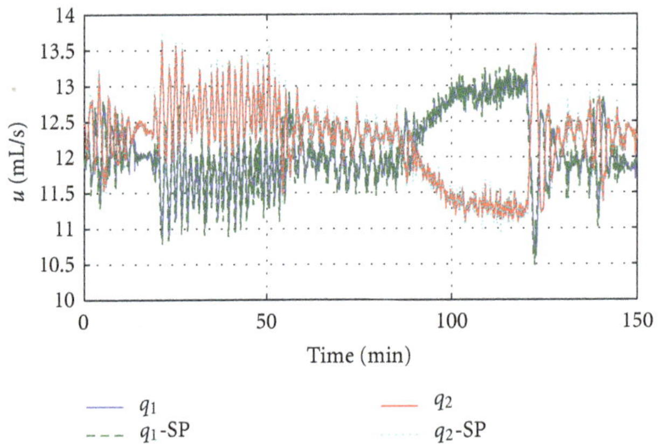

Figure 19: Experimental response: inputs.

easier, according to what the simulation curves with this solution indicate. If the system was linear, the relation given by pH $\times q_2/q_1$ plots would be a line whose approximation can be realized by the simulated curve of larger amount of buffer solution. Montandon [3] presented results for the closed-loop system using pH control of the acid flow with classical PID controller. The conclusion was that this implementation was not able to reject disturbances in a satisfactory manner. During the preparation of this work it has been studied certain PID testing for the MIMO case,

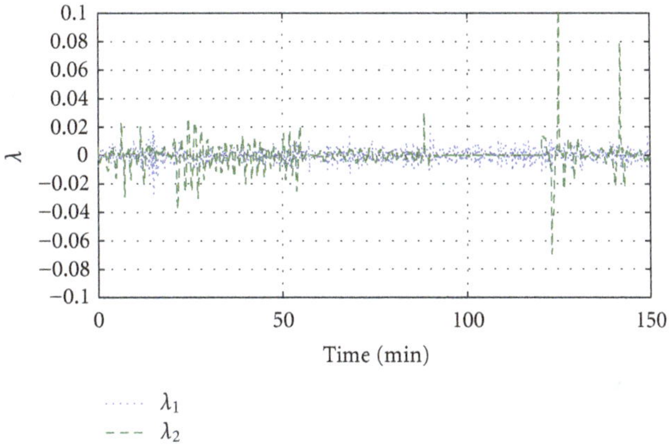

FIGURE 20: Experimental response: RSMPC—slack variables.

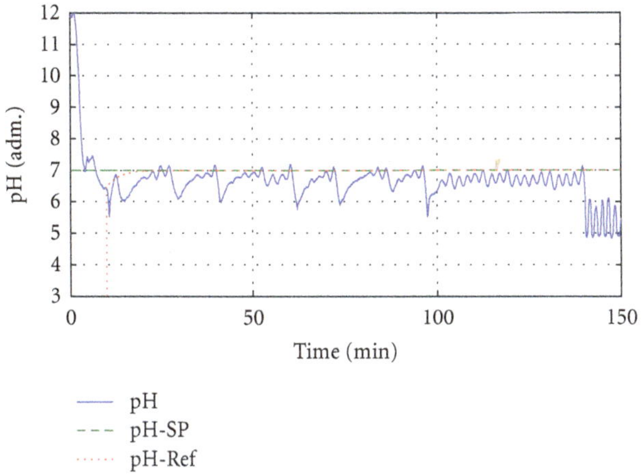

FIGURE 21: Experimental response RSMPC—pH (setpoint: SP and reference system: Ref).

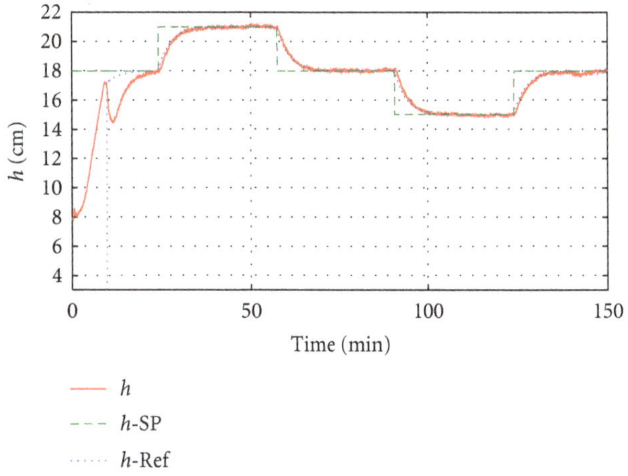

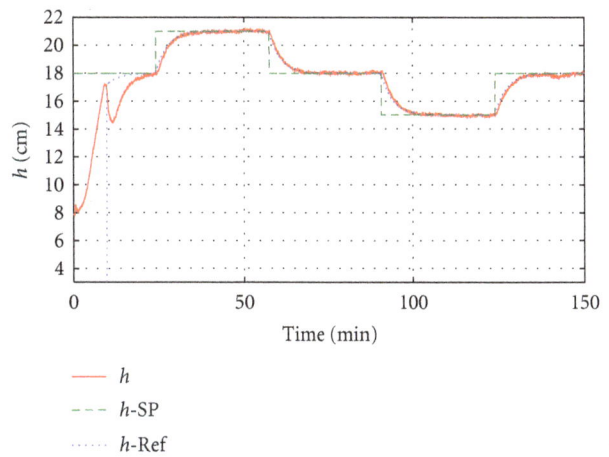

FIGURE 22: Experimental response: RSMPC—h (setpoint: SP and reference system: Ref).

but the answers proved to be mostly unstable and hard to tune.

3.1. Phenomenological Model. Hall [23] developed the physical model of this process, which is based on the hypotheses of the perfect mixture, constant density, and total solubility of the present ions. The chemical reactions involved in the acid-base neutralization (HNO_3–NaOH) are shown in (34)–(38). In order to have a complete model it will be considered the presence of a buffer ($NaHCO_3$). However, for the experimental results, this solution was not used:

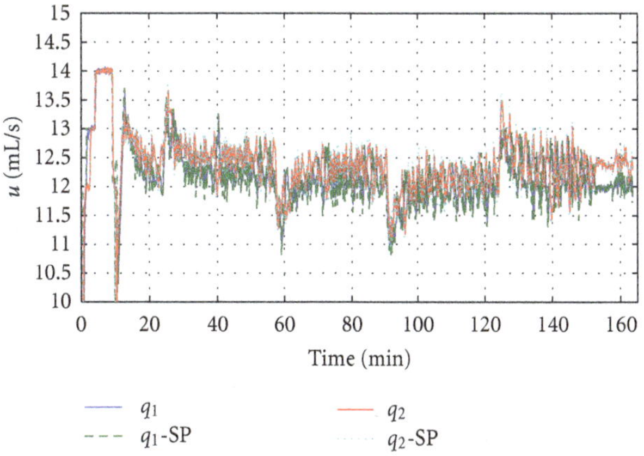

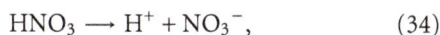

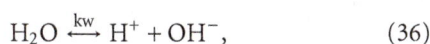

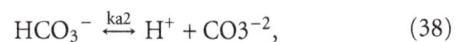

FIGURE 23: Experimental response: inputs.

$$HNO_3 \longrightarrow H^+ + NO_3{}^-, \quad (34)$$

$$NaOH \longrightarrow Na^+ + OH^-, \quad (35)$$

$$H_2O \overset{kw}{\longleftrightarrow} H^+ + OH^-, \quad (36)$$

$$\begin{aligned} NaHCO_3 + H_2O &\longrightarrow NaOH + H_2CO_3, \\ H_2CO_3 &\overset{ka1}{\longleftrightarrow} H^+ + HCO3^-, \end{aligned} \quad (37)$$

$$HCO_3{}^- \overset{ka2}{\longleftrightarrow} H^+ + CO3^{-2}, \quad (38)$$

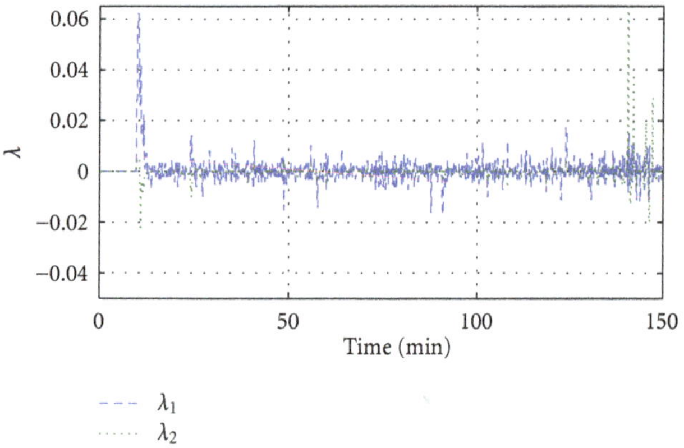

FIGURE 24: Experimental response: RSMPC—slack variables.

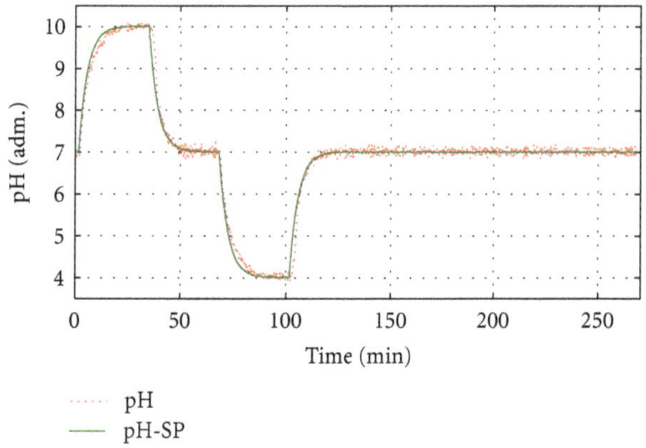

FIGURE 25: Closed-loop simulation: IHMPC-pH.

where

$$\mathrm{ka1} = \frac{\lfloor \mathrm{HCO_3}^- \rfloor \lfloor \mathrm{H}^+ \rfloor}{[\mathrm{H_2CO_3}]}, \quad (39)$$

$$\mathrm{ka2} = \frac{\lfloor \mathrm{CO_3}^- \rfloor \lfloor \mathrm{H}^+ \rfloor}{[\mathrm{HCO_3}^-]}. \quad (40)$$

The amounts of W_a and W_b are called as invariants, because their concentrations are not affected along the reaction. The reactions are fast enough to allow the system to be considered as in equilibrium. Then, the equilibrium reactions can be used to determine the concentration of the hydrogen (H) ions through the concentration of the reaction invariants. Equation (41) gives the equilibrium concentration. According to Gustafsson and Waller [24], there are two reaction invariants defined for the ith stream and presented in (42) and (43), which, when are combined,

TABLE 1: System nominal parameters.

Variables	Symbol	Nominal values
Level	H	25 cm
Area	Ar	$168.38\ \mathrm{cm}^2$
Volume	Vr	$4209.67\ \mathrm{cm}^3$
Acid flow	q_1	11.9130 mL/s
Base flow	q_2	11.8235 mL/s
Buffer solution flow	q_3	0.01 mL/s
pH	pH	7.0
Acid conc. in q_1	$[\mathrm{HNO_3}]$	3.510×10^{-03} M
Base conc. in q_2	[NaOH]	3.510×10^{-03} M

result in an implicit relation of H, W_a, and W_b, as (44) shows:

$$kw = \lfloor \mathrm{H}^+ \rfloor \lfloor \mathrm{OH}^- \rfloor, \quad (41)$$

$$W_{ai} = \lfloor \mathrm{H}^+ \rfloor_i - \lfloor \mathrm{OH}^- \rfloor_i - \lfloor \mathrm{HCO_3}^- \rfloor_i - 2\left\lfloor \mathrm{CO_3}^{-2} \right\rfloor_i, \quad (42)$$

$$W_{bi} = [\mathrm{H_2CO_3}]_i + [\mathrm{HCO_3}^-]_i - 2\left[\mathrm{CO_3}^{-2}\right]_i, \quad (43)$$

$$W_a = \mathrm{H} - \frac{\mathrm{kw}}{\mathrm{H}} - W_b \frac{(\mathrm{ka1/H}) + 2(\mathrm{ka1ka2})/\mathrm{H}^2}{1 + (\mathrm{ka1/H}) + \left(\mathrm{ka1ka2/H}^2\right)}. \quad (44)$$

Making the mass balance in the reactor, together with the invariant equations and considering that the density is constant by hypothesis, will result in a differential equation system (45)–(47):

$$\frac{dh}{dt} = \frac{1}{\mathrm{Ar}}(q_1 + q_2 + q_3 - q_4), \quad (45)$$

$$\frac{dW_a}{dt} = \frac{1}{\mathrm{Vr}}(q_1(W_{a1} - W_a) + q_2(W_{a2} - W_a) + q_3(W_{a3} - W_a)), \quad (46)$$

$$\frac{dW_b}{dt} = \frac{1}{\mathrm{Vr}}(q_1(W_{b1} - W_b) + q_2(W_{b2} - W_b), + q_3(W_{b3} - W_b)), \quad (47)$$

where Ar is the reactor area, Vr is the reaction volume; h is the solution height in the reactor; W_a is the acid reaction invariant; W_b is the base reaction invariant.

The flow output of the system is driven by gravity. The output flow (q_4) is done by a globe-type valve (48). This equation shows the relation between output flow and the reactor height and the parameters to be estimated (c_v and $p7$). The valve parameters were estimated by using the model and experimental data from open loop. After this process of defining the parameters of the valve it was locked so as to avoid modifying these parameters. Nominal operation conditions are presented in Table 1:

$$q_4 = c_v h^{p7}. \quad (48)$$

For estimation and validation of the model, the plant operation must be done through the inputs that excite all the dynamic modes and in a large frequency band, so that the

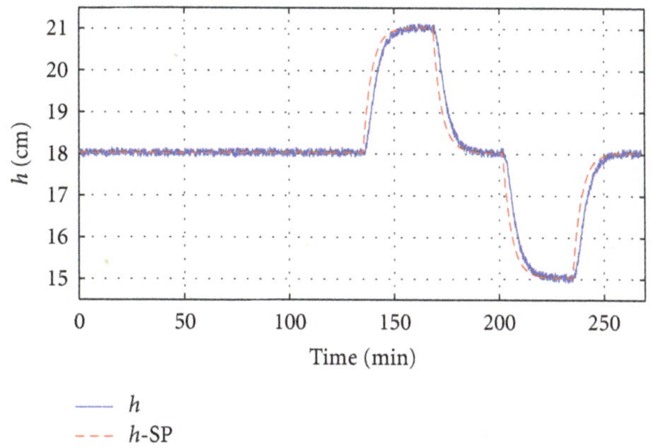

FIGURE 26: Closed-loop simulation: IHMPC-h.

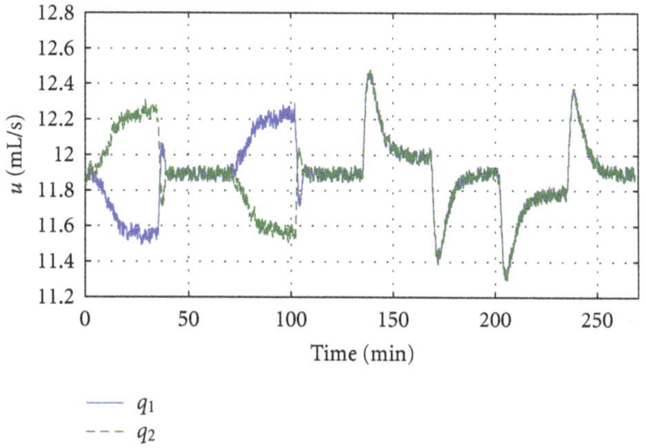

FIGURE 27: Closed-loop simulation: IHMPC-inputs.

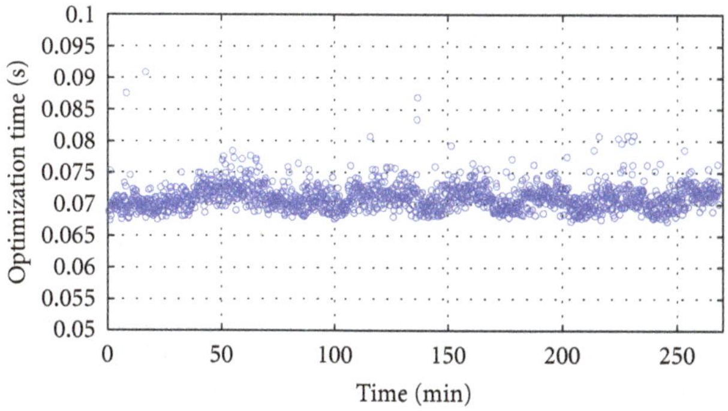

FIGURE 28: Closed-loop simulation: IHMPC-optimization time.

TABLE 2: Controller tuning.

Controller	Parameter
IHMPC	$N = 10$
	$\mathbf{R} = 10\mathbf{I} \rightarrow 100\mathbf{I}$
	$\mathbf{Q} = \mathbf{I}$
	$\mathbf{S} = 10\mathbf{I}$
	$\Delta\mathbf{u}_{max} = 0.25$ mL/s
RSMPC	$\mathbf{K} = 0.005\mathbf{I}$
	$\mathbf{R} = \mathbf{I} \rightarrow \mathbf{R} = 100\mathbf{I}$
	$\mathbf{S} = 10000\mathbf{I}$
	$N = 2$
	$P = 10$
	$\Delta\mathbf{u}_{max} = 0.5$ mL/s
General parameters	$\Delta t = 10$ s
	$\mathbf{u}_{max} = 38$ mL/s
	$\mathbf{u}_{min} = 4$ mL/s

outputs have enough information for that procedure. These are the most usual excitation types: ramp, step, and impulse. However, in order to gather information in a broader range of frequency, random inputs made of sequences of steps with variable amplitude and duration are applied, as it is indicated by Montandon [3]. The sampling time for the process was 10 s. The acid and base flows vary as $q_1 \in [9.856 \quad 13.310]$, $q_2 \in [9.977 \quad 13.215]$, with a step probability equal to 0,8, in which the duration of each step was 120 s. The flows in which the variations occurred were 12 mL/s each, in order to keep the flow sum inside the limit given by $q_1 + q_2 \in [21.5 \quad 24.5]$. The values presented were defined in such a way that, during the operation, the reactor did not get empty nor presents overflow. A set of experimental data composed by 6000

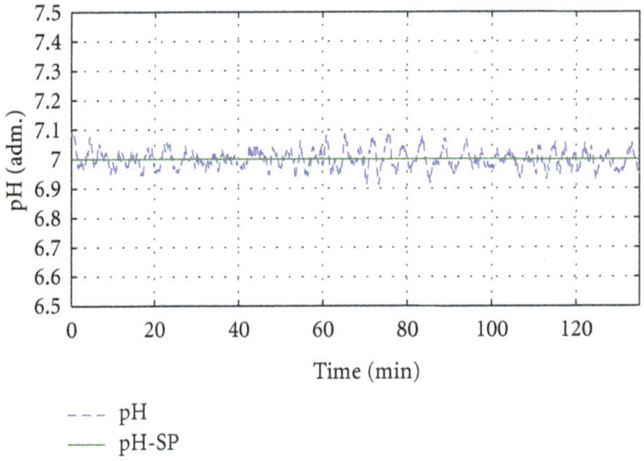

FIGURE 29: Experimental response: IHMPC-pH.

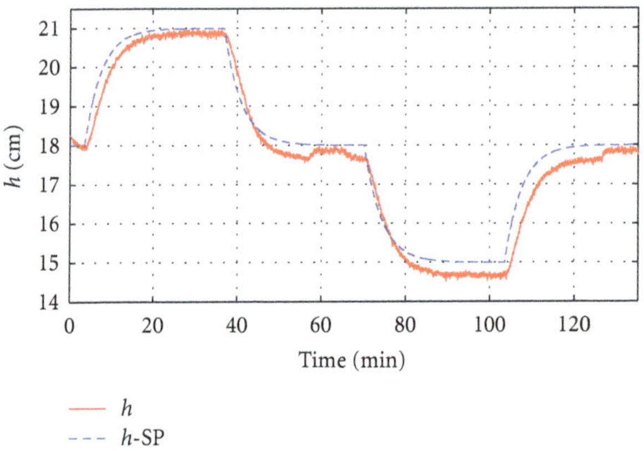

FIGURE 30: Experimental response: IHMPC-h.

acquisition points was used. The estimation procedure used half of the data and the validation employed the remaining data, in order to assure that the model would have the ability for predicting unknown data. The estimation is a nonlinear solution of the approximation curve that uses the least square method with a first trial given by a random value and a step equal to 15. The estimated values that resulted were $c_v = 20.4477$ and $p7 = 0.0523$. Figure 6 shows the error in predicting h. Figure 7 presents the set of input random data. Figure 8 presents the results of the simulation and the experimental ones for the closed-loop run, in which one can notice that the model is very representative of the process, with difficulties to predict around pH 7; this phenomenon was already expected, taking into account the titration curve of the system. Figure 9 shows the validation of the model by the prediction of the output that comes very close to the experimental data.

3.2. State-Space Model. The state-space approach resulting from the algebraic manipulation is given by (49)–(55). The terms $dWadz$ and $dWbdz$ are defined in (56) and (57). Equation (58) presents dydz relation:

$$\mathbf{x} = \begin{bmatrix} h & W_a & W_b \end{bmatrix}, \tag{49}$$

$$\mathbf{u} = \begin{bmatrix} q_1 & q_2 \end{bmatrix}, \tag{50}$$

$$\mathbf{y} = \begin{bmatrix} h & \text{pH} \end{bmatrix}, \tag{51}$$

$$\mathbf{d} = [q_3], \tag{52}$$

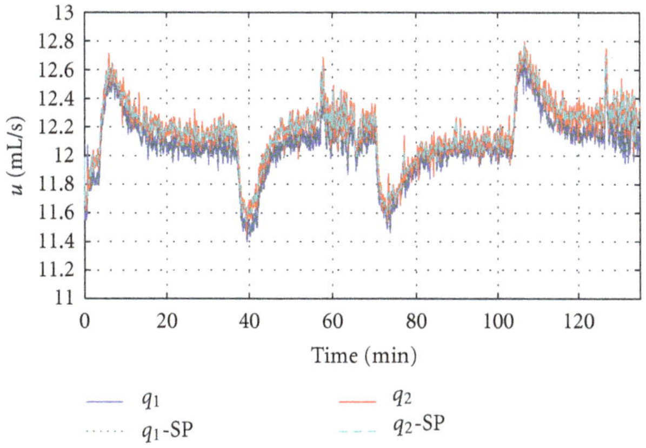

FIGURE 31: Experimental response: IHMPC-inputs.

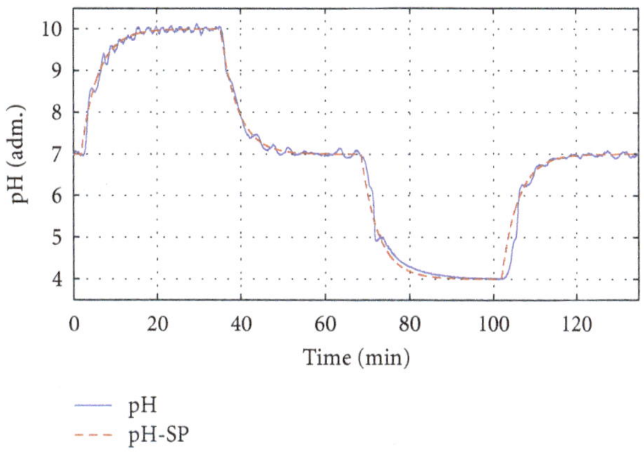

FIGURE 32: Experimental response: IHMPC-pH.

$$\mathbf{A} = \begin{bmatrix} -c_v p7 h^{p7-1} & 0 & 0 \\ -\dfrac{q_1(W_{a1}-W_a)+q_2(W_{a2}-W_a)+q_3(W_{a3}-W_a)}{\mathrm{Arh}^2} & \dfrac{-q_1-q_2-q_3}{\mathrm{Arh}} & 0 \\ -\dfrac{q_1(W_{b1}-W_b)+q_2(W_{b2}-W_b)+q_3(W_{b3}-W_b)}{\mathrm{Arh}^2} & 0 & \dfrac{-q_1-q_2-q_3}{\mathrm{Arh}} \end{bmatrix}, \tag{53}$$

$$\mathbf{B} = \begin{bmatrix} \dfrac{1}{\mathrm{Ar}} & \dfrac{1}{\mathrm{Ar}} \\ \dfrac{(W_{a1}-W_a)}{\mathrm{Arh}} & \dfrac{(W_{a2}-W_a)}{\mathrm{Arh}} \\ \dfrac{(W_{b1}-W_b)}{\mathrm{Arh}} & \dfrac{(W_{b2}-W_b)}{\mathrm{Arh}} \end{bmatrix}, \qquad \mathbf{G} = \begin{bmatrix} \dfrac{1}{\mathrm{Ar}} \\ \dfrac{(W_{a3}-W_a)}{\mathrm{Arh}} \\ \dfrac{(W_{b3}-W_b)}{\mathrm{Arh}} \end{bmatrix}, \tag{54}$$

$$\mathbf{C} = \begin{bmatrix} 1 & 0 & 0 \\ 0 & \dfrac{dydz}{dWadz} & \dfrac{dydz}{dWdz} \end{bmatrix}, \tag{55}$$

$$\begin{aligned} dWadz = {} & 1 + \frac{\mathrm{kw}}{\mathrm{H}^2} - \frac{W_b\left(-\left(\mathrm{ka1}/\mathrm{H}^2\right) - \left(4\mathrm{ka1}\cdot\mathrm{ka2}/\mathrm{H}^3\right)\right)}{1+(\mathrm{ka1}/\mathrm{H}) + \left(\mathrm{ka1}\cdot\mathrm{ka2}/\mathrm{H}^2\right)} \\ & + \frac{W_b\left((\mathrm{ka1}/\mathrm{H}) + \left(2\cdot\mathrm{ka1.ka2}/\mathrm{H}^2\right)\right)\left(-\left(\mathrm{ka1}/\mathrm{H}^2\right) + \left(2\cdot\mathrm{ka1}\cdot\mathrm{ka2}/\mathrm{H}^3\right)\right)}{\left(1+(\mathrm{ka1}/\mathrm{H}) + \left(\mathrm{ka1}\cdot\mathrm{ka2}/\mathrm{H}^2\right)\right)^2}, \end{aligned} \tag{56}$$

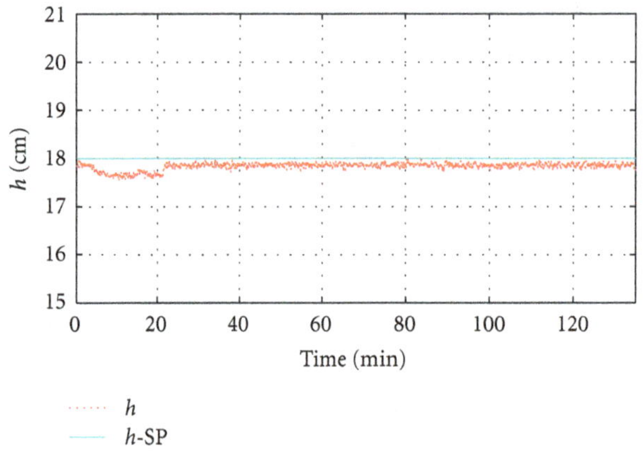

FIGURE 33: Experimental response: IHMPC-h.

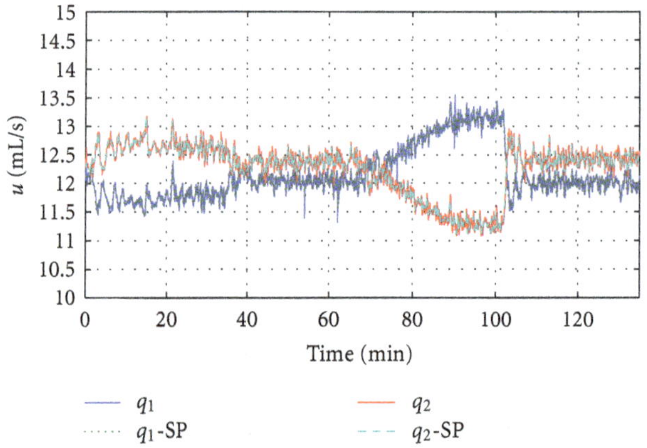

FIGURE 34: Experimental response: IHMPC-inputs.

$$
\begin{aligned}
dWdz = {} & \frac{\left(1+\left(\mathrm{kw}/\mathrm{H}^2\right)\right)\left(1+(\mathrm{ka1}/\mathrm{H})+\left(\mathrm{ka1}\cdot\mathrm{ka2}/\mathrm{H}^2\right)\right)}{(\mathrm{ka1}/\mathrm{H})+\left(2\cdot\mathrm{ka1}\cdot\mathrm{ka2}/\mathrm{H}^2\right)} \\
& - \frac{(\mathrm{H}-(\mathrm{kw}/\mathrm{H})-W_a)\left(1+(\mathrm{ka1}/\mathrm{H})+\left(\mathrm{ka1}\cdot\mathrm{ka2}/\mathrm{H}^2\right)\right)\left(-\left(\mathrm{ka1}/\mathrm{H}^2\right)-\left(4\cdot\mathrm{ka1}\cdot\mathrm{ka2}/\mathrm{H}^3\right)\right)}{\left((\mathrm{ka1}/\mathrm{H})+\left(2\cdot\mathrm{ka1}\cdot\mathrm{ka2}/\mathrm{H}^2\right)\right)^2} \\
& + \frac{(\mathrm{H}-(\mathrm{kw}/\mathrm{H})-W_a)\left(-\left(\mathrm{ka1}/\mathrm{H}^2\right)-\left(2\cdot\mathrm{ka1}\cdot\mathrm{ka2}/\mathrm{H}^3\right)\right)}{(\mathrm{ka1}/\mathrm{H})+\left(2\cdot\mathrm{ka1}\cdot\mathrm{ka2}/\mathrm{H}^2\right)},
\end{aligned}
\tag{57}
$$

$$
dydz = -\log_{10}\frac{e}{\mathrm{H}}. \tag{58}
$$

4. Results and Discussions

The real-time implementation was done using the successive linearization around the operating point. In this case, the linear form results (11).

A controller designed around a stationary state would not be able to control the system, because, for some regions, the stationary state would be very far from the desired point and so the plant model mismatch. Even with the successive linearization, the system control would not be successful, mainly in regions with high gain. There are several ways to deal with this problem. One of them is to expand the states, creating a new state matrix, and then the original structure of the optimization problem can be kept (59). The term related to the output linearization does originate problems for the controller, because it only maps state output and does not interfere in the systems dynamics as the term $\mathbf{f}_{k-1}$ does. Once this modification is done, it is possible to work on the controller previously defined. This option was chosen for real-time operation. Another form can be found in Reis [25]:

$$\frac{d\mathbf{x}}{dt} = \begin{bmatrix} \mathbf{A}_{k-1} & \mathbf{0} \\ \mathbf{0} & \mathbf{f}_{k-1} \end{bmatrix} \begin{bmatrix} \mathbf{x} \\ 1 \end{bmatrix} + \mathbf{B}_{k-1}\mathbf{u}. \quad (59)$$

For the real application in the experimental system, which has nonlinear characteristics, it is necessary to perform successive linearization of the process model in order to apply linear algorithms. The dynamic evolution of the process results in a time-varying system (LTV). The equation of the IHMPC and RSMPC in an LTV form can be easily extrapolated from the formulation presented herein, and for simplicity its description is omitted.

In the experimental system, there are the states h, W_a, and W_b. For this case, the last two states are not measured, as it was shown in the modeling section of the system, so you need to estimate them ($\hat{\mathbf{x}}$). To this end, we used a Kalman filter [26], (60). For the IHMPC controller, we also used an open-loop observer (61), informing the model that the error ($\mathbf{e}$) enters into the plant and model in order to minimize the interference of state estimation in the system behavior [27]:

$$\hat{\mathbf{x}}_{k+1|k} = \mathbf{A}\hat{\mathbf{x}}_{k|k-1} + \mathbf{B}\mathbf{u}_k + \mathbf{L}(\mathbf{y}_k - \mathbf{C}\hat{\mathbf{x}}_{k|k-1}), \quad (60)$$

$$\mathbf{e} = \mathbf{y}_{\text{Plant}} - \mathbf{y}_{\text{Model}}. \quad (61)$$

RSMPC and IHMPC were implemented in real-time operation of the experimental plant, using the interface through LabVIEW and the controller computation, using the routines implemented in Matlab. In the literature there are several techniques of controller tuning. In this work Henson and Seborg [11] and Montadon's [3] indications were adopted. Besides, a field refinement was done during the initial run in closed loop with each controller, making small adjustments in parameters, so that the system would deliver a better experimental performance in closed loop. Such results were omitted in this paper. Table 2 shows the parameters of each controller resulting from the simulation and used in the experimental run. The controllers were at first tuned with identity matrices and a control horizon equal to 10; besides, a tuning was done until achieving a satisfactory response. Figure 10 presents the structure of the experimental plant, and Figures 11 and 12 show the LabVIEW implementation.

Equation (62) shows that the calibration curves of the instruments. V_0, V_1, V_2, V_3, and V_4 are channel voltages. The regression coefficients are equal to 1 [4, 28]:

$$\begin{aligned} q_1 &= 5.2788V_0 - 9.4776, \\ q_2 &= 5.0997V_1 - 9.5541, \\ q_3 &= 0.2090V_2 - 0.3845, \\ \text{pH} &= 1.8982V_3 - 3.5583, \\ h &= 5.6347V_4 - 9.0838. \end{aligned} \quad (62)$$

Setpoint variations were done to evaluate the ability of the controllers to lead with transitions. Starting the system with a pH equal to 7 and a reactor height equal to 18 cm, variations were done according to the vector in (63), considering that each step took 2000 seconds. For all controller simulations a random noise of average equal to zero was added with about 10% of the outputs (h and pH):

$$\begin{bmatrix} h \\ \text{pH} \\ d \end{bmatrix} = \begin{bmatrix} 18 & 18 & 18 & 18 & 18 & 21 & 18 & 15 & 18 \\ 7 & 10 & 7 & 4 & 7 & 7 & 7 & 7 & 7 \\ 0 & 0 & 0 & 0 & 0 & 0 & 0 & 0 & 0 \end{bmatrix}. \quad (63)$$

Figures 13, 14, 15, and 16 present the closed-loop simulation that uses RSMPC. The performance shown is quite reasonable.

The experimental run was done in two steps to preserve and manipulate the data separately. In the first part was made a variation in the pH setpoint and kept the level constant. In the second part it was done otherwise. Figures 17, 18, 19, and 20 show the experimental run of the pH setpoint changes, and Figures 21, 22, 23, and 24 show the experimental run of level setpoint changes. The system was controlled in all runs. During the experiment, there was a need to increase the weight matrix of the control action (as indicated in Table 2); as the output response started to oscillate, it was not able to reach the desired setpoint. However, the overall response can be considered satisfactory.

The results for the IHMPC simulation are presented in Figures 25, 26, 27, and 28. The results are suitable to apply on line the control to the real process.

Like RSMPC, experimental runs with IHMPC were done in two steps. Figures 29, 30, and 31 show the experimental run of the pH setpoint changes, and Figures 32, 33, and 34 show the experimental run of level setpoint changes. This controller showed a performance superior to the RSMPC controller performance and with satisfactory and adequate response and performance.

5. Conclusions

Through this research it was able to better understand the use of predictive controllers in a real-time application, paving the way for research in the area. The experimental application led to an approximation of reality and industrial practice, experiencing some of the common problems and issues in the implementation of controllers. It was possible to verify the need for a mastery of techniques and concepts of process control, system modeling, parameter identification, scheduling, optimization, and, in addition, common sense engineering to solve the experimental problems.

The control of such class of nonlinear processes is a very challenging area with many possibilities for development and that undoubtedly has importance and influence on the performance of process and in consequence results in their organizations.

This work carried out the experimental application of two predictive controllers: IHMPC and RSMPC, both in simulation environment and in the experimental plant dealing with the level and pH control.

The experimental plant was modeled and has the required parameters identified through the run in open loop. The simulation results were satisfactory and indicated that the model is representative of the real process and suitable for the control purposes that were aimed. The simulation answers of the system outputs subjected to the controllers were adequate and satisfactory. The theoretical and experimental responses of the IHMPC runs were satisfactory.

The controller adjustment for real-time operations was suitable and feasible. The interference of matters related to leaking, process noises, and noises from electrical source and other problems associated to real-time application brought additional difficulties that demanded process knowledge of the process.

The experimental application of robust controllers associated or not to control systems tolerant to failures and the use of online identification employing neural networks will be presented elsewhere.

References

[1] G. Stephanopoulos, *Chemical Process Control: An Introduction to Theory and Practice*, Prentice Hall, Upper Saddle River, NJ, USA, 1984.

[2] K. R. Muske and J. B. Rawlings, "Model predictive control with linear models," *AIChE Journal*, vol. 39, no. 2, pp. 262–287, 1993.

[3] A. G. Montandon, *Controle preditivo em tempo real com trajetória de referência baseado em modelo neural para reator de neutralização*, M.S. thesis, Federal University of Uberlândia, 2005.

[4] C. H. F. Silva, *Uma contribuição ao estudo de controladores robustos*, Ph.D. thesis, Universidade Federal de Uberlândia , 2009.

[5] E. F. Camacho and C. Bordons, *Model Predictive Control*, Springer, Barcelona, Spain, 1999.

[6] D. Q. Mayne, J. B. Rawlings, C. V. Rao, and P. O. M. Scokaert, "Constrained model predictive control: stability and optimality," *Automatica*, vol. 36, no. 6, pp. 789–814, 2000.

[7] M. Nikolaou, "Model predictive controllers: a critical synthesis of theory and industrial needs," *Advances in Chemical Engineering*, vol. 26, pp. 131–204, 2001.

[8] M. Morari, "Model predictive control: multivariable control technique of choice in the 1990?" Tech. Rep. CIT/CDS 93-024, California Institute Of Technology, 1993.

[9] D. R. Saffer II and F. J. Doyle III, "Analysis of linear programming in model predictive control," *Computers and Chemical Engineering*, vol. 28, no. 12, pp. 2749–2763, 2004.

[10] J. B. Rawlings, "Tutorial: model predictive control technology," in *Proceedings of the American Control Conference (ACC'99)*, pp. 662–676, June 1999.

[11] M. A. Henson and D. E. Seborg, *Nonlinear Process Control*, Prentice Hall, Upper Saddle River, NJ, USA, 1997.

[12] S. de Oliveira and M. Morari, "Robust model predictive control for nonlinear systems with constraints," in *Advanced Control Of Chemical Processes*, pp. 295–300, Kyoto, Japan, 1994.

[13] J. H. Lee, "Recent advances in model predictive control and other related areas," *AIChE Symposium Series*, pp. 201–216, 1997.

[14] J. W. Eaton and J. B. Rawlings, "Model-predictive control of chemical processes," *Chemical Engineering Science*, vol. 47, no. 4, pp. 705–720, 1992.

[15] S. J. Qin and T. A. Badgwell, "A survey of industrial model predictive control technology," *Control Engineering Practice*, vol. 11, no. 7, pp. 733–764, 2003.

[16] J. Löfberg, *Minimax approaches to robust model predictive control*, Ph.D. thesis, Linköping, Sweden, 2003.

[17] N. L. Ricker, "Model predictive control with state estimation," *Industrial and Engineering Chemistry Research*, vol. 29, no. 3, pp. 374–382, 1990.

[18] J. Richalet, "Industrial applications of model based predictive control," *Automatica*, vol. 29, no. 5, pp. 1251–1274, 1993.

[19] A. Richards and J. How, "Robust model predictive control with imperfect information," in *Proceedings of the American Control Conference, (ACC'05)*, pp. 268–273, June 2005.

[20] Q. L. Zhang and Y. G. Xi, "An efficient model predictive controller with pole placement," *Information Sciences*, vol. 178, no. 3, pp. 920–930, 2008.

[21] M. Morari and J. H. Lee, "Model predictive control: past, present and future," *Computers and Chemical Engineering*, vol. 23, no. 4-5, pp. 667–682, 1999.

[22] H. M. Henrique, *Uma contribuição ao estudo de redes neuronais aplicadas ao controle de processos*, Ph.D. thesis, COPPE/UFRJ, 1999.

[23] R. C. Hall, *Development of a multivariable pH experiment*, M.S. thesis, University of California, Santa Barbara, Calif, USA, 1987.

[24] T. K. Gustafsson and K. V. Waller, "Dynamic modeling and reaction invariant control of pH," *Chemical Engineering Science*, vol. 38, no. 3, pp. 389–398, 1983.

[25] L. L. G. Reis, *Controle tolerante com reconfiguração estrutural acoplado a sistema de diagnóstico de falhas*, M.S. thesis, Federal University of Uberlândia, 2008.

[26] T. Kailath, *Linear Systems*, Prentice Hall, Englewood Cliffs, NJ, USA, 1980.

[27] J. M. Maciejowski, *Predictive Control With Constraints*, Pearson Education Limited, 2002.

[28] D. Gonçalves, *Controle preditivo em tempo real de uma planta piloto*, M. Sc qualification report, Federal University of Uberlândia, 2007.

Experimental Studies of Neural Network Control for One-Wheel Mobile Robot

P. K. Kim and S. Jung

Intelligent Systems and Emotional Engineering (I.S.E.E.) Laboratory, Department of Mechatronics Engineering, Chungnam National University, Daejeon 305-764, Republic of Korea

Correspondence should be addressed to S. Jung, jungs@cnu.ac.kr

Academic Editor: Haibo He

This paper presents development and control of a disc-typed one-wheel mobile robot, called GYROBO. Several models of the one-wheel mobile robot are designed, developed, and controlled. The current version of GYROBO is successfully balanced and controlled to follow the straight line. GYROBO has three actuators to balance and move. Two actuators are used for balancing control by virtue of gyro effect and one actuator for driving movements. Since the space is limited and weight balance is an important factor for the successful balancing control, careful mechanical design is considered. To compensate for uncertainties in robot dynamics, a neural network is added to the nonmodel-based PD-controlled system. The reference compensation technique (RCT) is used for the neural network controller to help GYROBO to improve balancing and tracking performances. Experimental studies of a self-balancing task and a line tracking task are conducted to demonstrate the control performances of GYROBO.

1. Introduction

Mobile robots are considered as quite a useful robot system for conducting conveying objects, conducting surveillance, and carrying objects to the desired destination. Service robots must have the mobility to serve human beings in many aspects. Most of mobile robots have a two-actuated wheel structure with three- or four-point contact on the ground to maintain stable pose on the plane.

Recently, the balancing mechanism becomes an important issue in the mobile robot research. Evolving from the inverted pendulum system, the mobile-inverted pendulum system (MIPS) has a combined structure of two systems: an inverted pendulum system and a mobile robot system. Relying on the balancing mechanism of the MIPS, a personal transportation vehicle has been introduced [1]. The MIPS has two-point contact to stabilize itself. Advantages of the MIPS include robust balancing against small obstacles on the ground and possible narrow turns while three- or four-point contact mobile robots are not able to do so.

In research on balancing robots, two-point contact mobile robots are designed and controlled [1–5]. A series of balancing robots has been implemented and demonstrated their performances [3–5]. Currently, successful navigation and balancing control performances with carrying a human operator as a transportation vehicle have been reported [5].

More challengingly, one-wheel mobile robots are developed. A one-wheel robot is a rolling disk that requires balancing while running on the ground [6]. Gyrover is a typical disc-typed mobile robot that has been developed and presented for many years [7–11]. Gyrover is a gyroscopically stabilized system that uses the gyroscopic motion to balance the lean angle against falling as shown in Figure 1. Three actuators are required to make Gyrover stable. Two actuators are used for balancing and one actuator for driving. Experimental results as well as dynamical modeling analysis on Gyrover are well summarized in the literature [12].

In other researches on one-wheel robots, different approaches of modelling dynamics of a one-wheel robot have been presented [13, 14]. Simulation studies of controlling

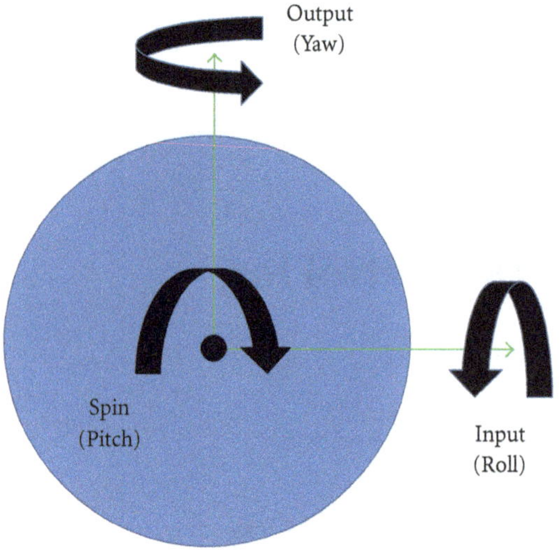

FIGURE 1: Gyro motion of one-wheel robot.

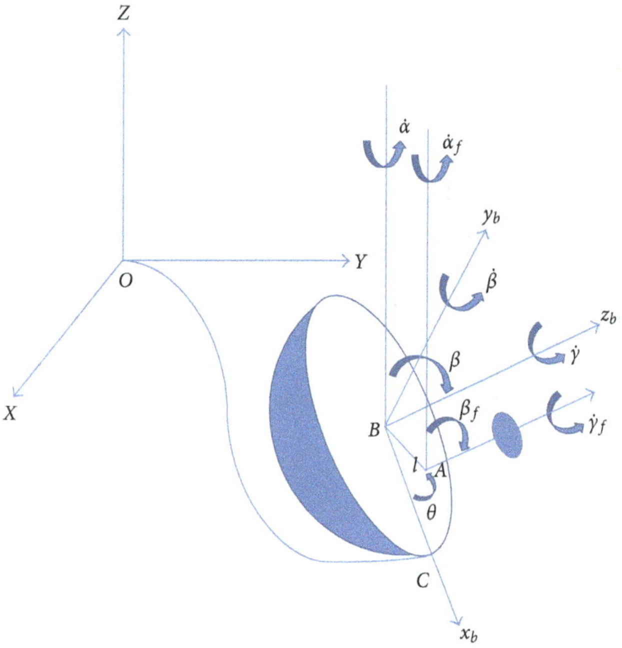

FIGURE 2: GYROBO Coordinates.

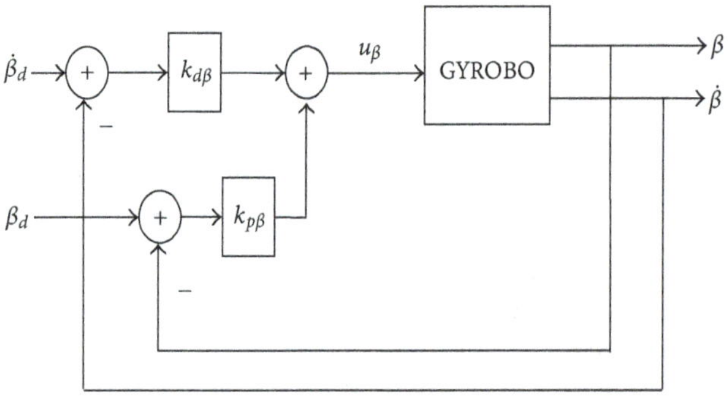

FIGURE 3: Lean angle control for balancing.

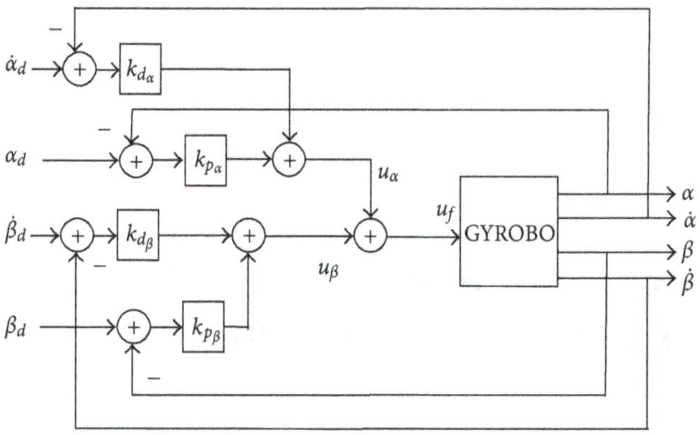

FIGURE 4: Straight line tracking control structure.

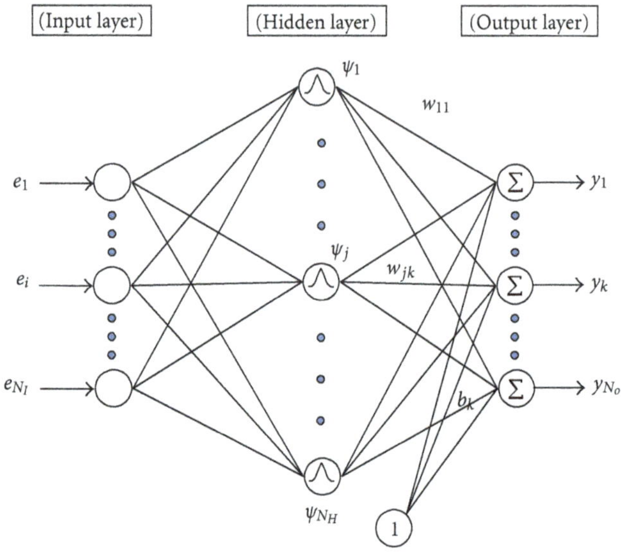

FIGURE 5: Neural network structure.

a Gyrobot along with design and fabrication are presented [15, 16]. An interesting design of a single spherical-wheel robot has been presented [17–19]. Successful balancing control of the spherical-wheel robot has been demonstrated.

In this paper, a one-wheel robot called GYROBO is designed, implemented, and controlled. GYROBO has three actuators to move forward and backward and to balance itself. Although aforementioned research results provide dynamics models for the one-wheel robot, in real physical system, it is quite difficult to control GYROBO based on dynamic models due to several reasons. First, modeling the system is not accurate. Second, it is hard to describe coupling effects between the body wheel and the flywheel since the system is nonlinearly coupled. Third, it is a nonholonomic system.

Therefore, we focus on optimal mechanical design rather than modeling the system since the dynamic model is quite complicated. Although we have dynamic models, other effects from uncertainties and nonmodel dynamics should be considered as well for the better performance.

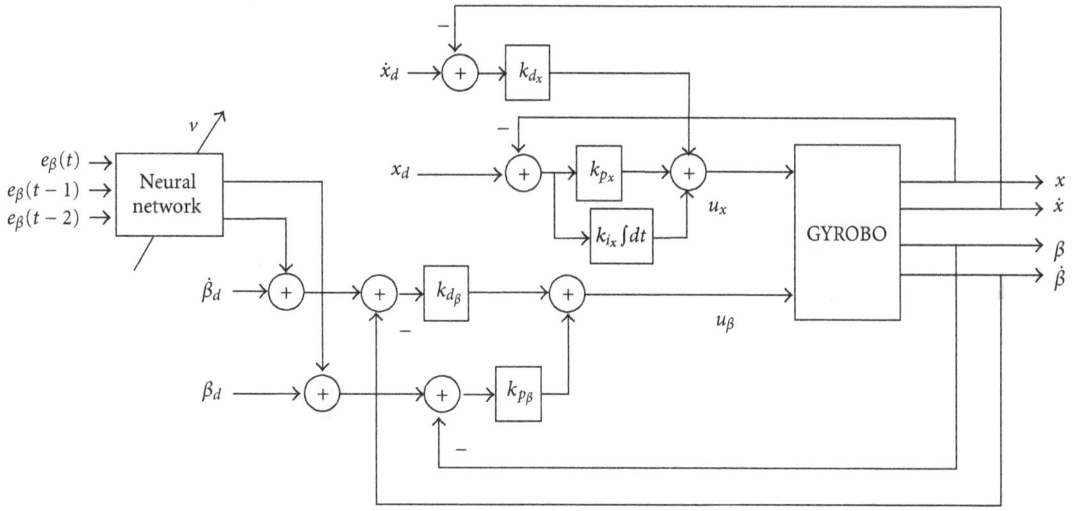

FIGURE 6: RCT neural network control structure.

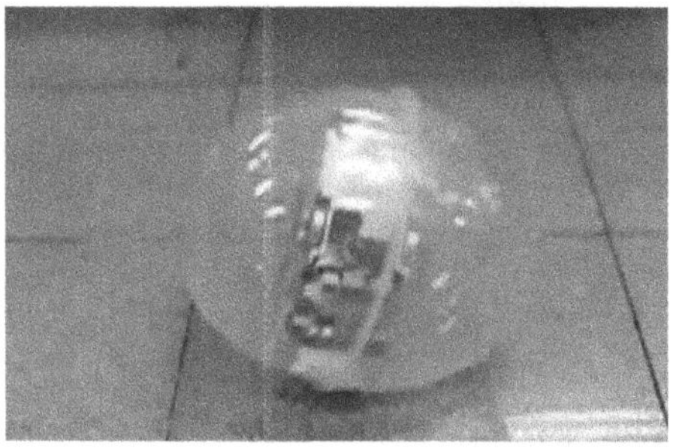

FIGURE 7: Sphere robot with links.

TABLE 1: Parameter definition.

α, α_f	Precession angle of a wheel and a flywheel
β	Lean angle of a wheel
β_f	Tilt angle of a flywheel
γ, γ_f	Spin angles of a wheel and a flywheel
θ	Angle between l and x_b axis
m_w, m_f	Masses of a wheel and a flywheel
m	Mass of whole body
R	Radius of a wheel
I_{xw}, I_{yw}, I_{zw}	Wheel moment of inertia about $X, Y,$ and Z axes
I_{xf}, I_{yf}, I_{zf}	Flywheel moment of inertia about $X, Y,$ and Z
g	Gravitational velocity
u_1, u_2	Drive torque and tilt torque
l	Distance between A and B

After several modifications of the design, the successful design and control of GYROBO are achieved. Since all of actuators have to be housed inside the one wheel, design and placement of each part become the most difficult problem.

After careful design of the system, a neural network control scheme is applied to the model-free-controlled system to improve balancing performance achieved by linear controllers. First, a simple PD control method is applied to the system and then a neural network controller is added to help the PD controller to improve tracking performance. Neural network has been known for its capabilities of learning and adaptation for model-free dynamical systems in online fashion [2, 20].

Experimental studies of a self-balancing task and a straight line tracking task are conducted. Performances by a PD controller and a neural controller are compared.

2. GYROBO Modelling

GYROBO is described as shown in Figure 2. Three angles such as spin, lean, and tilt angle are generated by three corresponding actuators. The gyro effect is created by the combination of two angle rotations, flywheel spin and tilt motion.

The dynamics of one-wheel robot system has been presented in [7, 11, 13]. Table 1 lists parameters of GYROBO. This paper is focused on the implementation and control of GYROBO rather than analyzing the dynamics model of the system since the modeling has been well presented in the literature [11].

Since the GYROBO is a nonholonomic system, there are kinematic constraints such that the robot cannot move in the lateral direction. GYROBO is an underactuated system that has three actuators to drive more states.

We follow the dynamic equation of the GYROBO described in [11]:

$$M(q)\ddot{q} + F(q,\dot{q}) = A^T\lambda + Bu, \tag{1}$$

(a) First model (b) Second model

(c) Third model (d) Fourth model

FIGURE 8: Models of GYROBO using gyro effects.

FIGURE 9: Real design of GYROBO I.

where

$$M(q) = \begin{bmatrix} m & 0 & 0 & 0 & 0 & 0 \\ 0 & m & 0 & 0 & 0 & 0 \\ 0 & 0 & M_{33} & 0 & I_{zw}S\beta & 0 \\ 0 & 0 & 0 & I_{xw} + I_{xf} + mR^2S^2\beta & 0 & I_{xf} \\ 0 & 0 & I_{zw}S\beta & 0 & I_{zw} & 0 \\ 0 & 0 & 0 & I_{xf} & 0 & I_{xf} \end{bmatrix},$$

$$M_{33} = I_{yw}C^2\beta + I_{zw}S^2\beta + I_{yf}C^2\left(\beta + \beta_f\right) + I_{zf}S^2\left(\beta + \beta_f\right),$$

$$F(q,\dot{q}) = \begin{bmatrix} 0 \\ 0 \\ F_3 \\ F_4 \\ I_{zw}C\beta\dot{\alpha}\dot{\beta} \\ F_6 \end{bmatrix}, \tag{2}$$

where

$$\begin{aligned} F_3 &= 2\left(I_{zw} - I_{yw}\right)C\beta S\beta\dot{\alpha}\dot{\beta} + I_{zw}C\beta\dot{\beta}\dot{\gamma} \\ &\quad + 2\left(I_{zf} - I_{yf}\right)C\left(\beta + \beta_f\right)S\left(\beta + \beta_f\right)\left(\dot{\beta} + \dot{\beta}_f\right)\dot{\alpha} \\ &\quad + I_{zf}C\left(\beta + \beta_f\right)\left(\dot{\beta} + \dot{\beta}_f\right)\gamma_f, \\ F_4 &= mR^2S\beta C\beta\dot{\beta}^2 + \left(I_{yw} - I_{zw}\right)C\beta S\beta\dot{\alpha}^2 - I_{zw}C\beta\dot{\gamma}\dot{\alpha} \\ &\quad + \left(I_{yf} - I_{zf}\right)C\left(\beta + \beta_f\right)S\left(\beta + \beta_f\right)\dot{\alpha}^2 \\ &\quad - I_{zf}C\left(\beta + \beta_f\right)\dot{\gamma}_f\dot{\alpha} - mgRS\beta, \\ F_6 &= \left(I_{yf} - I_{zf}\right)C\left(\beta + \beta_f\right)S\left(\beta + \beta_f\right)\dot{\alpha}^2 - I_{zf}C\left(\beta + \beta_f\right)\dot{\gamma}_f\dot{\alpha}, \end{aligned} \tag{3}$$

$$A = \begin{bmatrix} 1 & 0 & -RC\alpha C\beta & RS\alpha S\beta & -RC\alpha & 0 \\ 0 & 1 & -RC\beta C\alpha & -RC\alpha C\beta & -RS\alpha & 0 \end{bmatrix}, \tag{4}$$

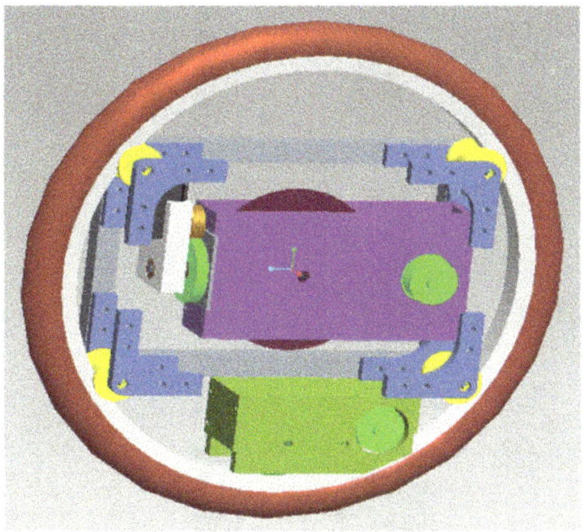

(a)

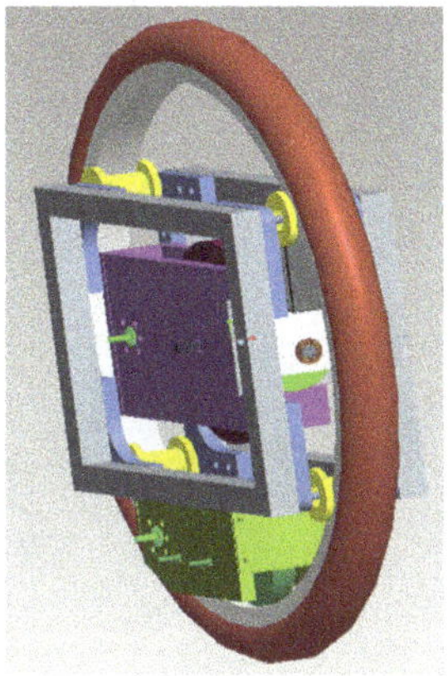

(b)

FIGURE 10: Design of GYROBO II.

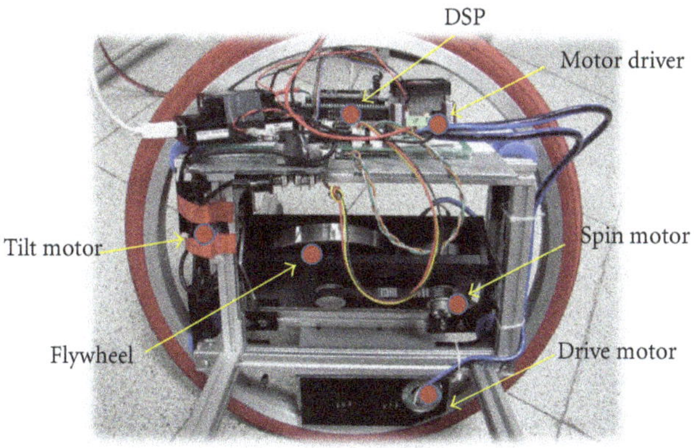

FIGURE 11: Real GYROBO II.

$$q = \begin{bmatrix} X \\ Y \\ \alpha \\ \beta \\ \gamma \\ \beta_f \end{bmatrix}, \quad \lambda = \begin{bmatrix} \lambda_1 \\ \lambda_2 \end{bmatrix}, \quad B = \begin{bmatrix} 0 & 0 \\ 0 & 0 \\ 0 & 0 \\ 0 & 0 \\ k_1 & 0 \\ 0 & 1 \end{bmatrix}, \quad u = \begin{bmatrix} u_1 \\ u_2 \end{bmatrix}, \tag{5}$$

where $C\alpha = \cos(\alpha)$, and $S\alpha = \sin(\alpha)$.

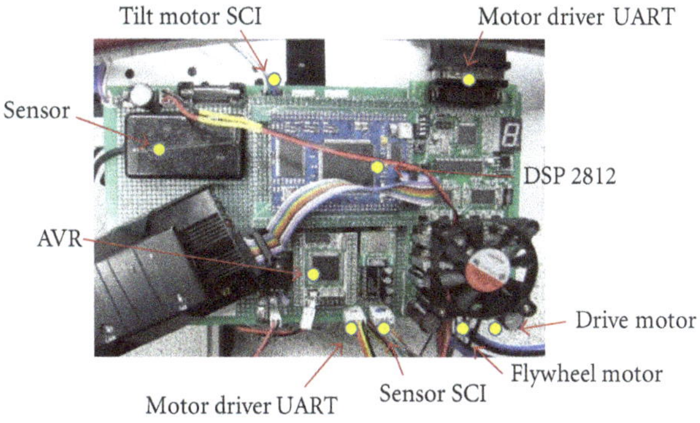

FIGURE 12: Control hardware.

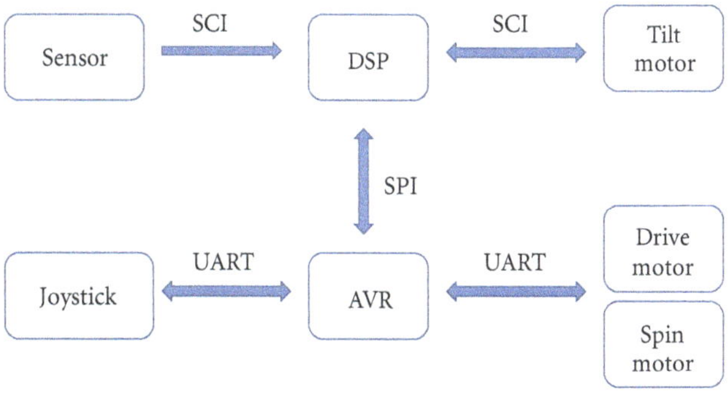

FIGURE 13: Control hardware block diagram.

3. Linear Control Schemes

The most important aspect of controlling GYROBO is the stabilization that prevents from falling in the lateral direction. Stabilization can be achieved by controlling the lean angle β. The lean angle in the y-axis can be separately controlled by controlling the flywheel while the flywheel rotates at a high constant speed. Spinning and tilting velocities of the flywheel induce the precession angle rate that enables GYROBO to stand up. Therefore, a linear controller is designed for the lean angle separately to generate suitable flywheel tilting motions.

3.1. PD Control for Balancing. The PD control method is used for balancing of GYROBO. The lean angle error is defined as

$$e_\beta = \beta_d - \beta, \tag{6}$$

where β_d is the desired lean angle value which is 0 for the balancing purpose and β is the actual lean angle. The angle error passes through the PD controller and generates the tilt torque u_β for the flywheel:

$$u_\beta = k_{p\beta} e_\beta + k_{d\beta} \dot{e}_\beta, \tag{7}$$

where $k_{p\beta}$ and $k_{d\beta}$ are controller gains. Figure 3 shows the PD control block diagram.

3.2. Straight Line Tracking Control. For the GYROBO to follow the straight line, a torque for the body wheel and a torque

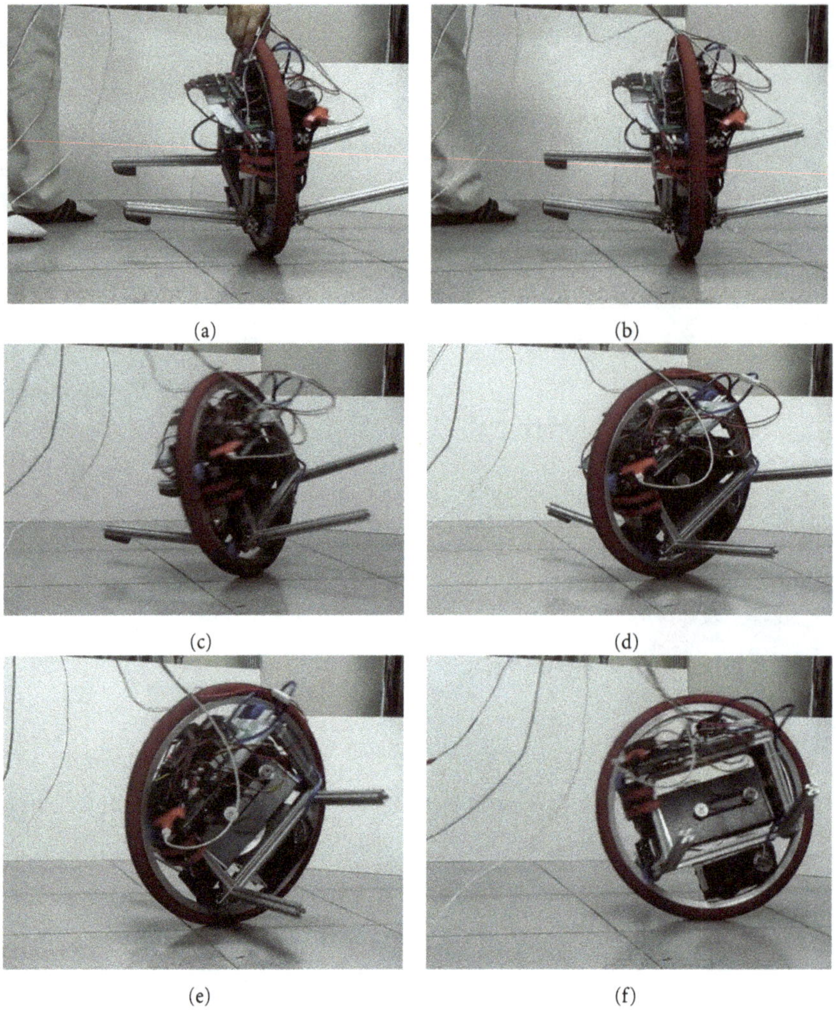

FIGURE 14: PD control: balancing task (a-b-c-d-e-f in order).

for the flywheel should be separately controlled. The body wheel rotates to follow the straight line while the lean angle is controlled to maintain balancing.

The position error detected by an encoder is defined by

$$e_p = x_d - x, \tag{8}$$

where x_d is the desired position value and x is the actual position. The detailed PID controller output becomes

$$u_x = k_{px}e_x + k_{dx}\dot{e}_x + k_{ix}\int e_x dt, \tag{9}$$

where k_{px}, k_{dx}, and k_{ix} are controller gains.

Thus line tracking control requires both position control and angle control. The detailed control block diagram for the straight line following is shown in Figure 4.

4. Neural Control Schemes

4.1. RBF Neural Network Structure. The purpose of using a neural network is to improve the performance controlled by linear controllers. Linear controllers for controlling GYROBO may have limited performances since GYROBO is a highly nonlinear and coupled system.

Neural networks have been known for their capabilities of learning and adaptation of nonlinear functions and used for nonlinear system control [20]. Successful balancing control performances of a two-wheel mobile robot have been presented [2].

One of the advantages of using a neural network as an auxiliary controller is that the dynamic model of the system is not required. The neural network can take care of nonlinear uncertainties in the system by an iterative adaptation process of internal weights.

Here the radial basis function (RBF) network is used for an auxiliary controller to compensate for uncertainties caused by nonlinear dynamics of GYROBO. Figure 5 shows the structure of the radial basis function.

The Gaussian function used in the hidden layer is

$$\psi_j(e) = \exp\left(-\frac{\left|e - \mu_j\right|^2}{2\sigma_j^2}\right), \tag{10}$$

where e is the input vector, $e = [e_1 e_2 \cdots e_{N_I}]^T$, μ_j is

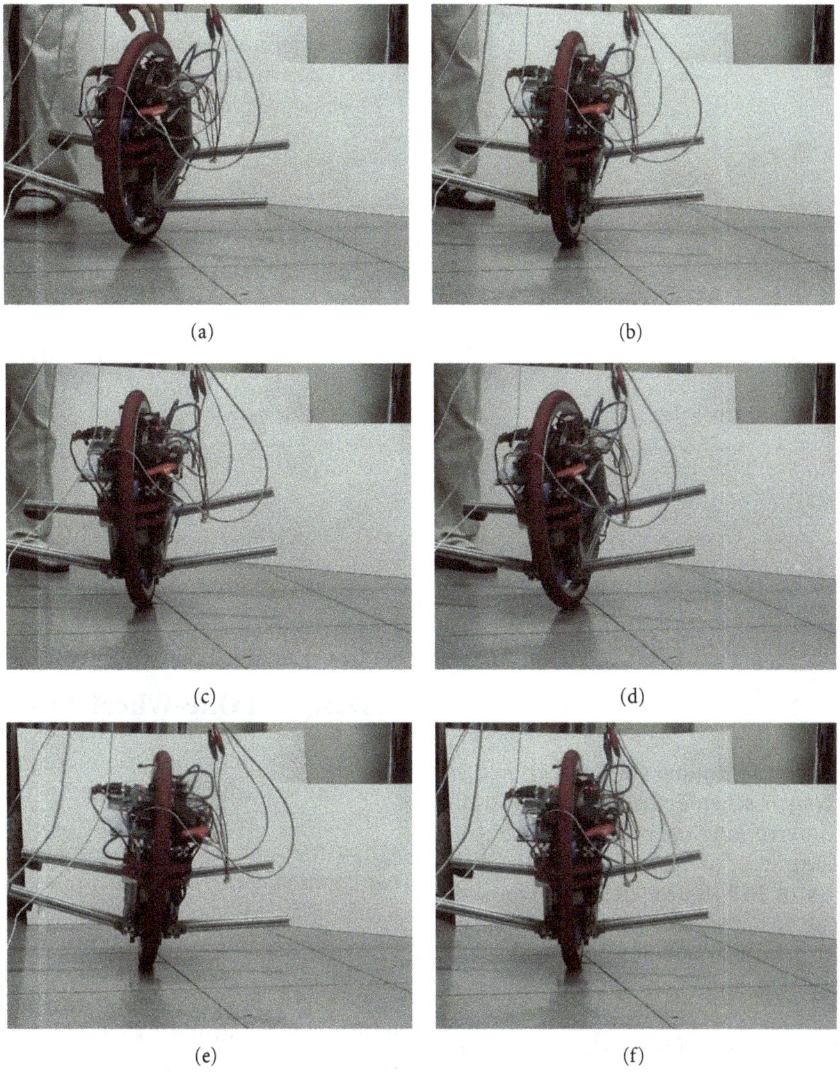

FIGURE 15: Neural network control: balancing task.

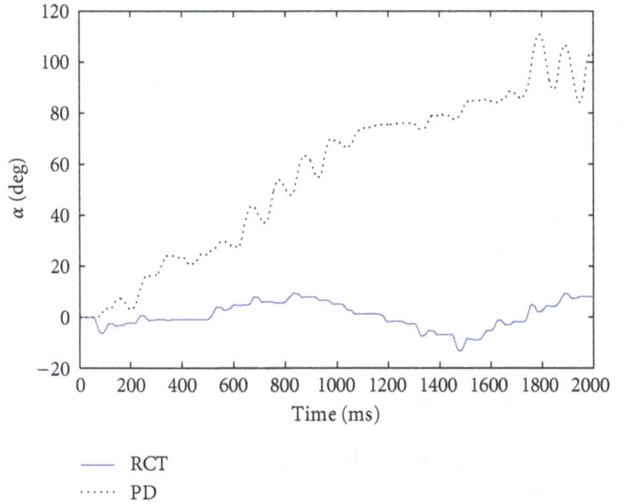

FIGURE 16: Heading angle comparison of PD (black dotted line) and RCT control (blue solid line).

the center value vector of the jth hidden unit, and σ_j is the width of the jth hidden unit.

The forward kth output in the output layer can be calculated as a sum of outputs from the hidden layer:

$$y_k = \sum_{j=1}^{N_H} \psi_j w_{jk} + b_k, \qquad (11)$$

where ψ_j is jth output of the hidden layer in (11), w_{jk} is the weight between the jth hidden unit and kth output, and b_k is the bias weight.

4.2. Neural Network Control. Neural network is utilized to generate compensating signals to help linear controllers by minimizing the output errors. Initial stability can be achieved by linear controllers and further tracking improvement can be done by neural network in online fashion.

Different compensation location of neural network yields different neural network control schemes, but they eventually perform the same goal to minimize the output error.

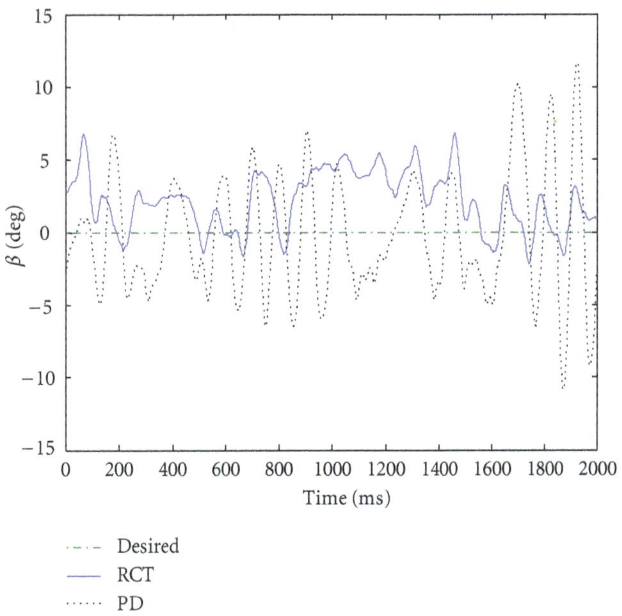

FIGURE 17: Lean angle comparison of PD (black dotted line) and RCT control (blue solid line).

The reference compensation technique (RCT) is known as one of neural network control schemes that provides a structural advantage of not interrupting the predesigned control structure [3, 4]. Since the output of neural network is added to the input trajectories of PD controllers as in Figure 4, tracking performance can be improved. This forms the totally separable control structure as an auxiliary controller shown in Figure 6:

$$u_\beta = k_{p\beta}\left(e_\beta + \phi_1\right) + k_{d\beta}\left(\dot{e}_\beta + \phi_2\right), \tag{12}$$

where ϕ_1 and ϕ_2 are neural network outputs.

Here, the training signal v is selected as a function of tracking errors such as a form of PD controller outputs:

$$v = k_{p\beta}e_\beta + k_{d\beta}\dot{e}_\beta. \tag{13}$$

Then (12) becomes

$$v = u_\beta - \left(k_{p\beta}\phi_1 + k_{d\beta}\phi_2\right). \tag{14}$$

To achieve the inverse dynamic control in (14) such as satisfying the relationship $u_\beta = \tau$ where τ is the dynamics of GYROBO, we need to drive the training signal of neural network to satisfy that $v \to 0$. Then the neural network outputs become equal to the dynamics of the GYROBO in the ideal condition. This is known as an inverse dynamics control scheme. Therefore learning algorithm is developed for the neural network to minimize output errors in the next section.

4.3. Neural Network Learning. Here neural network does not require any offline learning process. As an adaptive controller, internal weights in neural network are updated at each sampling time to minimize errors. Therefore, the selection of an appropriate training signal for the neural network becomes an important issue in the neural network control, even it determines the ultimate control performance.

The objective function is defined as

$$E = \frac{1}{2}v^2. \tag{15}$$

Differentiating (15) and combining with (14) yield the gradient function required in the back-propagation algorithm:

$$\frac{\partial E}{\partial w} = v\frac{\partial v}{\partial w} = -v\left(k_{p\beta}\frac{\partial \phi_1}{\partial w} + k_{d\beta}\frac{\partial \phi_2}{\partial w}\right). \tag{16}$$

The weights are updated as

$$w(t+1) = w(t) + \eta v\left(k_{p\beta}\frac{\partial \phi_1}{\partial w} + k_{d\beta}\frac{\partial \phi_2}{\partial w}\right), \tag{17}$$

where η is the learning rate.

5. Design of One-Wheel Robot

Design of the one-wheel robot becomes the most important issue due to the limitation of space. Our first model is shown in Figure 7. The original idea was to balance and control the sphere robot by differing rotating speeds of two links. Thus, two rotational links inside the sphere are supposed to balance itself but failed due to the irregular momentum induced by two links.

After that, the balancing concept has been changed to use a flywheel instead of links to generate gyro effect as shown in Figure 8. Controlling spin and tilt angles of the flywheel yields gyro effects to make the robot keep upright position.

We have designed and built several models as shown in Figure 8. However, all of models are not able to balance itself long enough. Through trial and error processes of designing models, we have a current version of GYROBO I and II as shown in Figures 9 and 10. Since the GYROBO I in Figure 9 has a limited space for housing, the second model of GYROBO has been redesigned as shown in Figure 11.

6. GYROBO Design

Figure 10 shows the CAD design of the GYROBO II. The real design consists of motors, a flywheel, and necessary hardware as shown in Figure 11.

GYROBO II is redesigned with several criteria. Locating motors appropriately in the limited space becomes quite an important problem to satisfy the mass balance. Thus the size of the body wheel is increased.

The placement of a flywheel effects the location of the center of gravity as well. The size of the flywheel is also critical in the design to generate enough force to make the whole body upright position. The size of the flywheel is increased. The frame of the flywheel system is redesigned and located at the center of the body wheel.

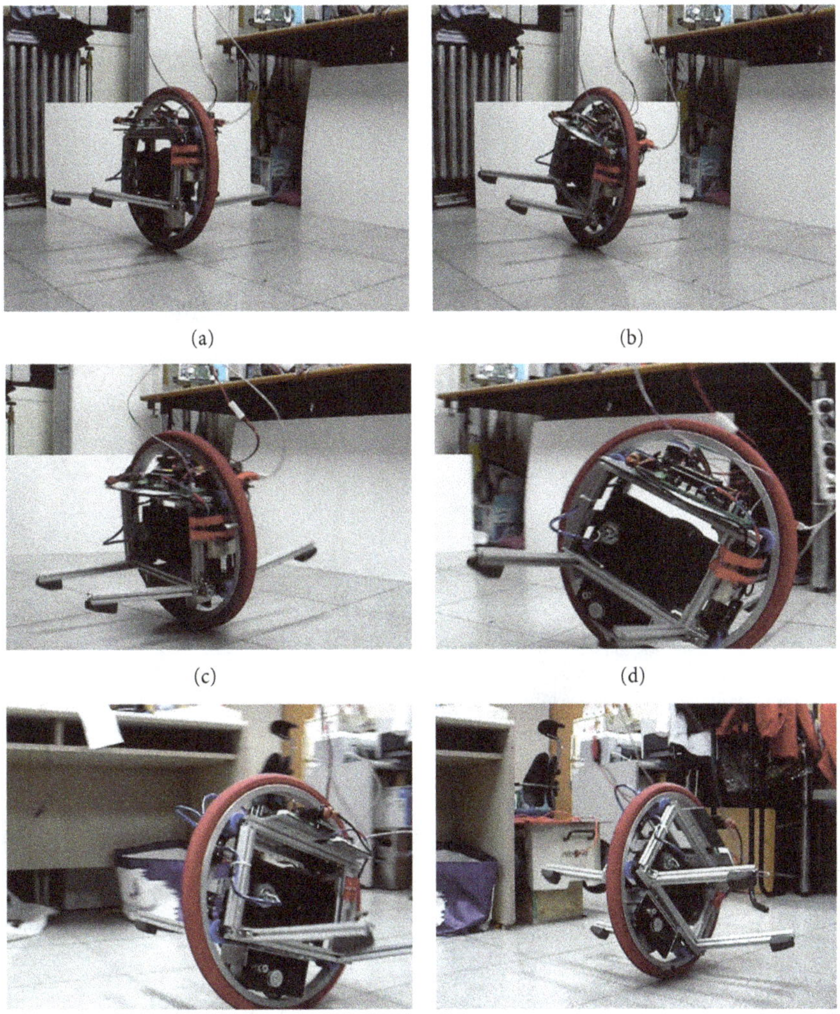

(a) (b) (c) (d) (e) (f)

FIGURE 18: PD control: line following task.

The drive motor is attached to the wheel so that it directly controls the movement of the wheel. A tilt motor is mounted on the center line to change the precession angle of the gyro effect. A high-speed spin motor is located to rotate the flywheel through a timing belt.

Control hardware includes a DSP2812 and an AVR as main processors. The AVR controls drive and spin motors while DSP controls a tilt motor. The detailed layout is shown in Figure 12. To detect the lean angle, a gyro sensor is used. Interface between hardware is shown in Figure 13.

7. Experimental Studies

7.1. Balancing Test at One Point

7.1.1. Scheme 1: PD Control. The speed of the flywheel is set to 7,000 rpm. The control frequency is 10 ms. When the PD controller is used, the GYROBO is able to balance itself as shown in Figure 14. However, the heading angle also rotates as it balances. Figures 14(d), 14(e), and 14(f) show the deviation of the heading angle controlled by a PD controller whose gains are selected as $k_{p\beta} = 10$, and $k_{d\beta} = 15$. PD gains are selected by empirical studies.

7.1.2. Scheme 2: Neural Network Compensation. The same balancing test is conducted with the help of the neural network controller. The learning rate is set to 0.0005, and 3 inputs, 4 hidden units, and 2 outputs are used for the neural network structure. Neural network parameters are found by empirical studies. The GYROBO balances itself and does not rotate much comparing with Figure 14. It stays still as shown in Figure 15. The heading angle is not changed much when neural network control is applied while it deviates for PD control. The larger heading angle error means rotation of GYROBO while balancing which is not desired.

Further clear comparisons of performances between PD control and neural network control are shown in Figures 16 and 17, which are the corresponding plots of Figures 14 and 15. We clearly see that the heading angle of GYROBO is

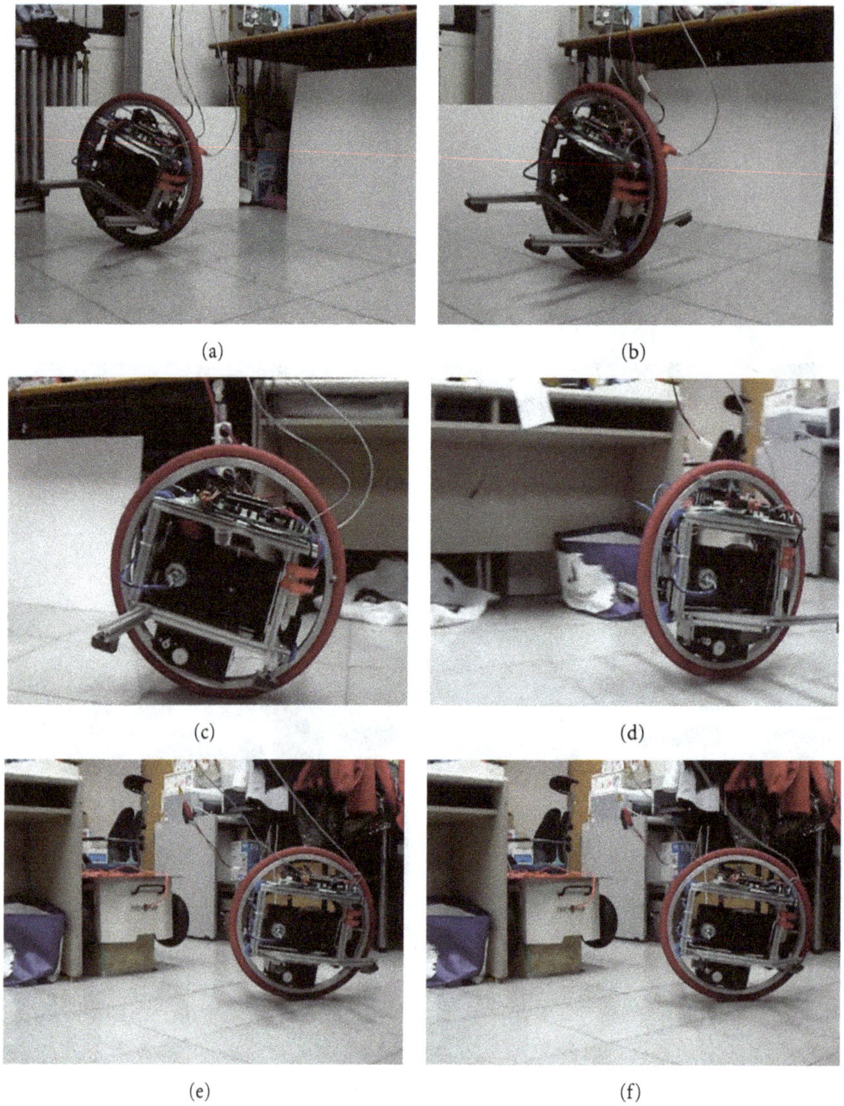

Figure 19: Neural network control: line following.

deviating in the PD control case. We clearly see the larger oscillation of PD control than that of RCT control.

7.2. Line Tracking Test

7.2.1. Scheme 1: PD Control. Next test is to follow the desired straight line. Figure 18 shows the line tracking performance by the PD controller. GYROBO moves forward about 2 m and then stops.

7.2.2. Scheme 2: Neural Network Compensation. Figure 19 shows the line tracking performance by the neural network controller.

We see that the GYROBO moves forward while balancing. Since the GYROBO has to carry a power line, navigation is stopped within a short distance about 2 m.

Clear distinctions between PD control and neural network control are shown in Figures 20 and 21. Both controllers maintain balance successfully. GYROBO controlled by a PD control method deviates from the desired straight line trajectory further as described in Figure 20. In addition, the oscillatory behaviour is reduced much by the neural network controller as shown in Figure 21.

8. Conclusion

A one-wheel robot GYROBO is designed and implemented. After several trial errors of body design, successful design is presented. An important key issue is the design to package all materials in one wheel. One important tip in controlling GYROBO is to reduce the weight so that the flywheel can generate enough gyro effect because it is not easy to find suitable motors. Although balancing and line tracking tasks are successful in this paper, one major problem has to be solved in the future. The power supply for the GYROBO should be independent. Since the stand-alone type of GYROBO is preferred, a battery should be mounted inside

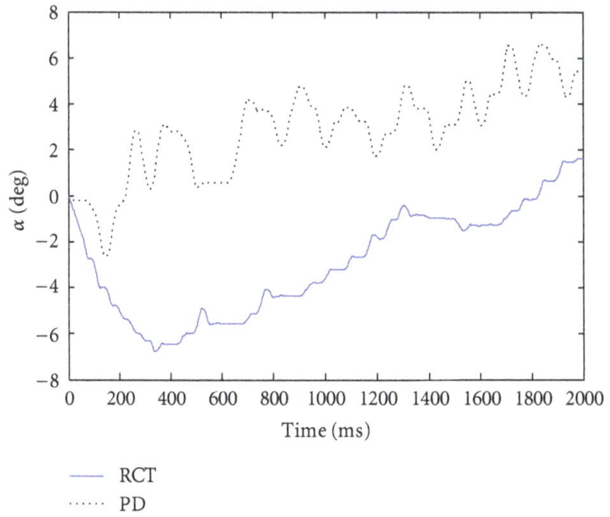

FIGURE 20: Heading angles of PD (black dotted line) and RCT control (blue solid line).

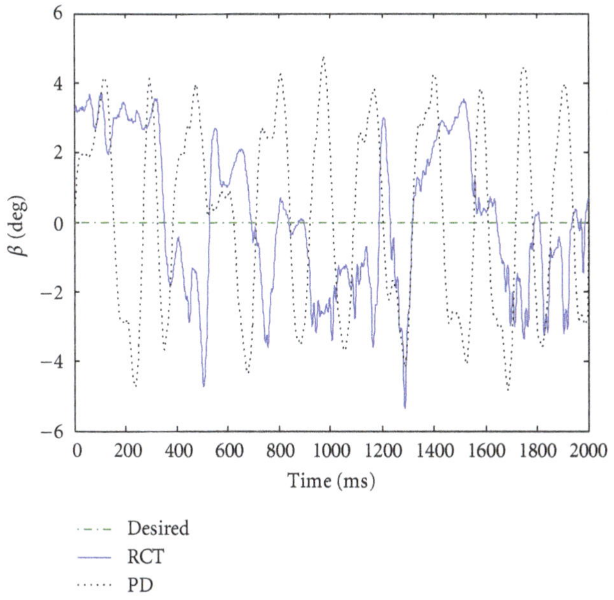

FIGURE 21: Lean angles of PD (black dotted line) and RCT control (blue solid line).

the wheel. Then a room for a battery requires modification of the body design again.

Acknowledgments

This work was financially supported by the basic research program (R01-2008-000-10992-0) of the Ministry of Education, Science and Technology (MEST) and by Center for Autonomous Intelligent Manipulation (AIM) under Human Resources Development Program for Convergence Robot Specialists (Ministry of Knowledge Economy), Republic of Korea.

References

[1] "Segway," http://www.segway.com/.

[2] S. H. Jeong and T. Takayuki, "Wheeled inverted pendulum type assistant robot: design concept and mobile control," in *Proceedings of the IEEE International Workshop on Intelligent Robots and Systems (IROS '07)*, p. 1937, 1932, 2007.

[3] S. S. Kim and S. Jung, "Control experiment of a wheel-driven mobile inverted pendulum using neural network," *IEEE Transactions on Control Systems Technology*, vol. 16, no. 2, pp. 297–303, 2008.

[4] J. S. Noh, G. H. Lee, H. J. Choi, and S. Jung, "Robust control of a mobile inverted pendulum robot using a RBF neural network controller," in *Proceedings of the IEEE International Conference on Robotics and Biomimetics, (ROBIO '08)*, pp. 1932–1937, February 2009.

[5] H. J. Lee and S. Jung, "Development of car like mobile inverted pendulum system : BalBOT VI," *The Korean Robotics Society*, vol. 4, no. 4, pp. 289–297, 2009.

[6] C. Rui and N. H. McClamroch, "Stabilization and asymptotic path tracking of a rolling disk," in *Proceedings of the 34th IEEE Conference on Decision and Control*, pp. 4294–4299, December 1995.

[7] G. C. Nandy and Y. Xu, "Dynamic model of a gyroscopic wheel," in *Proceedings of the IEEE International Conference on Robotics and Automation*, pp. 2683–2688, May 1998.

[8] Y. Xu, K. W. Au, G. C. Nandy, and H. B. Brown, "Analysis of actuation and dynamic balancing for a single wheel robot," in *Proceedings of the IEEE/RSJ International Conference on Intelligent Robots and Systems*, pp. 1789–1794, October 1998.

[9] Y. Xu, H. B. Brown, and K. W. Au, "Dynamic mobility with single-wheel configuration," *International Journal of Robotics Research*, vol. 18, no. 7, pp. 728–738, 1999.

[10] S. J. Tsai, E. D. Ferreira, and C. J. Raredis, "Control of the gyrover: a single-wheel gyroscopically stabilized robot," in *Proceedings of the IEEE International Workshop on Intelligent Robots and Systems (IROS '99)*, pp. 179–184, 1999.

[11] Y. Xu and S. K. W. Au, "Stabilization and path following of a single wheel robot," *IEEE/ASME Transactions on Mechatronics*, vol. 9, no. 2, pp. 407–419, 2004.

[12] Y. S. Xu and Y. S. Ou, *Control of One-Wheel Robots*, Springer, 2005.

[13] W. Nukulwuthiopas, S. Laowattana, and T. Maneewarn, "Dynamic modeling of a one-wheel robot by using Kane's method," in *Proceedings of the IEEE International Conference on Industrial Technology, (IEEE ICIT '02)*, pp. 524–529, 2002.

[14] A. Alasty and H. Pendar, "Equations of motion of a single-wheel robot in a rough terrain," in *Proceedings of the IEEE International Conference on Robotics and Automation*, pp. 879–884, April 2005.

[15] Z. Zhu, A. Al Mamun, P. Vadakkepat, and T. H. Lee, "Line tracking of the Gyrobot—a gyroscopically stabilized single-wheeled robot," in *Proceedings of the IEEE International Conference on Robotics and Biomimetics, (ROBIO '06)*, pp. 293–298, December 2006.

[16] Z. Zhu, M. P. Naing, and A. Al-Mamun, "Integrated ADAMS+MATLAB environment for design of an autonomous single wheel robot," in *Proceedings of the 35th Annual Conference of the IEEE Industrial Electronics Society, (IECON '09)*, pp. 2253–2258, November 2009.

[17] T. B. Lauwers, G. A. Kantor, and R. L. Hollis, "A dynamically stable single-wheeled mobile robot with inverse mouse-ball

drive," in *Proceedings of the IEEE International Conference on Robotics and Automation, (ICRA '06)*, pp. 2884–2889, May 2006.

[18] U. Nagarajan, A. Mampetta, G. A. Kantor, and R. L. Hollis, "State transition, balancing, station keeping, and yaw control for a dynamically stable single spherical wheel mobile robot," in *Proceedings of the IEEE International Conference on Robotics and Automation (ICRA'10)*, pp. 998–1003, 2009.

[19] U. Nagarajan, G. Kantor, and R. L. Hollis, "Trajectory planning and control of an underactuated dynamically stable single spherical wheeled mobile robot," in *Proceedings of the IEEE International Conference on Robotics and Automation, (ICRA '09)*, pp. 3743–3748, May 2009.

[20] P. K. Kim, J. H. Park, and S. Jung, "Experimental studies of balancing control for a disc-typed mobile robot using a neural controller: GYROBO," in *Proceedings of the IEEE International Symposium on Intelligent Control, (ISIC '10)*, pp. 1499–1503, September 2010.

5

Reinforcement Learning Ramp Metering without Complete Information

Xing-Ju Wang,[1,2] Xiao-Ming Xi,[1] and Gui-Feng Gao[1,2]

[1] *School of Traffic and Transportation, Shijiazhuang Tiedao University, Shijiazhuang, Hebei 050043, China*
[2] *Traffic Safety Engineering and Emergency Management Workgroup, Traffic Safety and Control Laboratory of Hebei Province, Shijiazhuang, Hebei 050043, China*

Correspondence should be addressed to Xing-Ju Wang, wangxingju@stdu.edu.cn

Academic Editor: Onur Toker

This paper develops a model of reinforcement learning ramp metering (RLRM) without complete information, which is applied to alleviate traffic congestions on ramps. RLRM consists of prediction tools depending on traffic flow simulation and optimal choice model based on reinforcement learning theories. Moreover, it is also a dynamic process with abilities of automaticity, memory and performance feedback. Numerical cases are given in this study to demonstrate RLRM such as calculating outflow rate, density, average speed, and travel time compared to no control and fixed-time control. Results indicate that the greater is the inflow, the more is the effect. In addition, the stability of RLRM is better than fixed-time control.

1. Introduction

Increasing dependence on car-based travel has led to the daily occurrence of recurrent and nonrecurrent freeway congestions not only in China but also around the world. Congestion on highways forms when the demand exceeds capacity. Recurrent congestion reduces substantially the available infrastructure capacity at rush hour, that is, at the time this capacity is most urgently needed. Moreover, congestion also causes delays, increases environmental pollution, and reduces traffic safety.

Ramp metering is essential to the efficient operation of highways, particularly when volumes are high. According to Papageorgiou and others, ramp metering is divided roughly into the reacted type and the preceded type [1]. DC (demand-capacity), OCC (occupancy), and ALNEA [2] are among the well-known local response type ramp metering [3]. In DC, the actual upstream volume is measured at regular short intervals and is then compared to the downstream capacity, which may be calculated by using downstream traffic conditions. OCC uses a predetermined relationship between occupancy rate and lane volume, developed from data previously collected at the highway adjacent to the ramp being considered. ALNEA is the ramp metering which sets up the private-use rate of an onramp based on the measured value of main line traffic. ALINEA has an example of application in some countries of Europe and is made highly validated compared to DC and OCC. Iwata, Tsubota, and Kawashima have proposed the ramp metering technique using the predicted value by a traffic simulator [4]. Reinforcement learning ramp metering based on traffic simulation model with desired speed was proposed by Wang et al. [5]. The aim of this study is to propose reinforcement learning ramp metering without complete information.

2. Methods

2.1. Traffic Flow Simulation Model. Figure 1 describes car-following behaviors. In a microsimulation model, a modeled fundamental behavior is the "car-following" which adjusts the driver's characteristics: the distance between two adjacent cars, the relative speed, and so forth.

In 1953, Pipes proposed the following basic differential equation model for car-following behavior:

$$\ddot{x}_{n+1}(t) = a[\dot{x}_n(t) - \dot{x}_{n+1}(t)], \tag{1}$$

where $\ddot{x}$, $\dot{x}$, and x denote the acceleration, speed, and distance from the reference point of vehicle n, respectively, and a

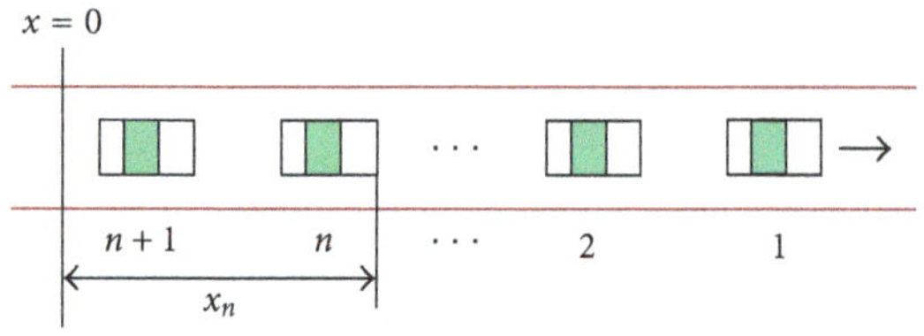

Figure 1: Car-following behavior.

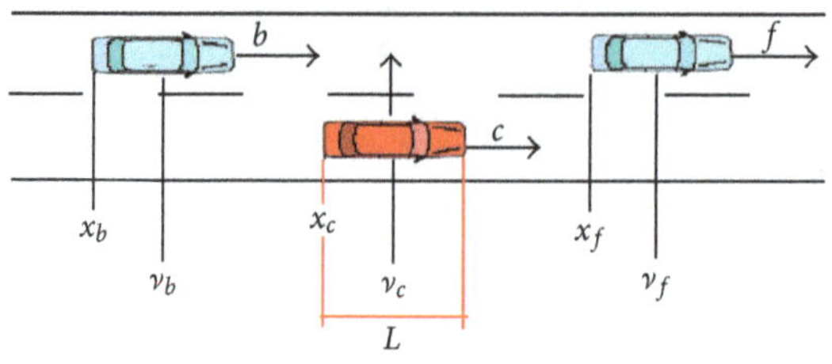

Figure 2: Lane change behavior.

is a constant. In the model, the acceleration of the vehicle which follows a leading vehicle is proportional to the speed difference between the vehicles. It is assumed that the delay of time in which the vehicle responds to the speed difference is so small that it can be neglected. To remove this drawback, Chandler introduced a reactive delay time T. Based on the rationale that the acceleration of the following car is also influenced by its speed and the distance between the vehicles, Gazis, Herman, and Rothery proposed the general type of car-following model:

$$\ddot{x}_{n+1}(t+T) = \frac{a[\dot{x}_{n+1}(t+T)]^m [\dot{x}_n(t) - \dot{x}_{n+1}(t)]}{[x_n(t) - x_{n+1}(t)]^l}. \quad (2)$$

Newell proposed the following model in which the acceleration is propositional to an exponential function of the distance between the vehicles, based on real data:

$$\ddot{x}_{n+1}(t+T) = a_1[\dot{x}_n(t) - \dot{x}_{n+1}(t)] \times e^{-(a_2/[x_n(t)-x_{n+1}(t)-a_3])}. \quad (3)$$

Although the above modifications have improved the reality of car-following model, they have the following two drawbacks. When the proceeding vehicle does not exist, this implies that a car will maintain an initial speed. On the other hand, when the speed difference is 0, the acceleration is 0. This implies the unrealistic phenomenon that the following car will not apply the brake even when the distance to the preceding car approaches 0 and will not accelerate even if the distance is very long. To solve the above-mentioned problems, Treiber and Helbing introduced the intelligent driver model [6], which introduces a desired speed and a shortest distance between cars. The IDM is given as

$$\dot{v}_n = a\left[1 - \left(\frac{v_n}{v_0}\right)^{\delta} - \left(\frac{s^*(v_n, \Delta v_n)}{s_n}\right)^2\right], \quad (4)$$

$$s^*(v, \Delta v) = s_0 + \max\left(Tv + \frac{v\Delta v}{2\sqrt{ab}}, 0\right), \quad (5)$$

$$s_n(t) = [x_{n-1} - x_n - l], \quad (6)$$

$$\Delta v_n(t) = [v_n(t) - v_{n-1}(t)], \quad (7)$$

where x is distance; n is the nth car; v is the speed; l is the length of car; s_0 is the desired minimum gap; a is the maximum acceleration; s^* is the effective gap; b is the comfortable deceleration ($a \le b$); δ is the parameter; T is the time gap; v_0 is the desired speed.

Figure 2 presents lane change behaviors. To simulate driver's behavior in the merging section on freeways and the merging behavior in the weave section, and so forth, the lane change model is needed [7]. We propose a new lane change model which describes driver's behavior depending on judgment functions [8, 9]. We focus on a vehicle approaching to a confluence point and describe its behavior with several variables: the relative speed between the car and cars in current lane, the locations of both the main line cars and the on-ramp cars, driver's judgment functions for changing his lane, and driver's desired speed. The driver's judgment function for the free merging is different from the judgment function for the forced merging. A free merging implies that a car on the ramp can merge into the main line without influences, and cars on the main line are not interfered. When forced merging models of psychological condition and physical condition are both satisfied, the driver conducts lane change behaviors. Otherwise, the driver continues the car-following behavior without lane change behaviors.

Physical condition presents the ability of lane change. The lane change model with driver's judgment function is expressed as follows:

$$h = \frac{x_f - x_c - L + (v_f - v_c)t + (-A + B)t^2}{2} + \delta\frac{(v_{0f} - v_f)}{v_{0f}}S + \zeta\frac{(v_{0c} - v_c)}{v_{0c}}S \ge S, \quad (8)$$

$$g = \frac{x_c - x_b - L + (v_c - v_b)t + (A - B)t^2}{2} - \theta\frac{(v_{b0} - v_f)}{v_{b0}}S - \xi\frac{(v_{c0} - v_c)}{v_{c0}}S \ge S, \quad (9)$$

$$0 \le A \le e, \quad (10)$$

$$0 \le B \le d, \quad (11)$$

where h, g are judgment function; x is the distance from reference point; v is the speed; L is the length of a vehicle; t is the judgment time; v_0 is the desired speed, subject to normal distribution; δ, ζ, θ, ξ (δ, ζ, θ, $\xi \in [0, 1]$) are the adjustment coefficients; A is the rapid acceleration with upper bound e; and B is the rapid deceleration with upper bound d. Parameters A and B are associated with vehicle c's judgment functions for lane change and decide the free merging or the forced merging. Since vehicle c judges to accelerate or decelerate to merge into the main line, two events are mutually exclusive.

The function h judges whether vehicle c accelerates or decelerates to merge according to the given space and speed conditions between vehicles f and c. Similarly, the function g is applied to judge in the relationship between vehicles c and b. If both A and B take 0, the distance between two vehicles f and b is large enough for vehicle c to be accommodated to enter into the main line, then the free merging occurs (no acceleration or deceleration behavior is required for vehicle c). Conversely, in the case of the forced merging, we need to examine whether the solution of inequality (8) to (11) exists. If A and B are mutually exclusive, then the following two conditions (1) and (2) are obtained.

(1) When a rapid brake event B does not exist, then $B = 0$, and only an event A could happen.

(2) When a rapid acceleration event A does not exist, then $A = 0$, and only an event B is approved.

The lane changing behavior of vehicle c could happen when a solution of (1) or (2) exists.

Psychological constraints describe driver's motivations on lane change. If the present car has not reached the desired speed and if the predicted speed of lane change is greater than that of no change, or gain speed advantage, a_1 and a_2 describe predicted acceleration of lane change and no lane change, respectively. a_1 and a_2 are given from the IDM. Then the psychological constraints can be given by

$$a_1 < a_2. \tag{12}$$

If (12) has a solution, the driver has maneuvers of changing the current lane to the target lane. Conversely, the driver does not conduct the lane changing maneuvers.

Lane change behaviors can be characterized as a sequence of three stages: the ability of lane change (physical condition); the motivation of lane change (psychological constraints); the execution of lane change. When lane change models of psychological condition and physical condition are both satisfied, the driver conducts the above-mentioned three stages. Otherwise, the driver continues the car-following behavior without lane change behaviors.

We develop a traffic flow simulation model consisting of car-following model and lane change model [10–12]. The basic concept of car-following theories is the relationship between stimuli and response. In the classic car-following theory, the stimuli are represented by the relative speed of following and leading vehicle, and the response is represented by the acceleration (or deceleration) rate of the following vehicle. The car-following model describes following behaviors that drivers follow each other in the traffic stream on only one lane. To reproduce the traffic flow in two or more lanes, lane change model which explores lane change behaviors is needed. By using the car-following model and lane change model, we express dynamic and complex traffic behaviors in two or more lanes. Moreover, traffic flow simulation models are applied to reproduce the traffic congestion represented by Helbing and Kerner [13–16].

2.2. The Reinforcement Learning Ramp Metering. Reinforcement learning is a kind of machine learning treating the problem at which the agent under a certain environment determines the action. And the action should observe and take the present state. An agent gets reward from environment by choosing actions. Reinforcement learning learns a policy from which most reward is obtained through a series of actions [17]. Reinforcement learning is a broad class of optimal control methods depending on estimating value functions from experience or simulations [18–21].

The model of reinforcement learning ramp metering (RLRM) is shown in Figure 3. qin is the inflow of the upstream of the main line; r is the metering rate; $qout$ is the outflow of the downstream of main line; dm is the density of the main line in merging section; dr is the density of onramp; vm is the average speed of the main line; vr is the average speed of onramp.

$$q = qin + r - qout. \tag{13}$$

According to the volume q in merging section, upstream traffic qin is updated by

$$qin_{t+1} \longleftarrow qin_{t+1} + q, \tag{14}$$

where qin called state variable can be collected by the control variable detector. r is set as a choosing action variable. Moreover, $qout$ is the reward based on the choosing action. ρ_L is the traffic density in the merging section of L long. ρ_L can be obtained by

$$\rho_L = \frac{qin_{t+1}}{L}. \tag{15}$$

According to Figure 4, the framework of RLRM is explained briefly. RLRM consists of metering rate choice model, outflow function, value function, and environmental model. The metering rate choice model is a rule to choose the optimal metering rate. Outflow function describes the data of downstream traffic which can be collected and calculated by detectors. Value function presents the total of volumes of downstream traffic. Environmental model predicts inflow and outflow in the next period of time depending on optimal metering rate and inflow.

2.3. RLRM with Complete Information. The RLRM with complete information faces a Markov decision problem (MDP). In addition, since inflow and metering rate's set denotes S, $A(qin)$ ($qin_t \in S$) is finite. We typically use a set of matrices

$$R^{r}_{qinqin'} = P_r\{qin_{t+1} = qin' \mid qin_t = qin, r_t = r\} \tag{16}$$

to describe the transition structure. Traffic outflow at time t is obtained by

$$R^{r}_{qinqin'} = E\{qout_{t+1} \mid qin_t = qin, r_t = r, qin_{t+1} = qin'\}, \tag{17}$$

$$\text{for all } qin \in S, \text{for all } r \in A(qin), \text{and for all } qin' \in S^{+}.$$

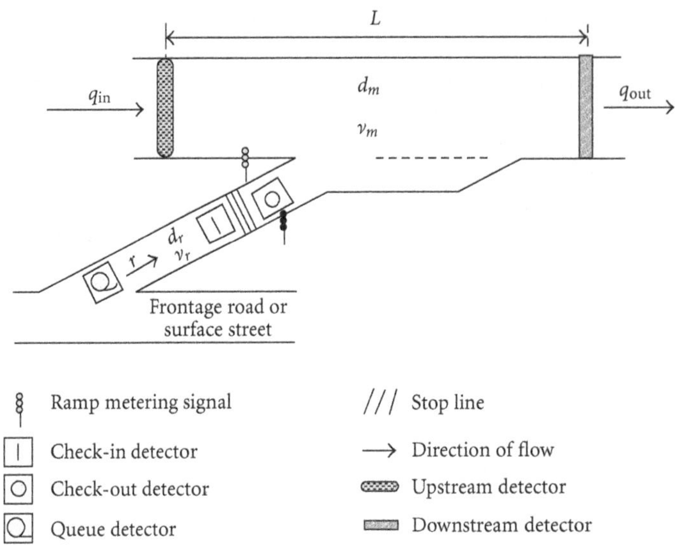

FIGURE 3: Reinforcement learning ramp metering model.

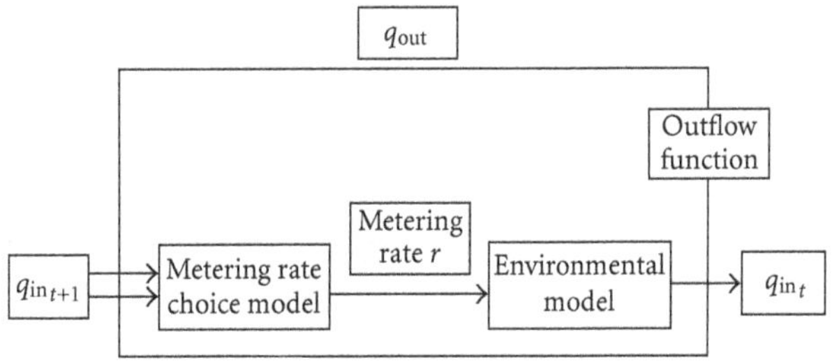

FIGURE 4: Block diagram for reinforcement learning ramp metering.

If maximum outflow V^* or Q^* is given by Bellman formula, we have

$$\begin{aligned} V^{\pi}(qin) &= \max_r E\{qout_{t+1} + \lambda V^*(qin_{t+1}) \mid qin_t = qin,\ r_t = r\} \\ &= \max_r \sum_{qin'} P^r_{qin\,qin'}\left[qout^r_{qin\,qin'} + \lambda V^{\pi}(qin')\right], \end{aligned} \tag{18}$$

or

$$\begin{aligned} Q^*(qin, r) &= E\Big\{qout_{t+1} + \lambda \max_{r'} Q^*(qin_{t+1}, r') \mid qin_t \\ &\qquad = qin, r_t = r\Big\} \\ &= \sum_{qin'} P^r_{qin\,qin'}\left[R^r_{qin\,qin'} + \lambda \max_{r'} Q^*(qin', r')\right]. \end{aligned} \tag{19}$$

We can obtain transit probability $P^r_{qin\,qin'}$ and next outflow $V^{\pi}(qin)$ with MDP's complete information. And we assume that traffic outflow is finite. Moreover, we can also compute traffic outflow.

2.4. RLRM without Complete Information. Supposed Markov decision process with complete information is given in Section 2.3. But this argument is untenable in fact. We can give ramp metering rate by using evaluation of the experience without complete information. Since transit probability is not necessary, we can rewrite (18) as

$$V^{\pi}(qin_t) = V^{\pi}(qin_t) + a_t[qout_t - V^{\pi}(qin_t)], \tag{20}$$

where $qout_t$ is real time outflow at time t, and constant a_t is transit probability function of t. Equation (19) can be replaced by

$$\begin{aligned} Q(qin_t, r_t) \longleftarrow Q(qin_t, r_t) + a_t[qout + \lambda E\{Q(qin_{t+1}, r_{t+1} \mid s_t)\} \\ - Q(qin_t, r_t)]. \end{aligned} \tag{21}$$

If expected value of metering rate is not given, we also replace

$$qout + \lambda E\{Q(qin_{t+1}, r_{t+1} s_t)\} - Q(qin_t, r_t) \tag{22}$$

by

$$qout + \lambda \sum_a \pi(qin_t, r_t) Q(qin_{t+1}, r_t) - Q(qin_t, r_t). \tag{23}$$

We get

$$\begin{aligned} Q(qin_t, r_t) &\longleftarrow Q(qin_t, r_t) \\ &\quad + a_t\left[qout + \lambda \sum_a \pi(qin_t, r_t) Q(qin_{t+1}, r_t) - Q(qin_t, r_t)\right]. \end{aligned} \tag{24}$$

We suppose that the probability of on-ramp control policy π can be obtained in (24). Here, it is difficult to satisfy the initial condition. The values $\sum_a \pi(qin_t, r_t) Q(qin_{t+1}, r_t)$ associated with an optimal on-ramp control policy are called the optimal ramp inflow and are often written as $\max Q(qin_{t+1}, r)$. We get

$$Q(qin_t, r_t) \longleftarrow Q(qin_t, r_t) + a_t[qout + \lambda \max Q(qin_{t+1}, r) - Q(qin_t, r_t)], \tag{25}$$

where

$$\sum_{t=1}^{\infty} a_t = \infty, \tag{26}$$

$$\sum_{t=1}^{\infty} a_t^2 < \infty. \tag{27}$$

In the (25), the action value function Q is gained by learning approximates Q^* (the optimal action value function) directly by using current policy. The state variable can be updated depending on the policy.

When the traffic reaches the jam density, it is possible to result in closure of the ramp for a long period of time, which must be taken into consideration. Maximum of waiting time ($T_{\max}$) and its metering rate (r_T) are given. When $\sum_{n=1}^{m} TS_n > T_{\max}$, the control (qin_t, r_T) is selected. In order to remove the curse of dimensionality, the discrete equation of the continuous variable r_t is represented. The average difference between 0 and $r_{\max}$ is divided by r_n. r_n is given by

$$N_r = \text{cell}\left(\frac{r_{\max}}{r_n}\right), \tag{28}$$

where N_r is the amount of the metering rate, and *cell* is the function of the bottom integral function. The metering rate is $\max(kr_n, r_{\max})$ for $k \in N$.

The algorithm of reinforcement learning on-ramp metering is shown in Figure 5.

(1) Initialize Q, $qout$, qin, and k.

(2) Determine cycle time of a traffic signal t.

(3) Update qin_t.

(4) Give metering rate by $r_t = k \times r_n$.

(5) Determine the traffic state (qin_t, r_t).

(6) Generate the density ρ_L by using traffic simulation and choose the metering rate.

(7) If $r_t < r_{\max}$, then update $k = k + 1$ and go to (4), and otherwise generate the optimal control (qin_t, r^*).

(8) If one closes the ramp, then update waiting time T by $T = T + t$, and otherwise initialize the waiting time T by $T = 0$. If $T > T_{\max}$, then update metering rate by $r_T \rightarrow r^*$.

(9) Operate the optimal control (qin_t, r^*) and update Q.

When the cycle time t is over, determine to continue the ramp metering. If yes, then collect the data of inflow qin_{t+1}, go to (3), and update qin_t, that is, $qin_{t+1} \rightarrow qin_t$; otherwise, complete the ramp metering.

Table 1: RLRM parameters.

$T_{\max}$ (min)	r_T	$r_{\max}$	r_n	N_r
5	200	1100	100	11

Table 2: Traffic inflow.

Case	A	B	C	D	E	F
Inflow of main line (pcu/hour)	1200	1500	1800	1800	2500	2500
Inflow of ramp (pcu/hour)	300	300	600	900	600	900
Total inflow	1500	1800	2400	2700	3100	3400

3. Data Combination and Reduction

Our aim is to design a reinforcement learning control law for the ramp metering controller without complete information. We need to control the inflow from the ramp into main line, and the metering rate should be given by traffic states. Traffic flow simulation is conducted to demonstrate this control of the ramp metering. In our simulation, we set the main line length on highways to 1000 m, ramp length to 200 m, and length in merging sections of the main line and ramp to100 m. Parameters of RLRM are shown in Table 1, and the metering rate matrix is $\{0, 100, 200, 300, \ldots\ldots, 900, 1000, 1100\}$.

Table 2 shows the inflow of cases A, B, C, D, E, and F. Inflow rate of the main line increases from 1200 pcu/hour of case A to 2500 pcu/hour of case F. Moreover, inflow rate of ramp rises from 300 pcu/hour of case A to 900 pcu/hour of case F. The cycle length of the fixed-time control is 20 s which consists of 15 s green time and 5s red time.

4. Result and Discussion

The results of no control, fixed-time control, and RLRM are shown in Figures 6–9. Total inflow increases from 1500 pcu/h in case A to 3400 pcu/h in case F. Figure 6 presents average speed and its rate compared to no control. The average speed of no control, about 108 km/h, is faster than fixedtime and RLRM in case A. The similar results are shown in case B. The average speed of no control, about 79 km/h, is faster than fixedtime and is slower than RLRM in case C. The average speed of no control, about 51 km/h, is slower than fixedtime and RLRM in case F. According to the average speed, rates of congestion reliefs of fixed-time control from case A to case F arrive at −7.80%, −6.65%, −3.77%, 0. 26%, 2.70%, and 8.26%, respectively. In addition, rates of congestion reliefs of RLRM from case A to case F arrive at −6.31%, −6.49%, 5.69%, 13.55%, 20.50%, and 18.18%, respectively.

Figure 7 describes density and its rate compared to no control. Densities of fixed-time control and RLRM are about 38 pcu/km, an about 60% increase, in case A. Densities of fixed-time control and RLRM are about 52 pcu/km and 45 pcu/km, about 11.46% and 22.60% decreases, in case C. Densities of fixed-time control, no control, and RLRM are about 120 pcu/km. According to densities, rates of congestion reliefs of fixed-time control from case A to case F

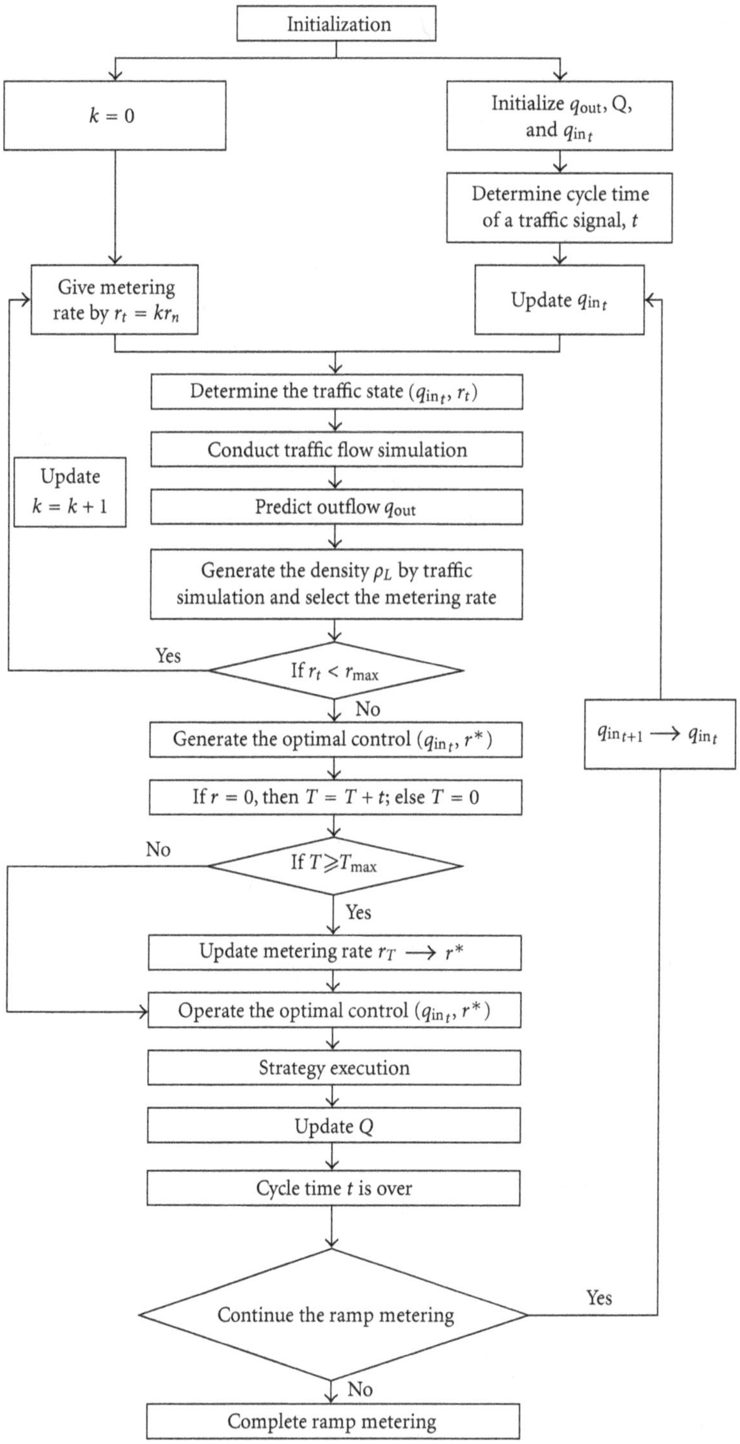

FIGURE 5: Algorithm of reinforcement learning ramp metering.

arrive at −57.55%, −19.92%, −11.46%, −21.35%, 7.6%, and 0.39%, respectively. In addition, rates of congestion reliefs of RLRM from case A to case F arrive at −59.59%, −22.05%, 22.60%, 8.18%, 9.65%, and 3.42%, respectively.

Figure 8 shows outflow and its rate compared to no control. Outflow rate rises from 1700 pcu/h without control to 2308 pcu/h with fixed-time control and 1800 pcu/h with RLRM in case A. Moreover, 3.82% and 7.85% increases are shown depending on outflow rate in case C. In addition, 18.97% and 30.65% increases are explored depending on outflow rate in case F. Rates of congestion reliefs of fixed-time control from case A to case F arrive at 35.76%, −14.25%, 7.82%, 7.93%, 12.51%, and 18.97%, respectively. On the other hand, rates of congestion reliefs of RLRM from case

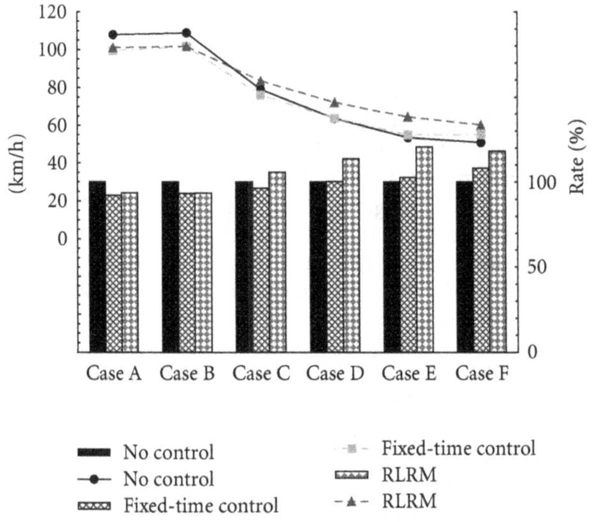

FIGURE 6: Average speed and its rate compared to no control.

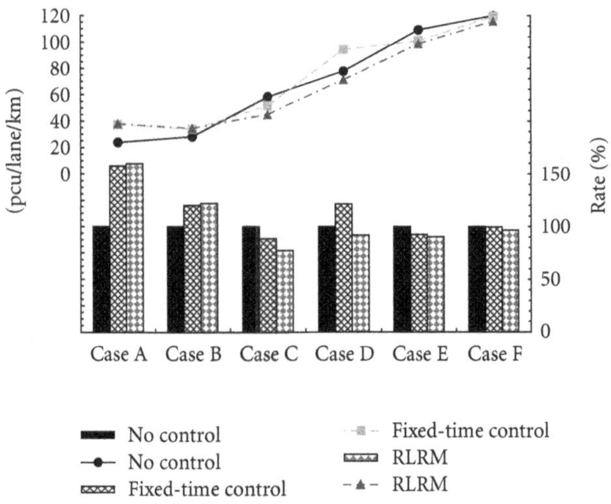

FIGURE 7: Density and its rate compared to no control.

A to case F arrive at 7.06%, 0.58%, 3.85%, 10.47%, 54.63%, and 30.65%, respectively.

Figure 9 represents travel time and its rate compared to no control. According to travel time, 6.25% and 9.38% increases are explored in case A. Travel time rises from 342 s without control to 370 s with fixed-time control and falls into 330 s with RLRM in case C. Travel time falls from 617 s to 469 s with fixed-time control and 343 s with RLRM in case F. Rates of congestion reliefs of fixed-time control from case A to case F arrive at −6.25%, −25.26%, −8.19%, 7.36%, 27.06%, and 23.99%, respectively. On the other hand, rates of congestion reliefs of RLRM from case A to case F arrive at −9.38%, −5.26%, 3.51%, 38.17%, 40.32%, and 44.41%, respectively.

According to Figures 6–9 when the traffic inflows are low, controls not efficient. Controls get efficient with the traffic inflows increasing. Controls are very efficient, and RLRM is optimal control when the traffic inflows are high. Moreover,

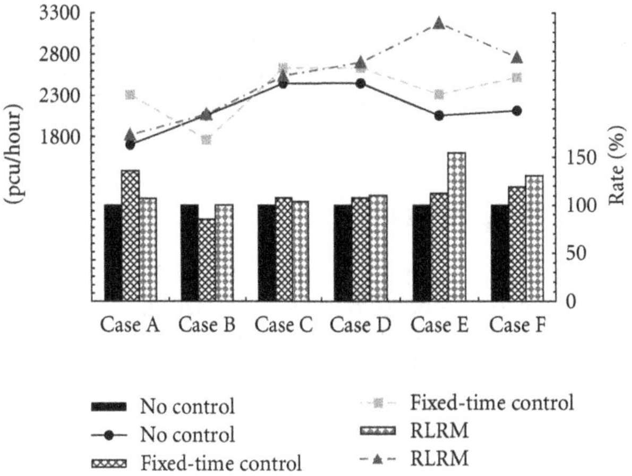

FIGURE 8: Outflow and its rate compared to no control.

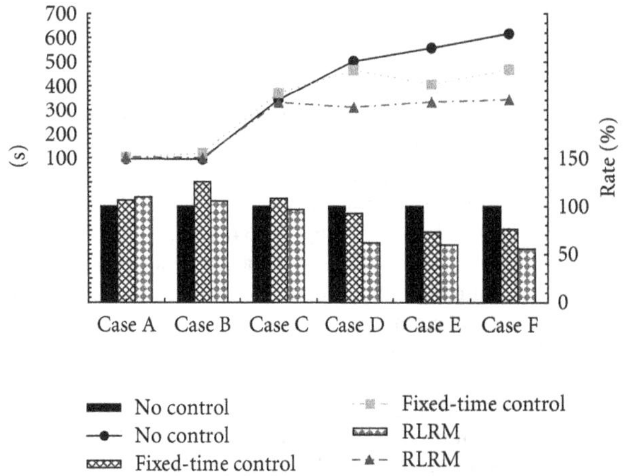

FIGURE 9: Travel time and its rate compared to no control.

based on curves of Figures 7–9 assessment indicators of fixed-time control fluctuate around indicators of no control. Fixed-time control shows instability compared to RLRM. Abilities of automaticity, memory, and performance feedback of RLRM are also shown.

5. Conclusion

The on-ramp metering ensures that traffic moves at a speed approximately equal to the optimum speed which results in maximum flow rates or travel time. This study develops an RLRM model without complete information, which consists of prediction tools depending on traffic flow simulation and optimal choice model based on reinforcement learning theories. Numerical cases are given to demonstrate RLRM compared to no control and fixed-time control. In addition, densities and outflow rates are calculated. Moreover, average speeds are computed, and travel times are assessed. According to cases A, B, C, D, E, and F, fixed-time control and RLRM are discussed depending on average speeds, densities, outflow rates, and travel times. When traffic inflow is low, controls are not efficient, and there are little differences among no control, fixed-time control, and RLRM. On the other hand, when traffic inflow is high, controls are very efficient, and RLRM is optimal control. Moreover, the greater is inflow, the more is the effect. In addition, the stability of RLRM is better than fixed-time control.

Acknowledgments

This research is founded by the National Natural Science Foundation of China (Grant no. 51008201). And this research is also sponsored by the Scientific Research Foundation for the Returned Overseas Chinese Scholars, State Education Ministry of China. Moreover, this research is also the key project supported by the Scientific Research Foundation, Education Department of Hebei Province of China (Grant no. GD2010235) and Society Science Development Program of Hebei Province of China (Grant no. 201004068).

References

[1] M. Papageorgiou and A. Kotsialos, "Freeway ramp metering: an overview," *IEEE Transactions on Intelligent Transportation Systems*, vol. 3, no. 4, pp. 271–281, 2002.

[2] M. Papageorgiou, H. H. Salem, and J. M. Blosseville, "A local feedback control law for on-ramp metering," *Transportation Research Record*, vol. 1320, pp. 58–64, 1991.

[3] M. Papageorgiou, H. Hadj-Salem, and F. Middelham, "ALINEA local ramp metering: summary of field results," *Transportation Research Record*, no. 1603, pp. 90–98, 1997.

[4] M. Iwata, "The comparative study about the ramp metering in high way," in *Proceedings of the Japan Society of Civil Engineers (JSTE '06)*, vol. 26, pp. 73–76, 2006.

[5] X. J. Wang, B. H. Liu, X. Q. Niu, and T. Miyagi, "Reinforcement learning control for on-ramp metering based on traffic simulation," in *Proceedings of the 9th International Conference of Chinese Transportation Professionals (ICCTP '09)*, vol. 358, pp. 2701–2707, Harbin, China, 2009.

[6] M. Treiber, A. Hennecke, and D. Helbing, "Congested traffic states in empirical observations and microscopic simulations," *Physical Review E*, vol. 62, no. 2, pp. 1805–1824, 2000.

[7] P. Hidas, "Modelling vehicle interactions in microscopic simulation of merging and weaving," *Transportation Research Part C*, vol. 13, no. 1, pp. 37–62, 2005.

[8] X. J. Wang, T. Miyagi, A. Takagi, and J. Q. Ying, "Analysis of the effects of acceleration lane length at merging sections by using micro-simulations," in *Proceedings of the 7th International Conference of Eastern Asia Society for Transportation Studies*, Dalian, China, 2007.

[9] X. J. Wang, T. Miyagi, and J. Q. Ying, "A simulation model for traffic behavior at merging sections in highways," in *Proceedings of the 2nd International Conference on Innovative Computing, Information and Control (ICICIC '07)*, Kumamoto, Japan, September 2007.

[10] X. J. Wang, G. F. Gao, J. J. Chen, and T. Miyagi, "Traffic flow simulation model based on desired speed," *Journal of Chang'an University*, vol. 30, no. 5, pp. 79–84, 2010.

[11] X. J. Wang, G. F. Gao, J. J. Chen, and T. Miyagi, "Traffic flow simulation model based on adaptive acceleration at merging sections in highways," *Highway*, vol. 10, pp. 128–132, 2010.

[12] X. J. Wang and T. Miyagi, "Reinforcement learning ramp metering," *Journal of Shijiazhuang Tiedao University*, vol. 2, pp. 104–108, 2010.

[13] D. Helbing, "High-fidelity macroscopic traffic equations," *Physical Review*, vol. 219, no. 3-4, pp. 391–407, 1995.

[14] M. Treiber, A. Hennecke, and D. Helbing, "Derivation, properties, and simulation of a gas-kinetic-based, nonlocal traffic model," *Physical Review E*, vol. 59, no. 1, pp. 239–253, 1999.

[15] D. Helbing, A. Hennecke, V. Shvetsov, and M. Treiber, "Micro-and macro-simulation of freeway traffic," *Mathematical and Computer Modelling*, vol. 35, no. 5-6, pp. 517–547, 2002.

[16] B. S. Kerner, "Synchronized flow as a new traffic phase and related problems for traffic flow modelling," *Mathematical and Computer Modelling*, vol. 35, no. 5-6, pp. 481–508, 2002.

[17] S. Richard and G. B. Andrew, *Reinforcement Learning: An Introduction*, MIT Press, London, UK, 1998.

[18] H. Boubertakh, M. Tadjine, P. Y. Glorennec, and S. Labiod, "Tuning fuzzy PD and PI controllers using reinforcement learning," *ISA Transactions*, vol. 49, no. 4, pp. 543–551, 2010.

[19] R. Razavi, S. Klein, and H. Claussen, "A fuzzy reinforcement learning approach for self-optimization of coverage in LTE networks," *Bell Labs Technical Journal*, vol. 15, no. 3, pp. 153–176, 2010.

[20] D. Vengerov, "A reinforcement learning framework for utility-based scheduling in resource-constrained systems," *Future Generation Computer Systems*, vol. 25, no. 7, pp. 728–736, 2009.

[21] P. Venkatraman, B. Hamdaoui, and M. Guizani, "Opportunistic bandwidth sharing through reinforcement learning," *IEEE Transactions on Vehicular Technology*, vol. 59, no. 6, Article ID 5452965, pp. 3148–3153, 2010.

6

Event-Based Stabilization over Networks with Transmission Delays

Xiangyu Meng and Tongwen Chen

Department of Electrical and Computer Engineering, University of Alberta, Edmonton, AB, Canada T6G 2V4

Correspondence should be addressed to Xiangyu Meng, xmeng2@ece.ualberta.ca

Academic Editor: Peter X. Liu

This paper investigates asymptotic stabilization for linear systems over networks based on event-driven communication. A new communication logic is proposed to reduce the feedback effort, which has some advantages over traditional ones with continuous feedback. Considering the effect of time-varying transmission delays, the criteria for the design of both the feedback gain and the event-triggering mechanism are derived to guarantee the stability and performance requirements. Finally, the proposed techniques are illustrated by an inverted pendulum system and a numerical example.

1. Introduction

Traditional control theory is built on the idea of perfect information flow from the sensor to the controller and from the controller to the actuator, that is, there is no delay and the transmitted signals are equal to received signals. However, this is not true for control loop closed over networks, where the actuators, sensors, and controllers are distributed in a wide geographical area, operating via some communication networks, such as DeviceNet, Ethernet, and FireWire, to name a few [1]. Because of the network uncertainties, data packets can be delayed, dropped, or reordered which make closed-loop control very difficult. Therefore, control over networks appears and has been drawing more and more attention in recent years from researchers working in the areas of systems and control [2–5]. A typical feature in the literature lies in the periodic execution of the control task due to the ease of analysis and design. However, the time synchronization problem presents a challenge in digital control applications when dealing with multiple sampling rates and systems with distributed computing devices; sampling jitter, time-varying delays, and coding errors introduced by networked distributed systems may degrade the performance or even cause closed-loop instability. On the other hand, periodic sampling only considers the system dynamics at every sampling instance triggered by a clock, and it does not take into account the constraints of both computer resources and communication bandwidth. Hence, the communication resources usage in this control scheme is inefficient.

To relax the periodicity assumption, event triggering tech-niques are proposed. Various terms are used to express event-based sampling strategy: the level crossing sampling [6], the magnitude-driven sampling, and, sometimes, sampling in the amplitude domain, Lebsegue sampling [7]. In the sensor network community, the magnitude-driven or level crossing sampling is known as send-on-delta [8] or deadbands [9]. By contrast, event-based communication mechanisms use resources more efficiently by invoking operation only when a specific event occurs in the system, which guarantees relatively little communication effort. Due to easy implementation, event-based control mechanisms have been used in industry for some time, ranging in sectors from oil and gas, power and utility, to manufacturing. Unfortunately there is little methodology available for event-based control at the beginning. This could be explained as the mathematical difficulties since event-based control integrates discrete logic functions with continuously evolving system dynamics. However, most existing control design methods focus only on dynamic systems, while ignoring logic constraints. Early results on discontinuous systems and impulsive control were used to solve problems concerning event-based control. Event-based control can also be viewed as a special case of hybrid systems from the viewpoint of the continuous variables and discrete transition associated with events. Recently, a few fundamental results have been reported for event-based control [10–18] and estimation [19]. Overall, the research of event-based control is still in its infancy, and the results obtained are still very limited, contrasting to its wider applications in practical control problems. The various benefits of event-based control necessitate overcoming the difficulties in the analysis and design of this type of control strategies.

In this paper, further results on event-based control recently dealt with in [10] are presented, where an event generation condition based on the control error is proposed rather than the state error considered in [10]. This is achieved by adopting the topology that the sensor, controller, and event detector reside on the same node in the network. The event detector contains sophisticated logic devices to trigger an event when the control error norm reaches a certain proportion of the state norm, and then send the current control signal to the actuator node. The effect of time-varying network transmission delays is considered instead of a constant computational delay as in [10]. The defined events guarantee that the controller designed can stabilize the event-based control system. The relationships between the parameters of the event detector, the upper bound of transmission delays, and the feedback gain are also established. Moreover, the feasibility of this event-based scheme is verified by estimating the lower bound of the difference between two consecutive event times. Two simulation examples are presented to illustrate the proposed approach.

The rest of this paper is organized as follows. Section 2 presents the event-triggered problem with consideration of time-varying transmission delays. Based on a control error event-triggered scheme, design for both the controller and the parameter of the event detector for event-based control systems is proposed in Section 3. Two simulation examples are given in Section 4 to demonstrate the advantage of the event-triggered algorithm. Finally, Section 5 concludes the paper.

2. Problem Statement

Consider the following continuous-time linear system described by

$$\dot{x}(t) = Ax(t) + Bu(t), \tag{1}$$

where $x(t) \in \mathbb{R}^n$ and $u(t) \in \mathbb{R}^m$ denote the system state vector, and control input, respectively. We make the assumption that all state variables are measurable. The parameter matrices A and B are known with appropriate dimensions.

Different from traditional control systems that the interconnection between the plant and the actuator is transparent, the actuator considered is connected to the system in (1) through a communication link. In this case, the usual assumption that the transmitted signals are equal to received signals is no longer applicable. As is well known, the periodic sampling mechanism has been used implicitly for a few decades due to the ease of analysis and design. However, the communication resources may be used unnecessarily in some situations. The event-based control scheme is used in this paper to avoid the overload of the network transmission and reduce the network bandwidth usage.

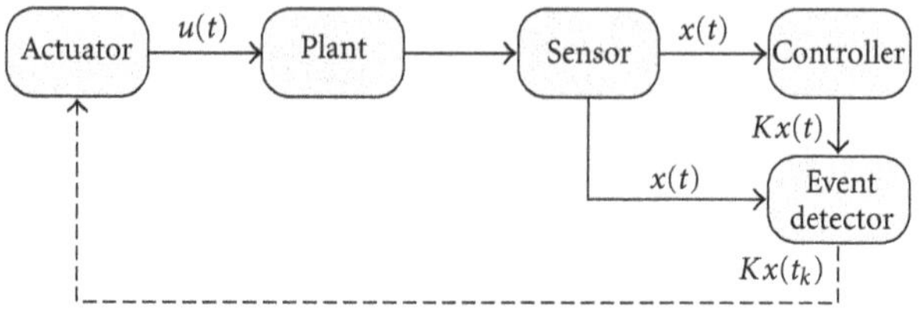

Figure 1: Event-based control loop.

The topological structure of the considered event-based control system consisting of a continuous-time linear time-invariant system, a sensor, a continuous-time state-feedback controller, an event detector, and an actuator is shown in Figure 1. The signal is transmitted continuously along the solid lines among the actuator, the plant, and the smart event detector; whereas the communication link denoted by the dash line is only used after an event has been generated. As Figure 1 suggests, the sensor, controller, and event detector reside on the same node in the network. The sensor samples the plant continuously, then the sampled signals are sent to the controller and the event detector. The event detector determines whether the control signal will be sent out through the network by using the event condition. Under the event condition, a sequence of time instants, $t_0, t_1, t_2, \ldots$, is determined as the event time, where $t_0 = 0$ is the initial time. The inter-event time is defined as $t_{k+1} - t_k$ which corresponds to the release period given by the event detector. Throughout this paper, it is assumed that the elapsed time between the sensor measurement and the event detector decision is negligible, the data is transmitted with a single packet, and packet loss does not occur in transmission. Thus the only effect considered for network uncertainties is the transmission delays on the system. Suppose the delay τ_k in the network communication is time varying and $\tau_k \in [0, \tau]$, where τ is a constant scalar representing the maximum delay. The control signal $u(t_0), u(t_1), u(t_2), \ldots$ will arrive at the actuator side at the instants $t_0 + \tau_0, t_1 + \tau_1, t_2 + \tau_2, \ldots$, respectively. Moreover, the actuator will hold the control value $u(t_k)$ and drive the controlled plant until a new message arrives. Therefore, the output of the actuator can be expressed as

$$u(t) = Kx(t_k), \qquad \text{for } t \in [t_k + \tau_k, t_{k+1} + \tau_{k+1}), \tag{2}$$

where K is a state feedback gain matrix of appropriate dimension to stabilize $A + BK$.

Based on the above analysis, the system model under the controller in (2) with event-based communication over networks can be described by

$$\dot{x}(t) = Ax(t) + BKx(t_k), \qquad \text{for } t \in [t_k + \tau_k, t_{k+1} + \tau_{k+1}). \tag{3}$$

The purpose of this paper is to design an event condition rendering the event-based control system in (3) asymptotically stable.

3. Main Results

Inspired by [10], the event detector has the following form:

$$\|e_k(t)\| < \sigma \|x(t)\|, \tag{4}$$

where $e_k(t)$ is defined as the control error between the currently computed control value and the previously submitted one

$$e_k(t) = K(x(t_k) - x(t)), \qquad \text{for } t \in [t_k, t_{k+1}) \tag{5}$$

and σ is a positive scalar to be determined later.

Remark 1. In the literature, event generation is usually based on the difference between the current plant state and the previously submitted plant state or the difference between the plant state and the state of a reference model. The results presented in this paper are based on the assumption that the event detector is implemented with respect to control errors. This does make sense in a networked control setting where the shared resource is the transmission medium.

Remark 2. To reduce communication bus load, the computed control signal satisfying the event condition in (4) will not be sent to update the actuator. Only the one that violates the inequality in (4) will be transmitted, but the communication will not be invoked when the system is in steady state. Intuitively, decreasing the value of σ has the effect of shrinking the average inter-event time since the ratio between the control error and the state will need less time to reach the threshold as the value of σ decreases. Particularly, the event-triggered scheme reduces to a continuous communication case when $\sigma = 0$.

Combining the definition of $e_k(t)$, the dynamic of the digitally implemented control system for $t \geq t_k + \tau_k$ can be described by

$$\begin{aligned}\dot{x}(t) &= Ax(t) + BKx(t_k) \\ &= (A + BK)x(t) + BK(x(t_k) - x(t)) \\ &= (A + BK)x(t) + Be_k(t).\end{aligned} \tag{6}$$

If we consider the control error as a perturbation, it is natural to apply the perturbation method [20]. The input-to-state stable Lyapunov function candidate $V(x,t) = x^T(t)Px(t)$ with respect to control errors $e(t)$ is used to investigate the stability for the event-based control system. The derivative of $V(x,t)$ along the trajectories of (6) is given by

$$\begin{aligned}\frac{\partial V(x,t)}{\partial t} &= \frac{\partial V(x,t)}{\partial x(t)}((A + BK)x(t) + Be_k(t)) \\ &= x^T(t)\left((A + BK)^T P + P(A + BK)\right)x(t) \\ &\quad + x^T(t)PBe_k(t) + e_k^T(t)B^T Px(t) \\ &= -x^T(t)Qx(t) + 2x^T(t)PBe_k(t),\end{aligned} \tag{7}$$

where Q is a symmetric matrix defined by

$$(A + BK)^T P + P(A + BK) + Q = 0. \tag{8}$$

From (7), it can be obtained that

$$\frac{\partial V(x,t)}{\partial t} \leq -\lambda_{\min}(Q)\|x\|^2 + 2\|PB\|\,\|x\|\,\|e_k\|. \tag{9}$$

Thus based on the Lasalle's invariance principle, a sufficient condition to guarantee the asymptotic stability is

$$\|e_k(t)\| < \sigma_1 \|x(t)\|, \tag{10}$$

where

$$\sigma_1 = \frac{\lambda_{\min}(Q)}{2\|PB\|}. \tag{11}$$

The next event instant is given by

$$t_{k+1} = \inf\{t > t_k \mid \|e_k(t)\| \geq \sigma \|x(t)\|\}. \tag{12}$$

Choose $\sigma \leq \sigma_1$, which implies the asymptotic stability of the system in (3) for $t \in [t_k + \tau_k, t_{k+1})$ since the inequality in (10) is satisfied.

Now consider the interval $[t_{k+1}, t_{k+1} + \tau_{k+1})$. First look at the dynamic of $\|e_k(t)\|/\|x(t)\|$:

$$\begin{aligned}\frac{d}{dt}\frac{\|e_k(t)\|}{\|x(t)\|} &= \frac{d}{dt}\frac{\left[e_k^T(t)e_k(t)\right]^{1/2}}{\left[x^T(t)x(t)\right]^{1/2}} \\ &= -\frac{e_k^T(t)K\dot{x}(t)}{\|e_k(t)\|\,\|x(t)\|} - \frac{x(t)^T\dot{x}(t)}{\|x(t)\|^2}\frac{\|e_k(t)\|}{\|x(t)\|} \\ &\leq \frac{\|e_k(t)\|\,\|K\|\,\|\dot{x}(t)\|}{\|e_k(t)\|\,\|x(t)\|} + \frac{\|x(t)\|\,\|\dot{x}(t)\|}{\|x(t)\|\,\|x(t)\|}\frac{\|e_k(t)\|}{\|x(t)\|} \\ &= \left(\|K\| + \frac{\|e_k(t)\|}{\|x(t)\|}\right)\frac{\|\dot{x}(t)\|}{\|x(t)\|} \\ &\leq \left(\|K\| + \frac{\|e_k(t)\|}{\|x(t)\|}\right) \\ &\quad \times \frac{\|A + BK\|\,\|x(t)\| + \|B\|\,\|e_k(t)\|}{\|x(t)\|} \\ &= \|K\|\,\|A + BK\| + (\|K\|\,\|B\| + \|A + BK\|) \\ &\quad \times \frac{\|e_k(t)\|}{\|x(t)\|} + \|B\|\left(\frac{\|e_k(t)\|}{\|x(t)\|}\right)^2.\end{aligned} \tag{13}$$

The comparison lemma in [20] as a tool to compute bounds on a solution without computing the solution itself can be used to estimate the bound on $\|e_k(t)\|/\|x(t)\|$. Consider the scalar differential equation

$$\dot{\varphi}(t) = \alpha\varphi^2(t) + (\alpha\beta + \gamma)\varphi(t) + \beta\gamma, \tag{14}$$

where

$$\alpha = \|B\|, \qquad \beta = \|K\|, \qquad \gamma = \|A + BK\|. \tag{15}$$

Let $[t_{k+1}, t_{k+1} + \tau)$ be the interval of existence of the solution $\varphi(t)$. Recall that (12) implies $e_k(t) = \sigma x(t)$ at event instant $t = t_{k+1}$ and thus $\|e_k(t_{k+1})\|/\|x(t_{k+1})\| = \sigma$. Let

$$\sigma \le \varphi(t_{k+1}) = \sigma_2 < \sigma_1. \tag{16}$$

Then $\|e_k(t)\|/\|x(t)\| \le \varphi(t)$ for all $t \in [t_{k+1}, t_{k+1}+\tau)$. Rewrite (14) as

$$\begin{aligned}\dot{\varphi}(t) &= \alpha\varphi^2(t) + (\alpha\beta + \gamma)\varphi(t) + \beta\gamma \\ &= \alpha\left(\varphi^2(t) + \frac{\alpha\beta + \gamma}{\alpha}\varphi(t) + \frac{\beta\gamma}{\alpha}\right) \\ &= \alpha\left[\left(\varphi(t) + \frac{\alpha\beta+\gamma}{2\alpha}\right)^2 - \frac{(\gamma - \alpha\beta)^2}{4\alpha^2}\right].\end{aligned} \tag{17}$$

Denote $(\gamma - \alpha\beta)^2/4\alpha^2$ as q^2, and take the transform

$$\varphi(t) + \frac{\alpha\beta + \gamma}{2\alpha} = s(t), \qquad d\varphi = ds, \tag{18}$$

then

$$\begin{aligned}\tau &= \int_{t_{k+1}}^{t_{k+1}+\tau} dt = \frac{1}{\alpha}\int_{\varphi(t_{k+1})}^{\varphi(t_{k+1}+\tau)} \frac{1}{\varphi^2 + ((\alpha\beta+\gamma)/\alpha)\varphi + \beta\gamma/\alpha} d\varphi \\ &= \frac{1}{\alpha}\int_{\varphi(t_{k+1})+((\alpha\beta+\gamma)/2\alpha)}^{\varphi(t_{k+1}+\tau)+((\alpha\beta+\gamma)/2\alpha)} \frac{1}{s^2 - q^2} ds \\ &= \frac{1}{2q\alpha}\ln\frac{s-q}{s+q}\Big|_{\varphi(t_{k+1})+((\alpha\beta+\gamma)/2\alpha)}^{\varphi(t_{k+1}+\tau)+((\alpha\beta+\gamma)/2\alpha)} \\ &= \frac{1}{\gamma - \alpha\beta}\ln\frac{(\varphi(t_{k+1}+\tau)+\beta)(\varphi(t_{k+1})+\gamma/\alpha)}{(\varphi(t_{k+1}+\tau)+\gamma/\alpha)(\varphi(t_{k+1})+\beta)}.\end{aligned} \tag{19}$$

The desired upper bound for σ is obtained by solving the last equation in (19) with

$$\varphi(t_{k+1} + \tau) = \sigma_1, \qquad \varphi(t_{k+1}) = \sigma_2, \tag{20}$$

that is,

$$\sigma_2 = \frac{\gamma(\sigma_1 + \beta) - \beta(\alpha\sigma_1 + \gamma)e^{(\gamma-\alpha\beta)\tau}}{(\alpha\sigma_1 + \gamma)e^{(\gamma-\alpha\beta)\tau} - \alpha(\sigma_1 + \beta)}. \tag{21}$$

The inequality $\sigma \le \sigma_2 < \sigma_1$ implies that (10) can be guaranteed for $t \in [t_{k+1}, t_{k+1} + \tau_{k+1})$ by generating an event at time instant t_{k+1}.

In addition, another constraint needs to be enforced on σ to guarantee that there is no event being generated for the time $t \in [t_{k+1}, t_{k+1} + \tau_{k+1})$, that is, $\|e_{k+1}(t)\|/\|x(t)\| < \sigma$. At $t = t_{k+1}$, an event occurs, and the control error changes from $e_k(t) = K(x(t_k) - x(t))$ to $e_{k+1}(t) = K(x(t_{k+1}) - x(t))$. To avoid the out-of-order transmission for $t \in [t_{k+1}, t_{k+1} + \tau_{k+1})$, the dynamic of $\|e_{k+1}(t)\|/\|x(t)\|$ should be bounded by σ. Follow the same arguments as (13) to get

$$\begin{aligned}\frac{d}{dt}\frac{\|e_{k+1}(t)\|}{\|x(t)\|} &= \frac{d}{dt}\frac{\left[e_{k+1}^T(t)e_{k+1}(t)\right]^{1/2}}{\left[x^T(t)x(t)\right]^{1/2}} \\ &= -\frac{e_{k+1}^T(t)K\dot{x}(t)}{\|e_{k+1}(t)\|\|x(t)\|} - \frac{x^T(t)\dot{x}(t)}{\|x(t)\|^2}\frac{\|e_{k+1}(t)\|}{\|x(t)\|} \\ &\le \frac{\|e_{k+1}(t)\|\|K\|\|\dot{x}(t)\|}{\|e_{k+1}(t)\|\|x(t)\|} \\ &\quad + \frac{\|x(t)\|\|\dot{x}(t)\|}{\|x(t)\|\|x(t)\|}\frac{\|e_{k+1}(t)\|}{\|x(t)\|} \\ &= \left(\|K\| + \frac{\|e_{k+1}(t)\|}{\|x(t)\|}\right)\frac{\|\dot{x}(t)\|}{\|x(t)\|} \\ &\le \left(\|K\| + \frac{\|e_{k+1}(t)\|}{\|x(t)\|}\right) \\ &\quad \times \frac{\|A+BK\|\|x(t)\| + \|B\|\|e_k(t)\|}{\|x(t)\|} \\ &= \left(\|K\| + \frac{\|e_{k+1}(t)\|}{\|x(t)\|}\right) \\ &\quad \times \left(\|A+BK\| + \|B\|\frac{\|e_k(t)\|}{\|x(t)\|}\right),\end{aligned} \tag{22}$$

where $\|e_k(t)\|/\|x(t)\|$ is bounded by $\varphi(t)$, which can be found by solving the last equation in (19) with $\varphi(t_{k+1}) = \sigma_2$. By the comparison principle, an upper bound for the evolution of the ratio $\|e_{k+1}(t)\|/\|x(t)\|$ can be immediately obtained by solving

$$\dot{\phi}(t) = (\|K\| + \phi(t))(\|A + BK\| + \|B\|\varphi(t)), \tag{23}$$

with $\|e_{k+1}(t)\|/\|x(t)\| \le \phi(t)$. Furthermore, it follows from (23) and $\phi(t_{k+1}) = 0$ that

$$\begin{aligned}&\phi(t_{k+1} + \tau) \\ &\quad = \|K\|\left(\exp\left(\int_{t_{k+1}}^{t_{k+1}+\tau}(\|A+BK\| + \|B\|\varphi(t))dt\right) - 1\right) \\ &\quad \le \|K\|\left(\exp\left(\int_{t_{k+1}}^{t_{k+1}+\tau}(\|A+BK\| + \|B\|\sigma_1)dt\right) - 1\right) \\ &\quad = \|K\|(\exp((\|A+BK\| + \|B\|\sigma_1)\tau) - 1) \\ &\quad = \sigma_3,\end{aligned} \tag{24}$$

which implies that there is no another event being triggered before the termination of the previous one if $\sigma \ge \sigma_3$. The lower bound of the difference between two consecutive event times is described by $\tau_k + \eta$, where η is the time for $\varphi(t)$ to evolve from σ_3 to σ. Substituting the corresponding values in (19), thus

$$\eta = \frac{1}{\gamma - \alpha\beta}\ln\frac{(\sigma + \beta)(\sigma_3 + \gamma/\alpha)}{(\sigma + \gamma/\alpha)(\sigma_3 + \beta)} \tag{25}$$

is obtained.

Hence, the following theorem can be concluded.

Theorem 3. *For a given parameter τ and any $k \in \mathbb{N}$, the event condition in (4) with*

$$\sigma_3 \leq \sigma \leq \sigma_2 \tag{26}$$

enforced for any $t \in [t_k, t_{k+1})$ and the control law in (2) with K given by (8) executed for any $t \in [t_k + \tau_k, t_{k+1} + \tau_{k+1})$ guarantee that the system in (1) under the event-based control is asymptotically stable, and the inter-event intervals are lower bounded by $\tau_k + \eta$, where η is given in (25).

Remark 4. Theorem 3 provides a useful way of design for both the feedback gain K and the trigger parameter σ. Moreover, the information of the transmission delays is also involved. Therefore, the method can be used to tackle the case with time-varying network transmission delays. For given upper bound τ on the transmission delays, by solving (8), (21), (24), the corresponding feedback gain and trigger parameter can be obtained, which can be used to guarantee the required performance even though the transmission delays exist in the network communication.

Remark 5. The maximum τ can be solved by maximizing σ_1 in terms of (8) and letting $\sigma_2 = \sigma_3$. However, how to find the optimal value of σ_1 is still open. Appropriately selecting the value of σ_1 will lead to a relatively larger value of the upper bound τ.

Note that if $\tau_k = 0$, that is, no transmission delay or the effect of the transmission delay can be omitted, Theorem 3 reduces to the result in the following corollary.

Corollary 6. *For any $k \in \mathbb{N}$, the event condition in (4) with*

$$\sigma \leq \sigma_1 \tag{27}$$

enforced and the control law in (2) with K given by (8) executed for any $t \in [t_k, t_{k+1})$ guarantee that the system in (1) under the event-based control is asymptotically stable, and the inter-event intervals are lower bounded by η, where η satisfies

$$\eta = \frac{1}{\gamma - \alpha\beta} \ln \frac{(\sigma + \beta)\gamma}{(\alpha\sigma + \gamma)\beta}. \tag{28}$$

Proof. The inequality $\sigma \leq \sigma_1$ implies that (10) can be guaranteed, thus the controller renders the closed-loop system asymptotically stable for any $t \in [t_k, t_{k+1})$ and for any $k \in \mathbb{N}$. To estimate the inter-event time, the dynamic of $\|e_k(t)\|/\|x(t)\|$ should be bounded by $\varphi(t)$. Following the same arguments for $t \in [t_k, t_{k+1})$, we have

$$\eta = \frac{1}{\gamma - \alpha\beta} \ln \frac{(\varphi(t_k + \eta) + \beta)(\varphi(t_k) + \gamma/\alpha)}{(\varphi(t_k + \eta) + \gamma/\alpha)(\varphi(t_k) + \beta)}. \tag{29}$$

Substituting the corresponding values $\varphi(t_k)$ and $\varphi(t_k + \eta)$ by 0 and σ, thus (28) is obtained. □

Remark 7. The lower bound of the inter-event intervals provided in (28) is always positive if $\gamma \neq \alpha\beta$. It can be shown in the following. Without loss of generality, suppose $\gamma < \alpha\beta$, then

$$\begin{aligned} \gamma < \alpha\beta &\Longleftrightarrow \sigma\gamma < \sigma\alpha\beta \Longleftrightarrow \sigma\gamma + \beta\gamma \\ &< \sigma\alpha\beta + \beta\gamma \Longleftrightarrow \frac{(\sigma + \beta)\gamma}{(\sigma\alpha + \gamma)\beta} < 1. \end{aligned} \tag{30}$$

Thus both terms $\gamma - \alpha\beta$ and $\ln((\sigma + \beta)\gamma/(\alpha\sigma + \gamma)\beta)$ are negative. Similarly, the positiveness can be proved for the case $\gamma > \alpha\beta$.

Remark 8. For the case $\gamma = \alpha\beta$, go back to (17), which can be written as

$$\dot{\varphi}(t) = \alpha[\varphi(t) + \beta]^2. \tag{31}$$

Take the transform

$$\varphi(t) + \beta = s(t), \qquad d\varphi = ds, \tag{32}$$

then

$$\begin{aligned} \eta &= \int_{t_k}^{t_k+\eta} dt = \frac{1}{\alpha} \int_{\varphi(t_k)}^{\varphi(t_k+\eta)} \frac{1}{[\varphi + \beta]^2} d\varphi \\ &= \frac{1}{\alpha} \int_{\varphi(t_k)+\beta}^{\varphi(t_k+\eta)+\beta} \frac{1}{s^2} ds \\ &= -\frac{1}{\alpha}\frac{1}{s} \Bigg|_{\varphi(t_k)+\beta}^{\varphi(t_k+\eta)+\beta} \\ &= \frac{1}{\alpha}\left[\frac{1}{\varphi(t_k) + \beta} - \frac{1}{\varphi(t_k + \eta) + \beta}\right]. \end{aligned} \tag{33}$$

The desired lower bound η for the inter-event times is obtained when $\varphi(t_k) = 0$ and $\varphi(t_k + \eta) = \sigma$, that is,

$$\eta = \frac{\sigma}{\alpha\beta(\sigma + \beta)}. \tag{34}$$

4. Simulation Examples

The event-based control strategy proposed in this paper is now applied to solve practical and numerical problems.

Example 9. Consider the unstable inverted pendulum system. This process, which was nonlinear, had been analyzed theoretically to obtain a linearized process model

$$\dot{x}(t) = Ax(t) + Bu(t) \tag{35}$$

with

$$A = \begin{bmatrix} 0 & 1 & 0 & 0 \\ 0 & 0 & -\dfrac{mg}{M} & 0 \\ 0 & 0 & 0 & 1 \\ 0 & 0 & \dfrac{g}{l} & 0 \end{bmatrix}, \qquad B = \begin{bmatrix} 0 \\ \dfrac{1}{M} \\ 0 \\ -\dfrac{1}{Ml} \end{bmatrix}, \tag{36}$$

where $M = 10$ is the cart mass, $m = 1$ is the mass of the pendulum bob, $l = 3$ is the length of the pendulum arm, and

TABLE 1: The comparison in terms of inter-event intervals for several values of σ.

Event condition parameter σ	0.01	0.02	0.03	0.04	0.05	0.06	0.07
The average inter-event interval	0.0073	0.0123	0.0183	0.0234	0.0282	0.0338	0.0389
The maximum inter-event interval	0.6569	0.9188	1.4787	1.2648	1.0860	1.4672	1.3987
The number of total events	5466	3250	2183	1709	1417	1183	1028

$g = 10$ is the gravitational acceleration. The states $x(t) = [x_1(t), x_2(t), x_3(t), x_4(t)]^T$ are the cart position and velocity, the pendulum angular position and velocity, respectively. The initial state of the system is chosen as $x_0 = [\,0.98\ 0\ 0.2\ 0\,]^T$. Since the eigenvalues of the matrix A are $-\sqrt{g/l}, 0, 0, \sqrt{g/l}$, the open-loop system is unstable. The following feedback gain

$$K = \begin{bmatrix} 2 & 12 & 378 & 210 \end{bmatrix} \quad (37)$$

is chosen in the design of the event-triggered scheme.

Consider the case when $\tau_k = 0$. Applying Corollary 6 with feedback gain in (37) and $Q = I$, the corresponding P is given by

$$P = 10^3 \times \begin{bmatrix} 0.0046 & 0.0089 & 0.0589 & 0.0343 \\ 0.0089 & 0.0251 & 0.1694 & 0.0988 \\ 0.0589 & 0.1694 & 1.1536 & 0.6726 \\ 0.0343 & 0.0988 & 0.6726 & 0.3926 \end{bmatrix}. \quad (38)$$

The value of σ_1 is solved by (11) as 0.0781. It can be concluded that system stability is guaranteed for any parameter $\sigma \leq 0.0781$. In this case, taking $\sigma = 0.0781$, and using the event triggering condition in (4), a simulation is conducted for $t \in [0, 40]$. It can be calculated that the event-based control algorithm in this paper leads to a maximum inter-event interval 1.0414, and the average inter-event time is 0.0428. Comparing with the average inter-event interval less than 10^{-5} in [21], the improvement over the result in [21] on the average inter-event time is obvious. Under the same conditions, the event instants and the inter-event intervals are shown in Figure 2, and the state responses of the system in (3) are shown in Figure 3. In addition, the comparison for different values of σ chosen from the feasible range is reported in Table 1. From Table 1, it can be found that increasing the value of σ has the effect of increasing the average inter-event interval but not the maximum inter-event interval. The intuition behind the statement is that the ratio between the norms of the control error and the state will need more time to reach the threshold as the value of σ increases.

Example 10. Consider a second-order linear control system described by

$$\begin{bmatrix} \dot{x}_1(t) \\ \dot{x}_2(t) \end{bmatrix} = \begin{bmatrix} 0 & 1 \\ -2 & 3 \end{bmatrix} \begin{bmatrix} x_1(t) \\ x_2(t) \end{bmatrix} + \begin{bmatrix} 0 \\ 1 \end{bmatrix} u(t), \quad (39)$$

and the controller

$$u(t) = x_1(t_k) - 4x_2(t_k) \quad (40)$$

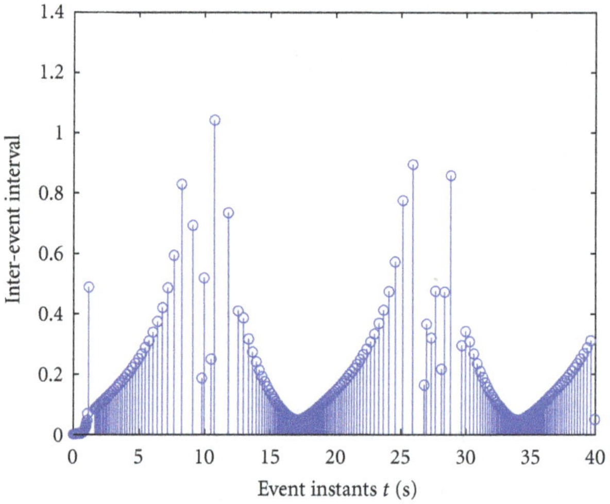

FIGURE 2: The event instants and inter-event intervals with event triggering condition in (4).

is designed to stabilize the closed-loop system. Choose

$$Q = \begin{bmatrix} \frac{1}{2} & \frac{1}{4} \\ \frac{1}{4} & \frac{1}{2} \end{bmatrix}, \quad (41)$$

then P is obtained via solving the Lyapunov equation:

$$P = \begin{bmatrix} 1 & \frac{1}{4} \\ \frac{1}{4} & 1 \end{bmatrix}. \quad (42)$$

Using $\sigma_2 = \sigma_3$ with $\alpha = 1$, $\beta = 4.1231$, $\gamma = 1.618$, and $\sigma_1 = 0.2139$, the upper bound for the random delays τ_k is computed as 0.014, that is, the closed-loop system can tolerate the transmission delays bounded by 0.014. For $\tau = 0.005$ s and according to Theorem 3, any σ satisfying $0.0379 \leq \sigma \leq 0.1748$ can be selected. The theoretical value for the minimum inter-event interval corresponding to $\sigma = 0.1064$ is 0.0097. The evolution of the Lyapunov function $V(t)$ based on the event condition in (4) is depicted in Figure 4. From the simulation result, it can be seen that the event-based control system is robust to time-varying transmission delays. Figure 5 shows the evolution of $\|e(t)\|$ based on the event condition in (4) in the presence of time delays. In this figure, an event is generated when the control error norm reaches the dash-dot line, and the control signal is transmitted to the actuator node via the network. Therefore, the error will never go beyond the dash line which

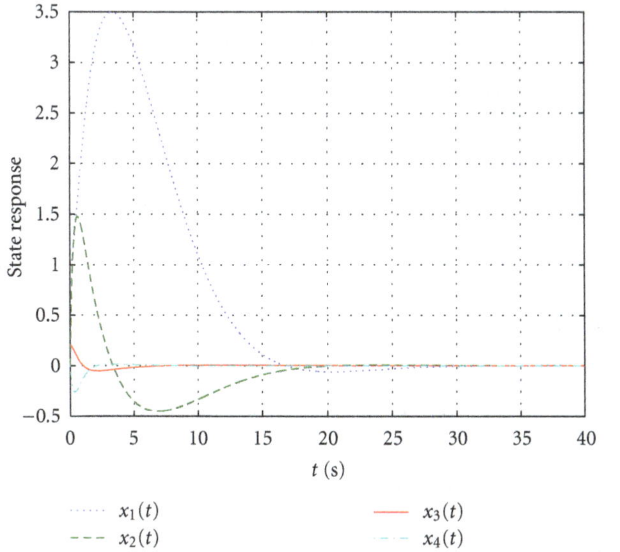

FIGURE 3: The state responses of the system in (3) with event triggering condition in (4).

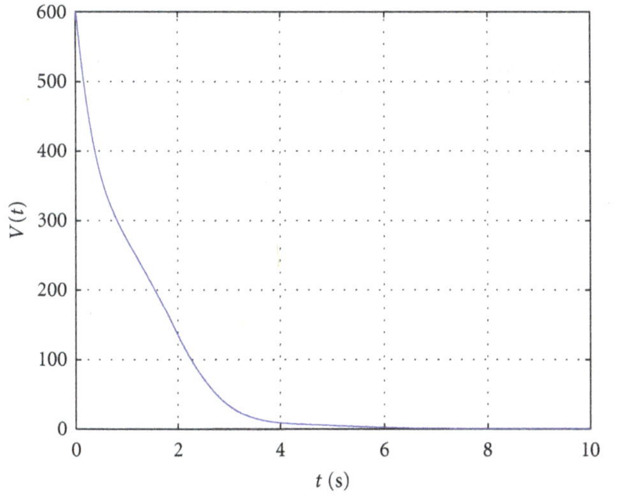

FIGURE 4: Evolution of $V(t)$ for $\sigma = 0.1064$ and initial condition $(x_1(0), x_2(0)) = (10, 20)$.

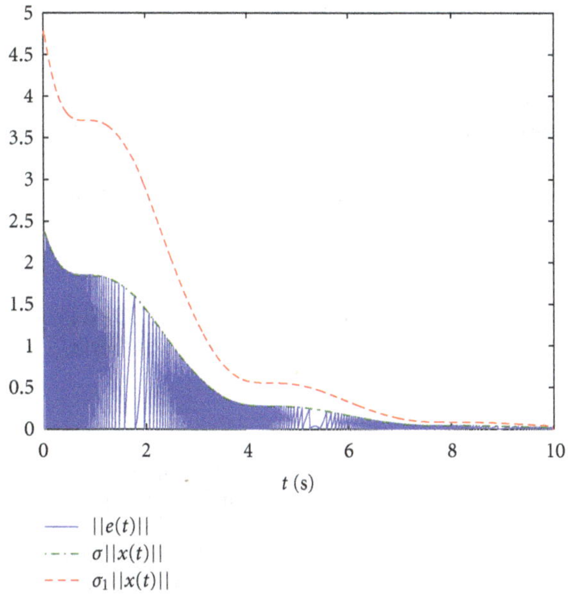

FIGURE 5: Evolution of $\|e\|$, $\sigma\|x\|$, $\sigma_1\|x\|$.

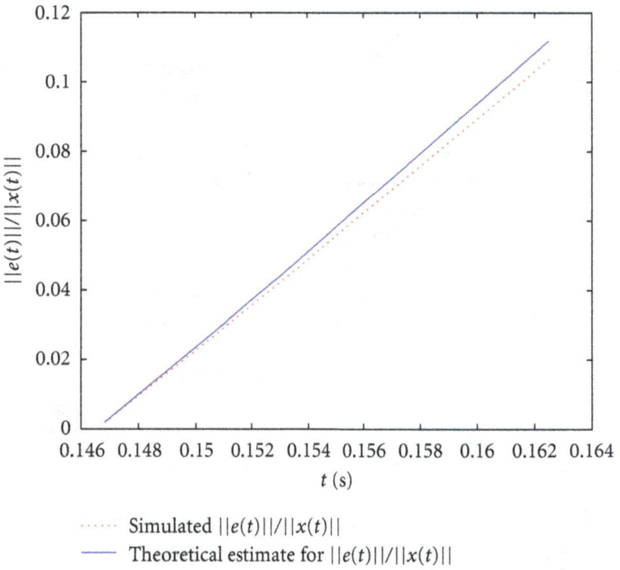

FIGURE 6: Evolution of $\|e(t)\|/\|x(t)\|$ obtained by simulation and its estimation according to (19) for $\sigma = 0.1064$.

guarantees the asymptotic stability. The comparison between the estimated and the simulated evolution of $\|e(t)\|/\|x(t)\|$ is provided over one of the shortest inter-event intervals $[0.1468, 0.1625]$ in Figure 6. The gap between the simulated and the estimated is relatively small. Thus, the equation developed in (25) guarantees a tight lower bound of the inter-event intervals.

5. Conclusions

To save communication bandwidth, a new event-triggered communication strategy has been developed for control over networks, which can be used to determine when the control signals will be transmitted. The event detector is based on the control error; the control is implemented via sample and hold devices. Notice that the results consider the effect of the random time-varying network-induced delays. Two examples detail the advantages of event-based implementation. However, this approach has some limitations. It can be seen that the constructed event detector requires delicate hardware to monitor the control signal and test the logic condition continuously. To overcome this disadvantage, the strategy of discrete detection will be proposed in our future work, where the event detector only needs a supervision of the event condition at discrete sampling instants. Moreover, the parameter of the event detector is chosen with the assumption that the controller gain is designed to guarantee the global asymptotic stability of the closed-loop system in advance without considering the effect of network transmission delays. The method of jointly designing the parameter of the event detector and the controller gain will also be considered in our future work to give higher resource utilization and better performance.

Acknowledgment

This work was supported by NSERC and an iCORE Ph.D. Recruitment Scholarship from the Province of Alberta.

References

[1] J. P. Hespanha, P. Naghshtabrizi, and Y. Xu, "A survey of recent results in networked control systems," *Proceedings of the IEEE*, vol. 95, no. 1, pp. 138–162, 2007.

[2] L. A. Montestruque and P. Antsaklis, "Stability of model-based networked control systems with time-varying transmission times," *IEEE Transactions on Automatic Control*, vol. 49, no. 9, pp. 1562–1572, 2004.

[3] L. Zhang, Y. Shi, T. Chen, and B. Huang, "A new method for stabilization of networked control systems with random delays," *IEEE Transactions on Automatic Control*, vol. 50, no. 8, pp. 1177–1181, 2005.

[4] L. Bakule and M. de la Sen, "Non-Fragile controllers for a class of time-delay nonlinear systems," *Kybernetika*, vol. 45, no. 1, pp. 15–32, 2009.

[5] L. Bakule and M. de la Sen, "Decentralized resilient H_∞ observer-based control for a class of uncertain interconnected networked systems," in *Proceedings of the American Control Conference (ACC '10)*, pp. 1338–1343, Baltimore, Md, USA, 2010.

[6] E. Kofman and J. H. Braslavsky, "Level crossing sampling in feedback stabilization under data-rate constraints," in *Proceedings of the 45th IEEE Conference on Decision and Control, (CDC '06)*, pp. 4423–4428, December 2006.

[7] K. J. Åström and B. M. Bernhardsson, "Comparison of Riemann and Lebesgue sampling for first order stochastic systems," in *Proceedings of the 41st IEEE Conference on Decision and Control*, pp. 2011–2016, December 2002.

[8] M. Miskowicz, "Send-on-delta concept: an event-based data reporting strategy," *Sensors*, vol. 6, no. 1, pp. 49–63, 2006.

[9] P. G. Otanez, J. R. Moyne, and D. M. Tilbury, "Using deadbands to reduce communication in networked control systems," in *Proceedings of the American Control Conference*, pp. 3015–3020, Anchorage, Alaska, USA, May 2002.

[10] P. Tabuada, "Event-triggered real-time scheduling of stabilizing control tasks," *IEEE Transactions on Automatic Control*, vol. 52, no. 9, pp. 1680–1685, 2007.

[11] W. P. M. H. Heemels, J. H. Sandee, and P. P. J. Van Den Bosch, "Analysis of event-driven controllers for linear systems," *International Journal of Control*, vol. 81, no. 4, pp. 571–590, 2008.

[12] M. Velasco, P. Martí, and E. Bini, "Control-driven tasks: modeling and analysis," in *Proceedings of the Real-Time Systems Symposium*, pp. 280–290, Barcelona, Spain, 2008.

[13] M. Mazo and P. Tabuada, "On event-triggered and self-triggered control over sensor/actuator networks," in *Proceedings of the 47th IEEE Conference on Decision and Control, (CDC '08)*, pp. 435–440, Cancun, Mexico, December 2008.

[14] J. Lunze and D. Lehmann, "A state-feedback approach to event-based control," *Automatica*, vol. 46, no. 1, pp. 211–215, 2010.

[15] D. V. Dimarogonas and K. H. Johansson, "Event-triggered cooperative control," in *Proceedings of the European Control Conference*, pp. 3015–3020, Budapest, Hungary, 2009.

[16] X. Wang and M. D. Lemmon, "Event-triggering in distributed networked systems with data dropouts and delays," in *Hybrid Systems: Computation and Control*, R. Majumdar and P. Tabuada, Eds., vol. 5469 of *Lecture Notes in Computer Science*, pp. 366–380, Springer, Berlin, Germany, 2009.

[17] X. Wang and M. Lemmon, "On event design in event-triggered feedback systems," *Automatica*, vol. 47, no. 10, pp. 2319–2322, 2011.

[18] D. Yue, E. Tian, and Q. Han, "A delay system method to design of event-triggered control of networked control systems," in *Proceedings of the 50th IEEE Conference on Decision and Control and European Control Conference*, pp. 1668–1673, Orlando, Fla, USA, 2011.

[19] J. Sijs and M. Lazar, "On event based state estimation," in *Hybrid Systems: Computation and Control*, R. Majumdar and P. Tabuada, Eds., vol. 5469 of *Lecture Notes in Computer Science*, pp. 336–350, Springer, Berlin, Germany, 2009.

[20] H. K. Khalil, *Nonlinear Systems*, Prentice Hall, 3rd edition, 2002.

[21] P. Tabuada and X. Wang, "Preliminary results on state-trigered scheduling of stabilizing control tasks," in *45th IEEE Conference on Decision and Control 2006, CDC*, pp. 282–287, San Diego, Calif, USA, December 2006.

7

The Switching Message Estimator for Network-Based Motion Control Systems

Chen-Chou Hsieh and Pau-Lo Hsu

Department of Electrical Engineering, National Chiao-Tung University, 1001 Ta Hsueh Road, Hsinchu 300, Taiwan

Correspondence should be addressed to Pau-Lo Hsu, plhsu@mail.nctu.edu.tw

Academic Editor: Ya-Jun Pan

Missing commands from the interpolator caused by the dropout effect of network transmission will cause motion error in motion plants implemented on network-based control systems (NCSs). Dropout data can be properly recovered by applying different message estimators to improve motion contouring accuracy. This study shows that the dropout rate and the distribution of missing commands dominate the motion error, and that more centralized missing commands result in a higher maximum contouring error. The short-window dropout quantity (SDQ) is proposed in this paper to estimate the network quality based on the dropout rate and its distribution of the missing data. Furthermore, according to the condition of missing data based on the SDQ, the switching least-square estimator (LSE) is proposed to compensate for missing motion commands. Simulation and experimental results on the two-axis AC servo motor NCS indicate that motion contouring accuracy is greatly improved by applying the proposed estimator.

1. Introduction

Recently, network-based control systems (NCSs) have been widely studied because of their advantages, such as lower cost, easier troubleshooting, and implementation flexibility [1, 2]. However, network-introduced time delay is unavoidable and the data dropout of NCS becomes significant as node and data length increase or as the system sampling time decreases with a limited network bandwidth. Recently, coping with the network-introduced delay on NCS has been widely studied using various approaches, such as robust H_∞ control [3], passive control [4], and predictive control [5]. These methods handle the stochastic network delay with the assumption that the delay is either relatively small or similar to the sampling time in NCS.

In real applications, however, the network-induced time delay may be significant enough to cause data traffic congestion and collision; thus, data dropout occurs and leads to the severe degradation of performance in NCS [7]. The linear quadratic Gaussian (LQG) control has been successfully used to solve random packet loss (e.g., TCP-like protocols) to compensate for the missing data in NCS under a relatively low dropout rate [8]. Optimal filtering based on the H_n-norm estimation error has been presented to handle multiple packet dropouts [9, 10]. The network predictor control (NPC), which consists of a control prediction generator (GPC), observer-based output predictor (OP), and network delay compensator (NDC), has been proposed to overcome network delay and data dropout rate [11]. A reliable estimator to restore the missing data in real motion NCS applications is still being pursued by automation and industry engineers.

The general NCS architecture in Figure 1 shows that all control and feedback signals are communicated through the network for multiaxis motion systems. However, as the number of motion axes increases, network traffic also increases. Therefore, in real control applications, the practical motion NCS architecture with more efficient transmission for multiaxis system is generally modified, as shown in Figure 2. Only command messages are transmitted from the master to the controller. These control and feedback signals are not sent through the network in multiaxis motion systems. Therefore, the transmission can meet the hard real-time requirement within a sampling period to avoid heavy traffic in networks.

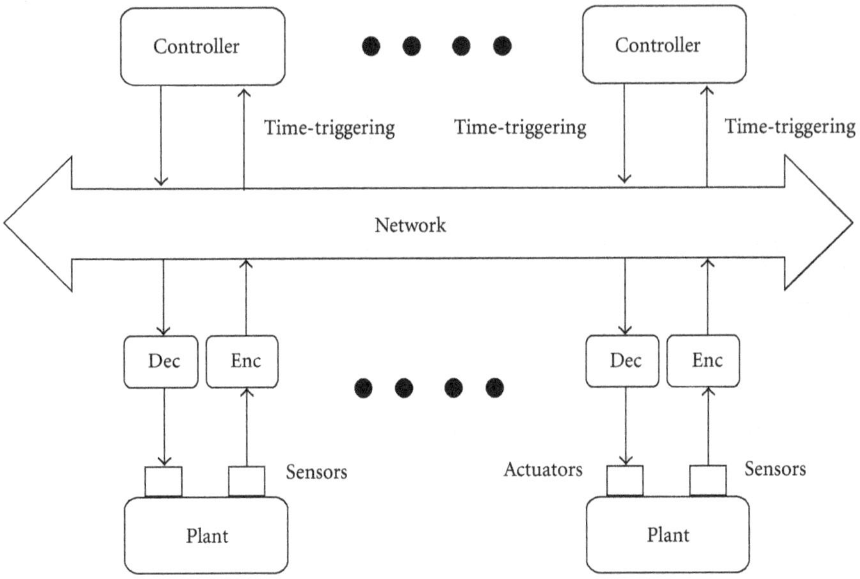

FIGURE 1: The general NCS architecture with multiple nodes.

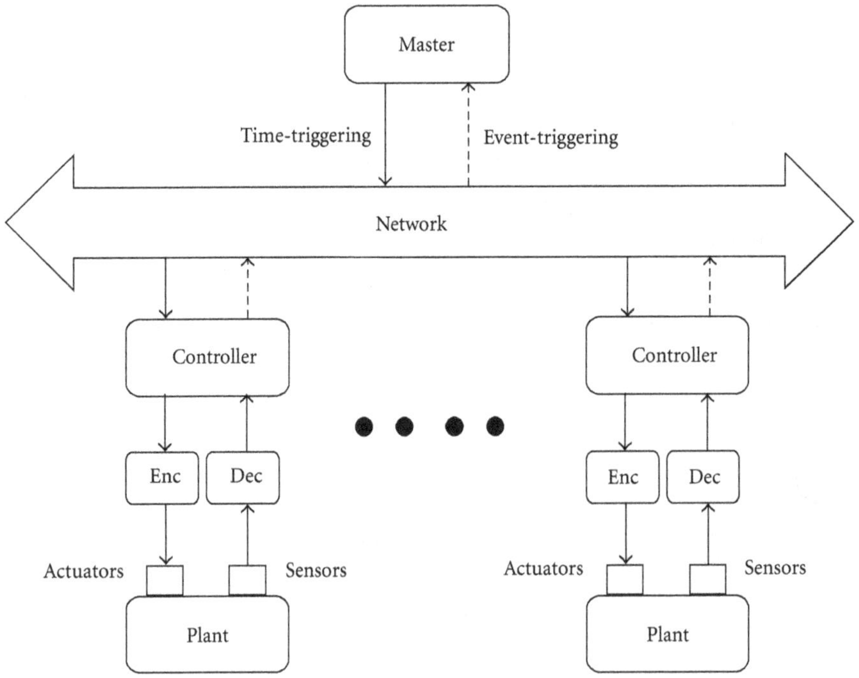

FIGURE 2: The modified NCS architecture with multiaxis motion control systems.

Feedback messages may still be transmitted according to monitoring functions by the event-triggering approach.

However, missing messages in motion NCS become unavoidable when the network delay is longer than the sampling period. Under such circumstances, a message estimator is required to estimate the missing commands and to compensate for their effect in motion accuracy. Various message estimators have been proposed to cope with the dropout effect for motion NCS under different conditions. The 1-delay message estimator is implemented by estimating the missing message using previously received data [12]. The nonlinear NCS has been modeled as a Markovian jump linear system, and the finite loss history estimator (FLHE) has been proposed to improve data dropout effects when the dropout rate is accurately known [13]. However, these methods require an accurate plant/network model. Recently, model-free strategies for control packet dropout compensators, such as the proportional plus derivative (PD) predictor with a different order of derivatives, have been proposed [14]. The Taylor estimator has also been proposed to significantly improve the control performance of motion control NCS [6, 15]. These methods are preferred as dropout data over

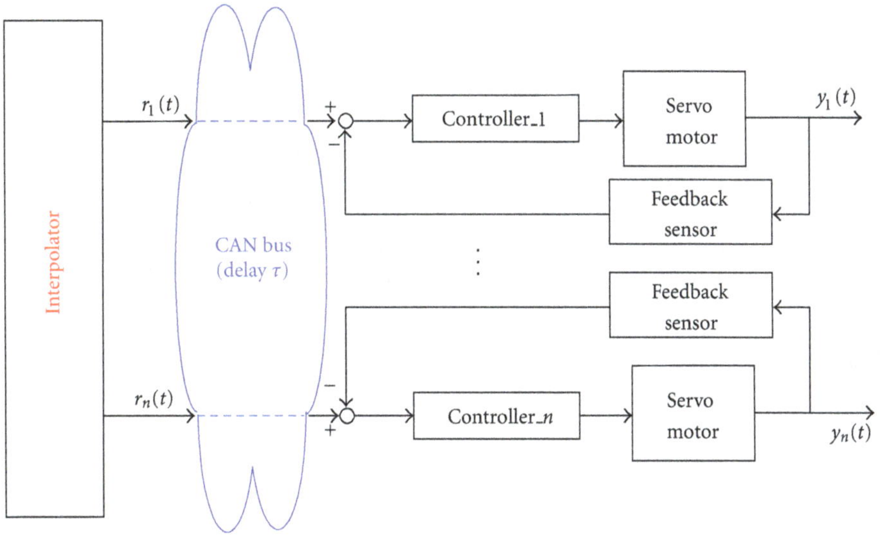

FIGURE 3: The multiaxis motion NCS.

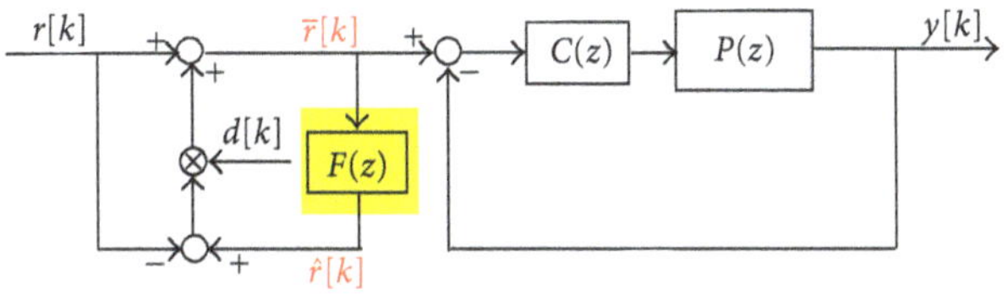

FIGURE 4: The motion NCS with the message estimator [6].

the network are evenly distributed. Missing data that occur in a continuous format tend to lead to a more significant maximum contouring error. Therefore, motion NCS require a suitable index to express the data dropout quantity and to imply its distribution.

This paper proposes the use of short-window dropout quantities (SDQs) to indicate the network communication quality for motion NCS based on both the dropout rate and distribution of the missing data. This paper also proposes an intelligent message estimator (IME) with a switching mechanism based on real-time measured SDQ to obtain a significantly-reduced motion error compared with the 1-delay and Taylor message estimators in the literature [6, 12–15].

Data dropout also leads to the problem of asynchronization among different axes. Both simulation and experimental results, with the nonuniform rational B-spline (NURBS) motion commands [16, 17], have been verified using the proposed message estimator. Results indicate that the proposed estimator maintains the lowest transmission error and the least motion contouring error as missing messages become more severe. The CAN-based two-axis AC servo motor control system has also been implemented with the proposed estimator. Although a high dropout rate degrades NCS performance, the contouring error in the motion control is also closely related to centralized missing commands. Therefore, an effective approach to estimate the quantity of network communication by considering both the dropout rate and the distribution of missing motion commands is crucial to obtain a reliable motion NCS.

2. Data Dropout Quantity in Motion NCS

In motion NCS, the control messages for each motion axis must be transmitted on time through the network protocol to meet control design specifications, as shown in Figure 3. Since the network-induced time delay exists in stochastic and time-varying natures, the transmitted messages may miss the hard real-time deadline because of the limited network bandwidth and the missing data thus occurs.

2.1. Modeling of the Data Dropout. Concerning the missing data in motion NCS, d represents a binary process with probability $P(d[k] = 1) = \varepsilon$ and $P(d[k] = 0) = 1 - \varepsilon$, and the data dropout occurs when $d[k] = 1$, where $C(z)$ and $P(z)$ represent the controller and the plant, respectively, as shown in Figure 4 [6]. During the network transmission, the general message estimator $F(z)$ is activated to compensate for the lost data packets with the estimated motion commands $\hat{r}$ as [9]:

$$\begin{aligned} \bar{r}[k] &= r[k], \quad \text{if } d[k] = 0, \\ \bar{r}[k] &= \hat{r}[k], \quad \text{if } d[k] = 1, \iff \text{dropout}. \end{aligned} \tag{1}$$

2.2. The Distribution Effect of Data Dropout. Traditionally, the data dropout rate ε is recognized as the quality of service

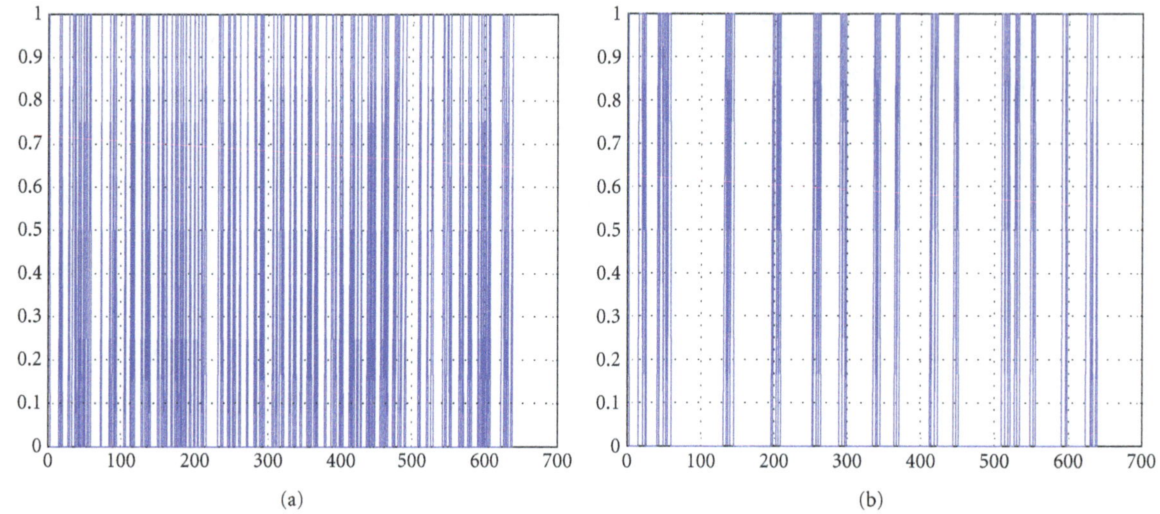

FIGURE 5: (a) Distributed-dropout signals and (b) centralized-dropout signals with the same dropout rate 20%.

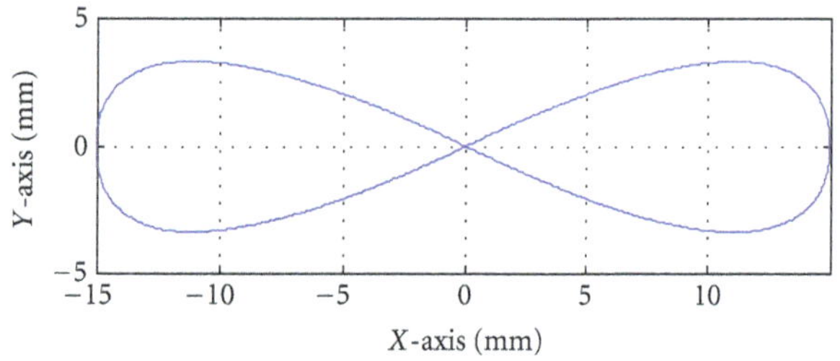

FIGURE 6: NURBS position commands.

(QoS) for most NCSs. However, in motion NCS, centralized missing data will cause a more serious motion error compared with evenly distributed missing data, as shown in Figure 5. Note that Figures 5(a) and 5(b) show two signals with the same data dropout rate of 20% applied to the butterfly profile for the fifth-order NURBS commands, as shown in Figures 6-7 [6]. By applying the same second-order Taylor estimator for the missing motion commands, simulation results show that the transmission error is more significant when the data dropout is more centralized, as shown in Figure 7(b). These results indicate that both data dropout and its distribution should be evaluated together particularly applied in motion NCS.

2.3. The SDQ Index. The dropout data of different network infrastructures possess different distributions [18]. It is mentioned before that the distribution of data dropout in NCS is a crucial factor to determine motion accuracy as well as the data dropout. This paper proposes the short-windows dropout quantities (SDQ) for on-line evaluating the transmission performance of motion NCS. The SDQ is activated only when a missing data occurs, and the total missing data in the past short window is accumulated as in the following:

$$\text{SDQ}\begin{cases}\text{nonactivated}, & \text{if } d[k]=0,\\ \text{activated}, & \text{if } d[k]=1.\end{cases} \tag{2}$$

For example, if the data length of the short window is determined as 5, six possible receiving/missing states, $0 \sim 5$, is recorded in a buffer during the network communication to indicate its micro QoS. The average of the total SDQ can be recognized and the macro-QoS of the NCS.

2.4. An Illustrative Example of the SDQ. Examples illustrated with three missing data among eighteen transmitted messages with different distribution as shown in Figures 8(a) and 8(b), separately. Even their data dropout rates are the same as 3/16, their average SDQ are different, and the more distributed missing data series presents a lower average value of SDQ as shown in Figure 8(b).

As shown in Figures 5(a) and 5(b) with the same data dropout rate of 20%, their average values of SDQ are very different as 0.9 and 2.1, for more distributed and more centralized dropouts, respectively. Simulation results indicate that as the 3rd-order Taylor estimator is applied, the contouring error increases when the averaging SDQ increases even with the same data dropout rate, as shown in Figures 6-7. Results also indicate that the index of SDQ is more appropriate to imply the distribution of the dropout data in motion NCS.

3. The Switching Least-Square Estimator

In real motion NCS, the position commands are generally in the form of smooth curves. At the curvature varying significantly along the contour, and missing data with higher

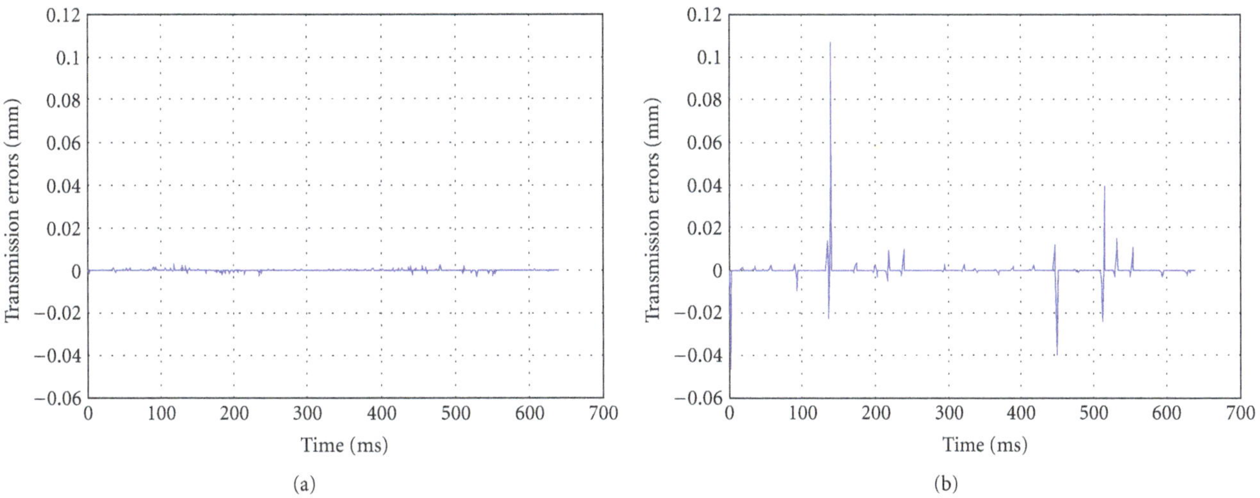

FIGURE 7: y-axis racking errors due to (a) distributed and (b) centralized missing data (20% dropout rate).

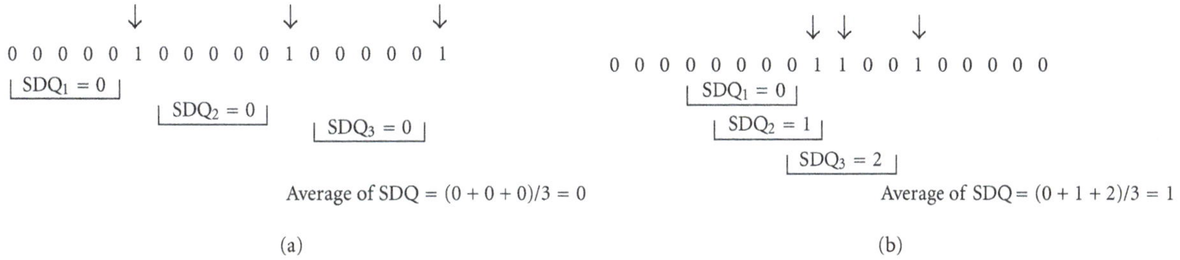

FIGURE 8: Real-time recorded SDQ with the same dropout rate 3/16 (a) distributed dropout and (b) concentrated dropout.

SDQ generally causes a more serious contouring error. To estimate the missing messages in NCS, the basic 1-delay estimator simply adopts the last received message as the current missing message. Moreover, the Taylor estimator was proposed to estimate the current missing message from past several received signals [15]. However, if the past received signal is also missing, the estimated message by the Taylor estimators thus becomes unreliable.

In this paper, the estimator with a switching mechanism is proposed based on the online measured SDQ with the suitable order of the least-square estimator (LSE). However, as the messages dropout becomes more serious, estimation based on the previous data is no longer reliable and the 1-delay estimator will be adopted.

3.1. The Least-Square Estimator. Since the online estimation processing is time consuming and the time-trigger NCS commands are simply in a time series, all parameters of the real-time least-square estimation (LSE) can be obtained in advance. Thus, an online estimation and compensation algorithm for the missing messages in motion NCS are proposed by applying the least-square approach on the past data within a short window. For a general time sequence $x[1], x[2], \ldots, x[M]$, a polynomial sequence can be suitably described as

$$x[k] = c_0 + c_1 k + c_2 k^2 + \cdots + c_N k^N. \tag{3}$$

Thus,

$$\begin{aligned} x[1] &= c_0 + c_1 + c_2 + \cdots + c_N, \\ x[2] &= c_0 + c_1 2 + c_2 2^2 + \cdots + c_N 2^N, \\ &\vdots \\ x[M] &= c_0 + c_1 M + c_2 M^2 + \cdots + c_N M^N. \end{aligned} \tag{4}$$

By rearranging (4) as

$$\begin{bmatrix} x[1] \\ x[2] \\ \vdots \\ x[M] \end{bmatrix} = \begin{bmatrix} 1 & 1 & \cdots & 1 \\ 2^0 & 2 & \cdots & 2^N \\ \vdots & \vdots & & \vdots \\ M^0 & M & \cdots & M^N \end{bmatrix} \begin{bmatrix} c_0 \\ c_1 \\ \vdots \\ c_N \end{bmatrix} \equiv x = A \cdot c. \tag{5}$$

The normal equation from the least-square approach can be applied to obtain the coefficient vector c as

$$c = \left(A^T A\right)^{-1} A^T x. \tag{6}$$

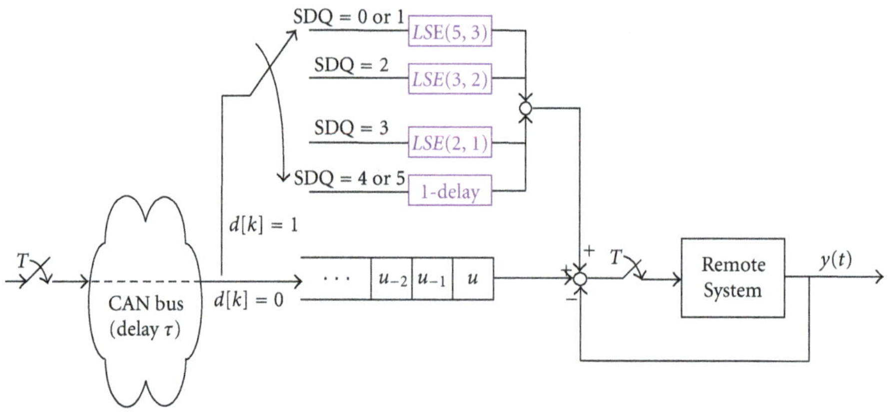

FIGURE 9: The architecture of the proposed estimator with a switching mechanism.

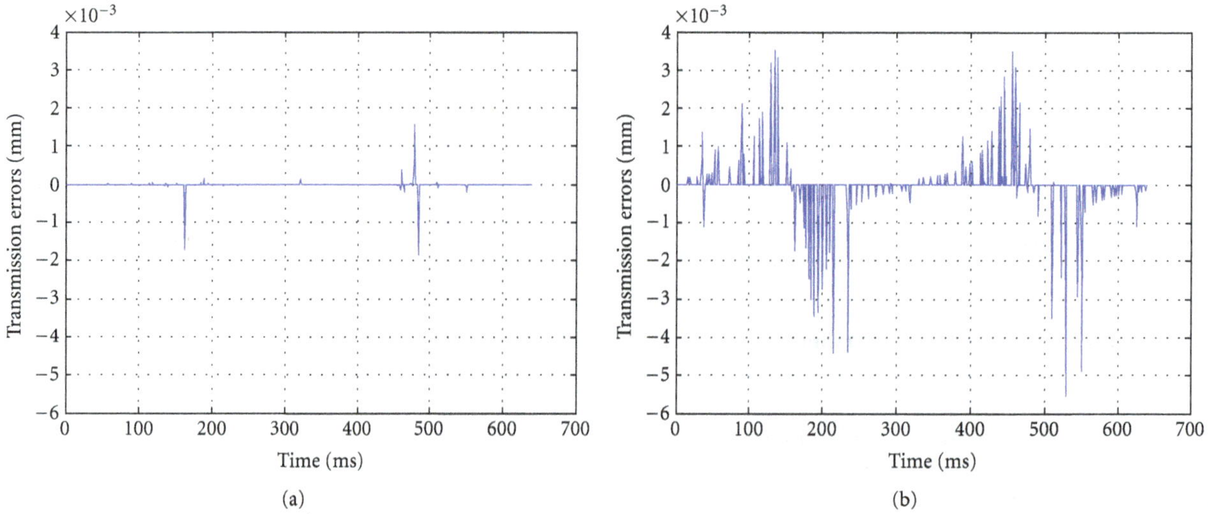

FIGURE 10: Transmission errors with (a) LSE(5, 3) and (b) LSE(2, 1) with the average SDQ = 0.9.

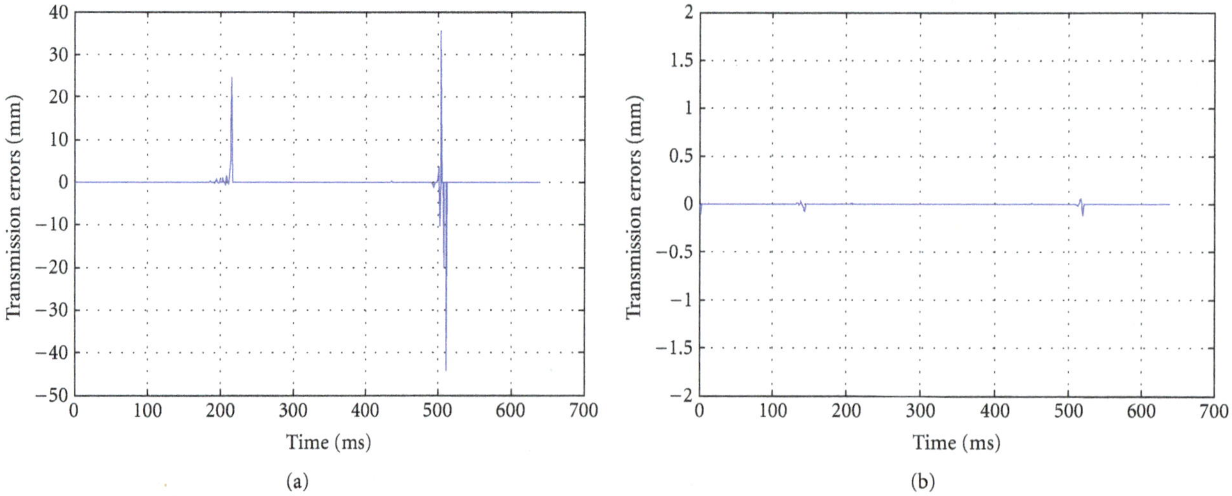

FIGURE 11: Transmission errors with (a) LSE(5, 3) and (b) LSE(2, 1) with the average SDQ = 1.9.

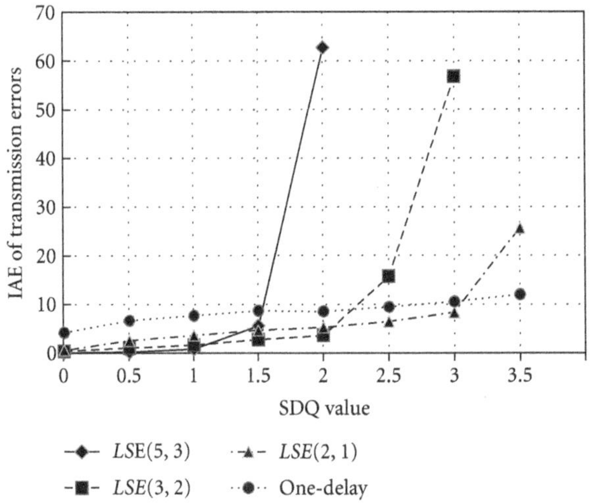

Figure 12: Analysis of compensation effects with averaging SDQ.

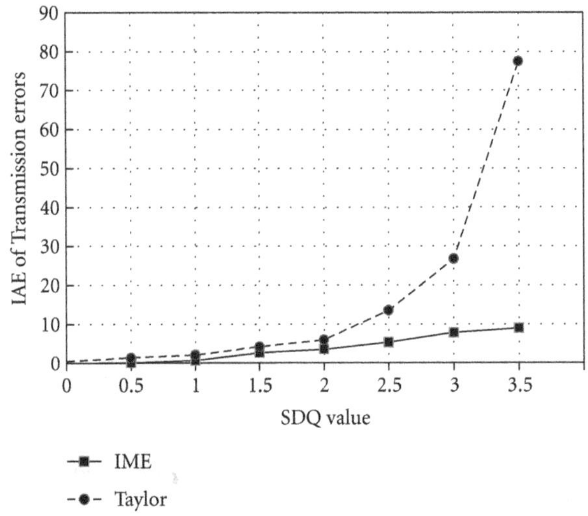

Figure 13: Simulation results with different estimators.

Thus, the missing value can be estimated based on the past M signals with the Nth order of the least-square estimator as

$$\begin{aligned} x[M+1] &= c_0 + c_1(M+1) + c_2(M+1)^2 + \cdots \\ &\quad + c_N(M+1)^N \\ &= \left[(M+1)^0 \ (M+1)^1 \ \cdots \ (M+1)^N\right] \cdot c \\ &= \left[(M+1)^0 \ (M+1)^1 \ \cdots \ (M+1)^N\right] \\ &\quad \cdot \left(A^T A\right)^{-1} A^T x \\ &\equiv \mathrm{LSE}(M,N) \cdot x, \end{aligned} \tag{7}$$

and the estimator matrix $\mathrm{LSE}(M,N)$ can thus be preobtained for real-time implementation, where M also indicates the data number in a window to be counted and N is the order of polynomial functions with $N+1$ coefficients.

3.2. Coefficients of the Least-Square Estimators. To achieve an online estimation for motion NCS, the order and the number of data within the window of the least-square estimator should be determined based on the motion commands. For example, the NURBS signal can be approximated by a third-order polynomial equation obtained from the LSE [19]. Practically, the window length can be properly chosen as large as five to estimate the missing NURBS commands or other motion contours. In practice, three useful $\mathrm{LSE}(M,N)$ are precalculated for real-time applications based on the quality of communication of NCS as follows.

3.2.1. LSE(5,3) for Low Dropout Cases. The third-order $\mathrm{LSE}(5,3)$ can properly estimate the motion trajectory concerning its position, velocity, acceleration, and even the change of acceleration as the jerk. In this transmission case, all data within the window length 5 are properly received, or at most, only one missing data is estimated within the window. $\mathrm{LSE}(5,3)$ is chosen to estimate a cubic-curve motion command with the order of 3 by using all five previous data, which may include at most one estimated data. The parameters of $\mathrm{LSE}(5,3)$ are determined from (5) as follows:

$$A = \begin{bmatrix} 1 & 1 & 1 & 1 \\ 1 & 2 & 4 & 8 \\ 1 & 3 & 9 & 27 \\ 1 & 4 & 16 & 64 \\ 1 & 5 & 25 & 125 \end{bmatrix}, \tag{8}$$

and by (6)-(7),

$$\begin{bmatrix} 1 & 6 & 36 & 216 \end{bmatrix} \left(A^T A\right)^{-1} A^T = \begin{bmatrix} -0.8 & 2.2 & -0.8 & -2.8 & 3.2 \end{bmatrix}. \tag{9}$$

The $\mathrm{LSE}(5,3)$ is thus obtained as

$$\mathrm{LSE}(5,3) = 3.2z^{-1} - 2.8z^{-2} - 0.8z^{-3} + 2.2z^{-4} - 0.8z^{-5}. \tag{10}$$

3.2.2. LSE(3,2) for Medium Dropout Cases. In this case, the medium data dropout condition occurs, and the missing data within the short window length 5 are as large as 2. In other words, at most only three reliable data are accountable within the window to correctly estimate the missing data. Therefore, $\mathrm{LSE}(3,2)$ is chosen to estimate the quadric-curve trajectory with the order of 2 by using three previous data through considering both its position, velocity and acceleration from (7) as

$$\mathrm{LSE}(3,2) = 3z^{-1} - 3z^{-2} + z^{-3}. \tag{11}$$

3.2.3. LSE(2,1) for Heavy Dropout Cases. In this situation, only two data within the window are received, and $\mathrm{LSE}(2,1)$ is chosen to estimate the motion trajectory concerning as high as its velocity only by applying previous two data as

$$\mathrm{LSE}(2,1) = 2z^{-1} - z^{-2}. \tag{12}$$

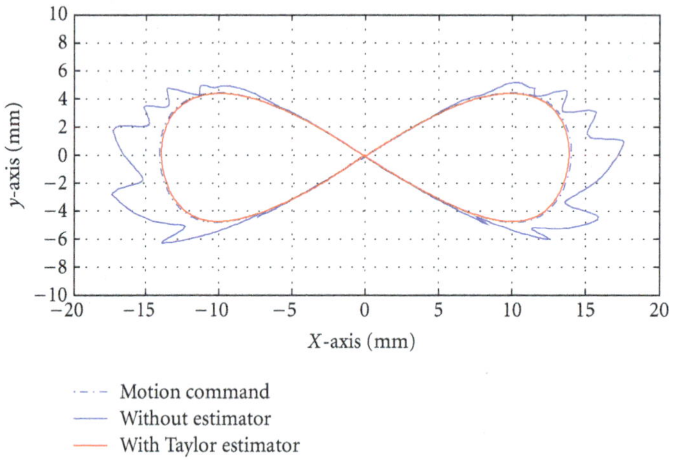

Figure 14: Contouring of motion NCS without/with the Taylor estimator (average SDQ = 0.8).

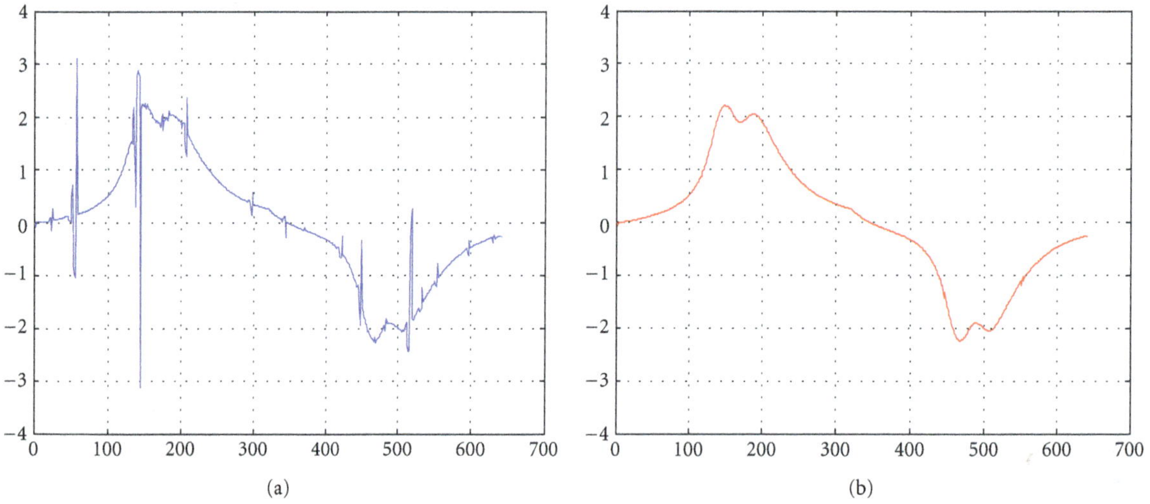

Figure 15: Contouring errors with (a) the Taylor estimator (b) IME (average SDQ = 0.8).

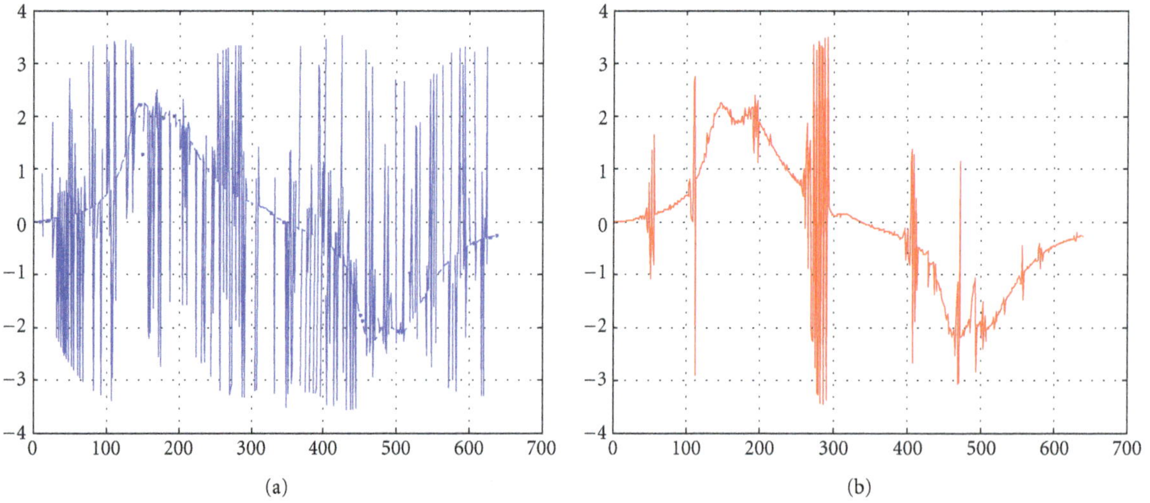

Figure 16: Contouring errors with (a) the Taylor estimator and (b) IME (average SDQ = 2.6).

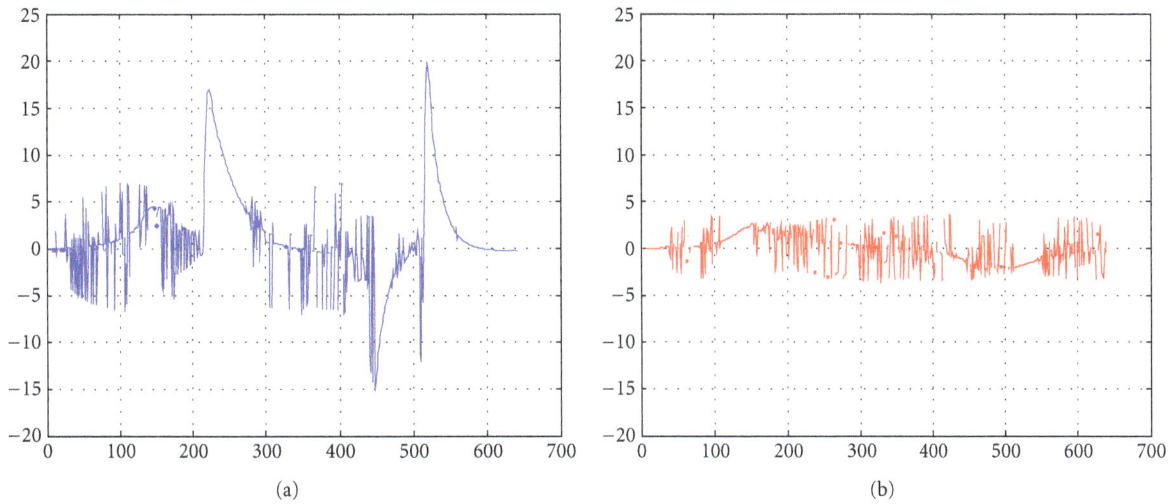

FIGURE 17: Contouring errors with (a) the Taylor estimator and (b) IME (average SDQ = 3.1).

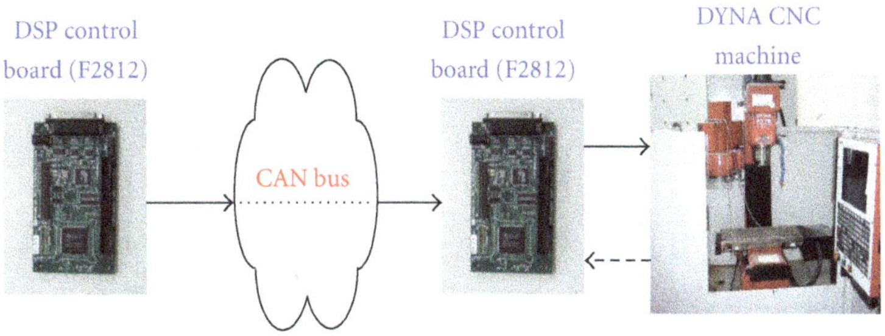

FIGURE 18: Experimental setup.

3.2.4. The 1-Delay Estimator for Serious Dropout Cases. In this situation, network communication presents such a heavy data dropout rate, either only one data is received within the previous window of length 5, or all five previous adopted data are obtained through the estimation process. Therefore, estimation results based on the least-square approach by adopting all data within the window length 5 is not reliable anymore, and the 1-delay estimator is determined under such circumstance to estimate the position only by directly adopting the previous data as [6, 12]

$$\text{1-delay estimator} = z^{-1}. \tag{13}$$

In summary, different estimators are applied to different transmission conditions based on the SDQ, as shown in Figure 9 as:

$$\text{SDQ} = \begin{cases} 0 \text{ or } 1, & \text{LSE}(5,3) \text{ is adopted}, \\ 2, & \text{LSE}(3,2) \text{ is adopted}, \\ 3, & \text{LSE}(2,1) \text{ is adopted}, \\ 4 \text{ or } 5, & \text{1-delay estimator is adopted}. \end{cases} \tag{14}$$

4. Analysis of the SDQ on Motion NSC

Figure 10 shows the transmission error obtained by applying the estimators of LSE(5, 3) and LSE(2, 1) to the motion NCS as the simulation shown in Figure 7. The case of SDQ = 0.9 implies that one data (close to 0.9 in average) is missing in average within the window length 5. Simulation results show that LSE(5, 3) renders a better compensation effect as compared to LSE(2, 1) which should be applied to a more serious data dropout case. However, as the SDQ increases to 1.9, of which implies that there are about two missing messages among the five transmitted messages, the transmission error increases and LSE(5, 3) is not suitable anymore. Figure 11 shows that the compensation results applying LSE(2, 1) render better performance.

Furthermore, the least-square approach with a different M applied to a different averaging SDQ shows that applying LSE(5, 3) to compensate for the low missing data rate presents the best motion accuracy as the averaging SDQ is less than one, but it becomes the worst as the averaging SDQ increases larger than 1.5, as shown in Figure 12. Moreover, LSE(3, 2) is more suitable for the situation as the averaging SDQ is between 1 and 2. Note that LSE(2, 1) is most suitable for the situation as the averaging SDQ is between

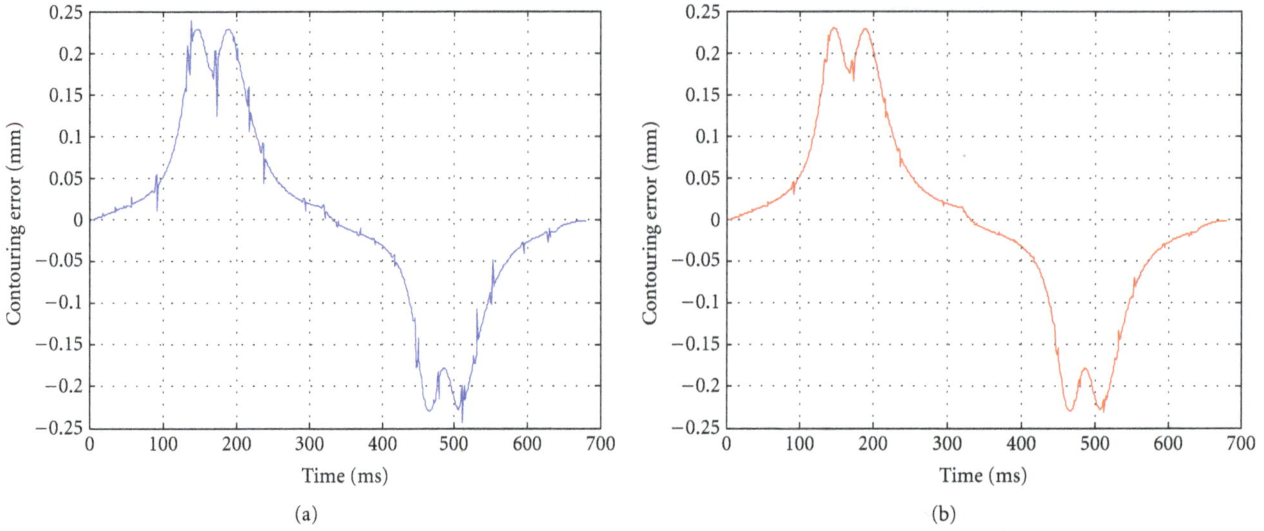

FIGURE 19: Contouring error (a) Taylor estimator and (b) IME (T_s = 0.5 ms and average SDQ = 1.8).

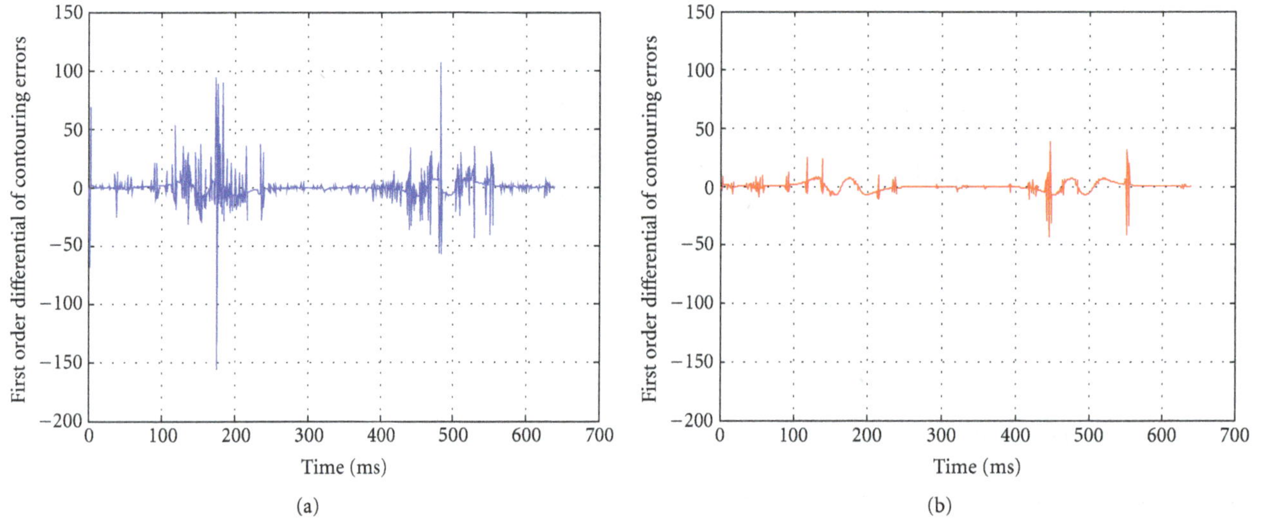

FIGURE 20: First-order differential of the contouring errors as shown in Figure 19.

2 and 3. In addition, the 1-delay estimator possesses the best compensation effect when the averaging SDQ is larger than 3. All switching laws according to (14) based on the SDQ thus agree with both the simulation results and the theoretical analysis, as shown in Figure 12. Although the 3rd-Taylor estimator has been proven to render more accurate results than the 1-delay estimator [6], Figure 13 further indicates that the proposed intelligent message estimator (IME) renders much better performance under different SDQ, especially as the missing data becomes more serious as shown in Figure 9.

5. Contouring Accuracy of Motion NCS

5.1. Simulation Results. Applications of the present estimator based on the SDQ have been applied to the two-axis motion NCS, as shown in Figure 2. The NURBS commands and the system response with averaging SDQ = 0.8 of the network are shown in Figure 14. Results show that both the 3rd-Taylor estimator and the proposed estimator can reduce the effects of data dropout at a lower averaging SDQ. Figure 15 also shows that the contouring error obtained by applying the proposed estimator is significantly reduced to achieve better contouring accuracy.

Figure 16 shows that the present estimator even renders a much better contouring accuracy when the averaging SDQ increases to 2.6. Furthermore, when the value of the averaging SDQ increases to 3.1, the Taylor estimator will lead to an unstable motion as shown in Figure 17. Under such circumstances, the proposed estimator still results in a stable motion and maintains the contouring accuracy well.

5.2. Experimental Results. The proposed estimator was applied to two Tamagawa motors with NCS realization, as

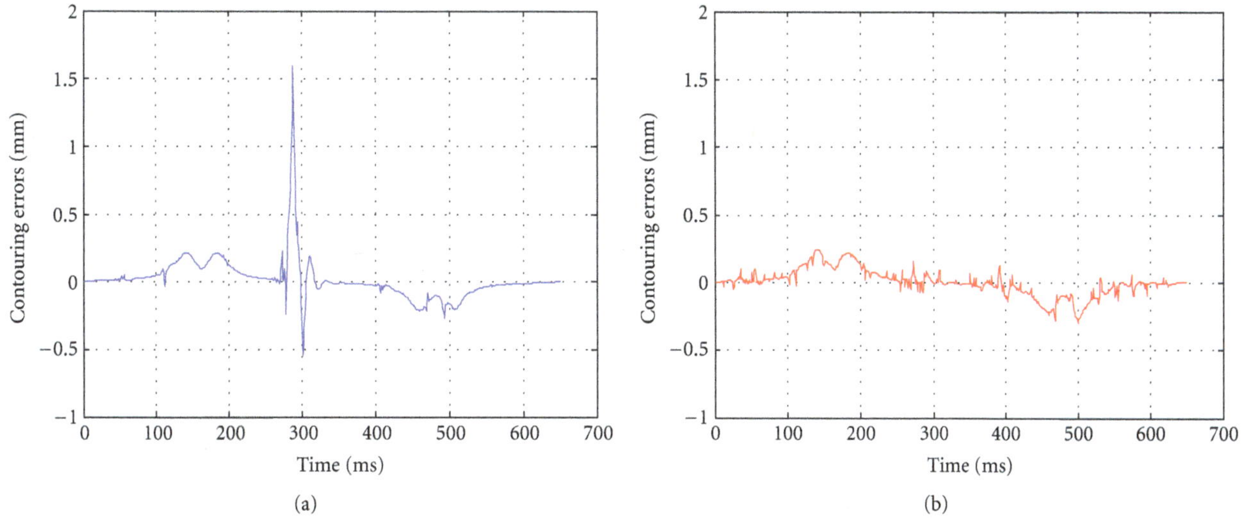

Figure 21: Contouring error with (a) Taylor estimator and (b) IME (T_s = 0.4 ms and average SDQ = 2.7).

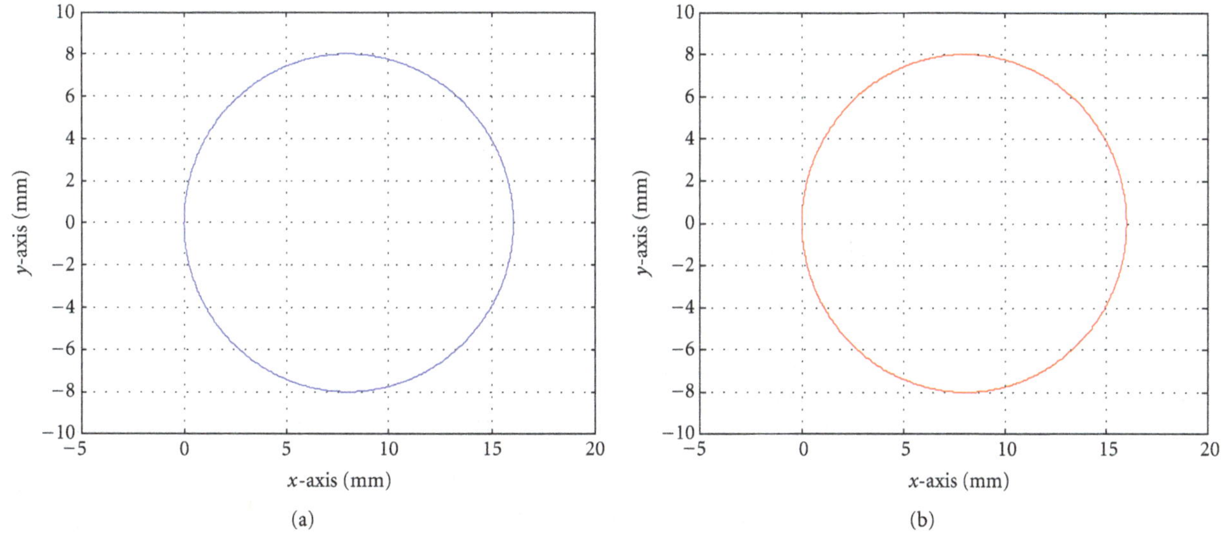

Figure 22: Circular response of (a) the Taylor estimator and (b) IME (T_s = 0.5 ms and average SDQ = 1.8).

shown in Figure 18. The butterfly NURBS profile for both the x-axis and y-axis position amplitudes is 30 mm under the feed rate 3,000 mm/min for experiments. Furthermore, the averaging SDQ is measured as 1.8 and 2.7 for the present CAN-bus implementation with the baud rate 1 M bit/sec under different sampling periods as 0.5 ms and 0.4 ms, respectively. Experimental results indicate that increasing the sampling rate will result in more serious missing data due to the saturation of network bandwidth.

Figure 19 also shows the contouring error when the averaging SDQ is 1.8. The first-order differential results of the measured contouring error with less oscillation are also shown in Figure 20. All results indicate that the proposed estimator renders a more stable and reliable motion than the Taylor estimator. By observing the contouring error as shown in Figure 21 with a more serious data dropout (average SDQ 2.7), results also show that the proposed estimator is more effective in reducing the asynchronization effect than the Taylor estimator. Similar results provided by the circular NURBS profile for the motion NCS obtained as shown in Figures 22 and 23 also indicate applicability of the proposed estimator to different motion profiles.

6. Conclusions

As the online measured time delay is crucial for the NCS controller design [20], this paper successfully measures transmitted signals with different dropout rates and proposes the SDQ, which is a suitable index for the dropout rate and concentration of missing commands in motion NCS. The missing commands can also be estimated by applying the proper order of LSE. Motion NCS with the proposed estimator render significant improvements in accuracy.

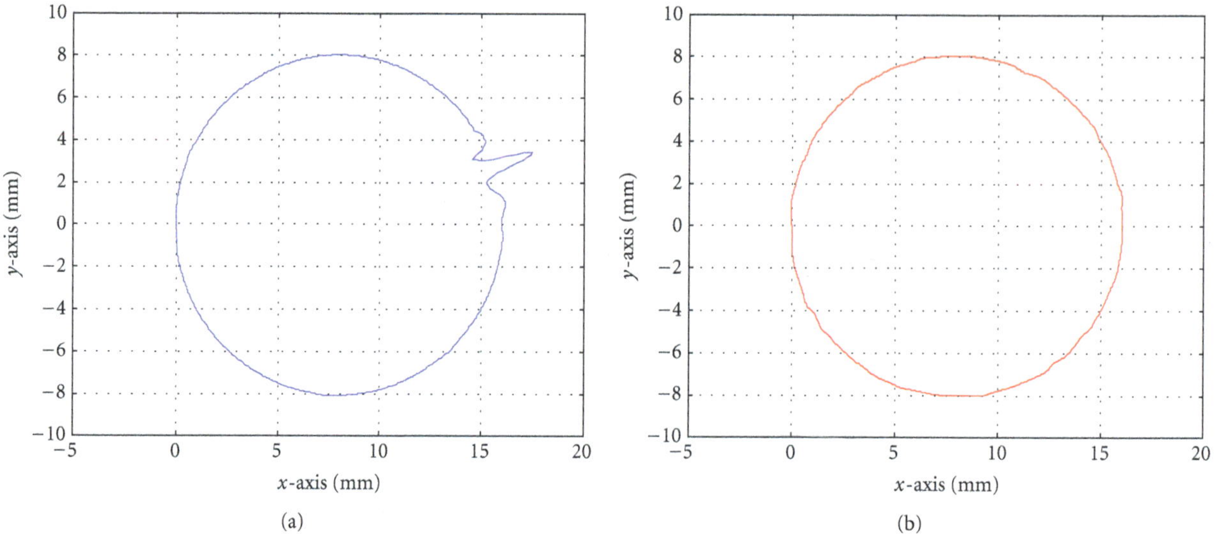

Figure 23: Circular response of (a) Taylor estimator and (b) the proposed estimator ($T_s = 0.4$ ms and average SDQ = 2.7).

In a real motion system, commands usually consider position, velocity, and acceleration. Thus, $N = 3$ is a suitable order for the least square estimator LSE(M, N), and 5 is chosen as the window length M, with more accurate estimation results if there are no or if there is only one missing data within the window length. A lower order of LSE will be determined if the QoS for the network worsens. The 1-delay estimator can be adopted only for the worst communication cases, as in (14).

The proposed estimator with a switching mechanism based on SDQ leads to the lowest contouring error. Results indicate that with the proposed estimator in motion NCS, contouring accuracy can be maintained even under severe missing commands compared with the 1-delay and the Taylor estimators. Experimental results have also proven the feasibility and applicability of the proposed approach in motion NCS.

Acknowledgment

This work was supported by the National Science Council, Taiwan, under Grant NSC 98-2218-E-009-005.

References

[1] S. Zampieri, "Trends in networked control systems," in *Proceedings of the 17th World Congress of the International Federation of Automatic Control (IFAC '08)*, pp. 2886–2894, Seoul, Korea, July 2008.

[2] J. P. Hespanha, P. Naghshtabrizi, and Y. Xu, "A survey of recent results in networked control systems," *Proceedings of the IEEE*, vol. 95, no. 1, pp. 138–162, 2007.

[3] F. Abdollahi and K. Khorasani, "A novel H∞ control strategy for design of a robust dynamic routing algorithm in traffic networks," *IEEE Journal on Selected Areas in Communications*, vol. 26, no. 4, pp. 706–718, 2008.

[4] T. Matiakis, S. Hirche, and M. Buss, "Control of networked systems using the scattering transformation," *IEEE Transactions on Control Systems Technology*, vol. 17, no. 1, pp. 60–67, 2009.

[5] G. P. Liu, Y. Xia, J. Chen, D. Rees, and W. Hu, "Networked predictive control of systems with random network delays in both forward and feedback channels," *IEEE Transactions on Industrial Electronics*, vol. 54, no. 3, pp. 1282–1297, 2007.

[6] C. C. Hsieh, P. L. Hsu, and B. C. Wang, "The motion message estimator in real-time network control systems," in *Proceedings of the 32nd Annual Conference on IEEE Industrial Electronics (IECON '06)*, pp. 4632–4637, November 2006.

[7] C. C. Hsieh, P. L. Hsu, and B. C. Wang, "The motion message estimator in networked control systems," in *Proceedings of the 17th World Congress of the International Federation of Automatic Control (IFAC '08)*, pp. 11606–11611, Seoul, Korea, July 2008.

[8] B. Sinopoli, L. Schenato, M. Franceschetti, K. Poolla, and S. Sastry, "Optimal linear LQG control over lossy networks without packet acknowledgment," *Asian Journal of Control*, vol. 10, no. 1, pp. 3–13, 2008.

[9] M. Sahebsara, T. Chen, and S. L. Shah, "Optimal H_2 filtering in networked control systems with multiple packet dropout," *IEEE Transactions on Automatic Control*, vol. 52, no. 8, pp. 1508–1513, 2007.

[10] S. Sun, L. Xie, W. Xiao, and N. Xiao, "Optimal filtering for systems with multiple packet dropouts," *IEEE Transactions on Circuits and Systems II*, vol. 55, no. 7, pp. 695–699, 2008.

[11] S. Chai, G. P. Liu, D. Rees, and Y. Xia, "Design and practical implementation of internet-based predictive control of a servo system," *IEEE Transactions on Control Systems Technology*, vol. 16, no. 1, pp. 158–168, 2008.

[12] Q. Ling and M. D. Lemmon, "Robust performance of soft real-time networked control systems with data dropouts," in *Proceedings of the 41st IEEE Conference on Decision and Control*, pp. 1225–1230, December 2002.

[13] S. C. Smith and P. Seiler, "Estimation with lossy measurements: jump estimators for jump systems," *IEEE Transactions on Automatic Control*, vol. 48, no. 12, pp. 2163–2171, 2003.

[14] Y. C. Tian and D. Levy, "Compensation for control packet dropout in networked control systems," *Information Sciences*, vol. 178, no. 5, pp. 1263–1278, 2008.

[15] C. C. Hsieh and P. L. Hsu, "Analysis and applications of the motion message estimator for network control systems," *Asian Journal of Control*, vol. 10, no. 1, pp. 45–54, 2008.

[16] L. Piegl and W. Tiller, *The NURBS Book*, Springer, New York, NY, USA, 1995.

[17] M. Gopi and S. Manohar, "A unified architecture for the computation of B-spline curves and surfaces," *IEEE Transactions on Parallel and Distributed Systems*, vol. 8, no. 12, pp. 1275–1287, 1997.

[18] F. L. Lian, J. R. Moyne, and D. M. Tilbury, "Performance evaluation of control networks: Ethernet, ControlNet, and DeviceNet," *IEEE Control Systems Magazine*, vol. 21, no. 1, pp. 66–83, 2001.

[19] H. W. Sorenson, "Least-square estimation: from Gauss to Kalman," *IEEE Spectrum*, vol. 7, no. 7, pp. 63–68, 1970.

[20] C. L. Lai and P. L. Hsu, "Design the remote control system with the time-delay estimator and the adaptive smith predictor," *IEEE Transactions on Industrial Informatics*, vol. 6, no. 1, pp. 73–80, 2010.

8

Robust MPC with Output Feedback of Integrating Systems

J. M. Perez,[1] D. Odloak,[2] and E. L. Lima[3]

[1] *Cenpes, 21941-915 Rio de Janeiro, RJ, Brazil*
[2] *Chemical Engineering Department, University of São Paulo, CP 61548, 05424-970 São Paulo, SP, Brazil*
[3] *Chemical Engineering Program/COPPE Federal University of Rio de Janeiro, CP 68502, 21941-970 Rio de Janeiro, RJ, Brazil*

Correspondence should be addressed to J. M. Perez, perez@petrobras.com.br

Academic Editor: Mohamed Zribi

In this work, it is presented a new contribution to the design of a robust MPC with output feedback, input constraints, and uncertain model. Multivariable predictive controllers have been used in industry to reduce the variability of the process output and to allow the operation of the system near to the constraints, where it is usually located the optimum operating point. For this reason, new controllers have been developed with the objective of achieving better performance, simpler control structure, and robustness with respect to model uncertainty. In this work, it is proposed a model predictive controller based on a nonminimal state space model where the state is perfectly known. It is an infinite prediction horizon controller, and it is assumed that there is uncertainty in the stable part of the model, which may also include integrating modes that are frequently present in the process plants. The method is illustrated with a simulation example of the process industry using linear models based on a real process.

1. Introduction

Model predictive control has achieved a remarkable popularity in the process industry with thousands of practical applications [1]. One of the reasons for this industrial acceptance is the ability of MPC to incorporate constraints in the control problem. However, one additional desirable characteristic, still not attended by commercial MPC packages, is closed-loop stability in the presence of model uncertainty, which is usually related to the nonlinearity of the real system. When model uncertainty is considered in the synthesis of the model predictive control, the majority of existing robust algorithms usually demand a computer effort that is prohibitive for practical implementation [2]. From the application viewpoint, we could not find in the control literature a satisfactory solution to the robust MPC problem with output feedback and input constraints. In Rodrigues and Odloak [3], it is presented a formulation to the robust unconstrained MPC with output feedback where stability is achieved through the explicit inclusion of a Lyapunov inequality into the control optimization problem. Later, in Rodrigues and Odloak [4] the method was extended to allow the switching of active input constraints during transient conditions. In Perez [5] the non-minimal state space model (realigned model) proposed in Maciejowski [6] was extended to the incremental form, and this model was used in the controller proposed in Rodrigues and Odloak [4] to produce a robust controller with reduced computational effort by avoiding the use of a state observer, since the state of the realigned model is directly read from the plant. In this approach, the MPC solves the optimization problem that produces the control law in two stages. The first stage is solved offline and computes a family of linear controllers in which the Lyapunov inequalities are added to the control problem so as to ensure state contraction (stability). These linear controllers match all possible configurations of saturation of manipulated variables for a given set of variables that need to be controlled. The second stage is solved online and computes a convex combination of these linear controllers through an optimization problem that includes all the input bounds. These combinations span all possible control configurations in terms of controlled outputs and manipulated inputs. For each control configuration, the linear controller is robust to all process models defining the multiplant uncertainty. The main disadvantage of this approach is that the number of possible control configurations corresponding to the possible combinations of saturated inputs may become very high in systems with large dimension. Then, the off-line step of the control problem may be computationally very expensive or there may not be a feasible solution. The realigned model was applied in González et al. [7] to the development of an infinite horizon MPC nominally stable to open-loop stable and integrating systems. The same non-minimal model was applied in González and Odloak [8] to the development of a robust MPC to uncertain stable systems.

The main objective of this work is to extend the controller proposed in González and Odloak [8] to consider systems with stable and integrating outputs. The method proposed here is based on the non-minimal state space model formulation presented in Maciejowski [6], and the control problem considers that one may have uncertainties in the stable part of the model, while the model for the integrating part is perfectly known. In Section 2, the non-minimal (realigned) model in the incremental form is developed for systems with stable and integrating outputs. The realigned model is split in two parts, one related to the stable outputs and the other related to the integrating outputs. Model uncertainty is only related to the stable outputs, and it is characterized by a family of possible linear time-invariant models. It is assumed that the real model of the system is unknown, but at any time instant the system can be represented by one of the models of this family of models. In practice the models of the integrating outputs are less affected by uncertainty, and here, this uncertainty is neglected. In Section 3, the optimization problem that produces the robust controller is presented. Since the inputs affect both the stable and integrating outputs, the cost function of the robust MPC proposed here includes a combination of the weighted predicted errors in all the outputs. Several results concerning the robustness of the convergence and stability of the resulting closed-loop system are provided. Next, in Section 4, a simulation example, based on linear models of a process system from the oil refining industry, is used to illustrate the application of the new method and finally, in Section 5, the paper is concluded.

2. The Nonminimal State Space Model with Measured State

Consider the following discrete time-invariant model:

$$y(k) + \sum_{i=1}^{na} A_i y(k-i) = \sum_{i=1}^{nb} B_i u(k-i), \tag{1}$$

where na and nb are the orders of the discrete model, $A_i \in \Re^{ny\times ny}$ and $B_i \in \Re^{ny\times nu}$ are the coefficients that correspond to the parameters of the model, $y(k) \in \Re^{ny}$ is the system output, and $u(k) \in \Re^{nu}$ is the system input. Here, it assumed that in the model represented above there are no zero-pole cancellations. In Maciejowski [6], it is shown that this model can be written in a state space form where the state is built with the measurements of the plant inputs and outputs at different time instants. In González et al. [7], it is adopted the same model written in the input incremental form:

$$\begin{bmatrix} x_y(k+1) \\ x_{\Delta u}(k+1) \end{bmatrix} = \begin{bmatrix} A_y & A_{\Delta u} \\ \underline{0} & \underline{I} \end{bmatrix} \begin{bmatrix} x_y(k) \\ x_{\Delta u}(k) \end{bmatrix} + \begin{bmatrix} B_{\Delta u} \\ \bar{I} \end{bmatrix} \Delta u(k),$$
$$y(k) = \begin{bmatrix} C_y & C_{\Delta u} \end{bmatrix} \begin{bmatrix} x_y(k) \\ x_{\Delta u}(k) \end{bmatrix}, \tag{2}$$

where

$$\begin{aligned} x_y(k) &= \begin{bmatrix} y(k)^T & y(k-1)^T & \cdots & y(k-na+1)^T & y(k-na)^T \end{bmatrix}^T \\ &\in \mathbb{R}^{(na+1)ny}, \\ x_{\Delta u}(k) &= \begin{bmatrix} \Delta u(k-1)^T & \Delta u(k-2)^T & \cdots & \Delta u(k-nb+1)^T \end{bmatrix}^T \\ &\in \mathbb{R}^{(nb-1)nu}, \end{aligned} \tag{3}$$

k is the present sampling instant, $\Delta u(k) = u(k) - u(k-1)$ is the input increment, $\Delta u \in \Re^{nu}$, $y \in \Re^{ny}$ is the output. The matrices involved in this model are (see more details in Maciejowski [6])

$$A_y = \begin{bmatrix} I-A_1 & A_1-A_2 & A_2-A_3 & A_3-A_4 & \cdots & A_{na-1}-A_{na} & A_{na} \\ I_{ny\times ny} & 0 & 0 & 0 & \cdots & 0 & 0 \\ 0 & I_{ny\times ny} & 0 & 0 & \cdots & 0 & 0 \\ 0 & 0 & I_{ny\times ny} & 0 & \cdots & 0 & 0 \\ 0 & 0 & 0 & I_{ny\times ny} & \cdots & 0 & 0 \\ \vdots & \vdots & \vdots & \vdots & \ddots & \vdots & \vdots \\ 0 & 0 & 0 & 0 & \cdots & I_{ny\times ny} & 0 \end{bmatrix}, \quad A_y \in \mathbb{R}^{(na+1)ny\times(na+1)ny},$$

$$A_{\Delta u} = \begin{bmatrix} B_2 & \cdots & B_{nb} \\ 0_{ny\times nu} & \vdots & 0_{ny\times nu} \\ 0_{ny\times nu} & \vdots & 0_{ny\times nu} \\ \vdots & \ddots & \vdots \\ 0_{ny\times nu} & \cdots & 0_{ny\times nu} \end{bmatrix}, \quad A_{\Delta u} \in \mathbb{R}^{(na+1)\cdot ny\times(nb-1)\cdot nu},$$

$$\underline{I} = \begin{bmatrix} 0_{nu} & \cdots & 0_{nu} \\ I_{nu} & \cdots & 0_{nu} \\ \vdots & \ddots & \vdots \\ 0_{nu} & I_{nu} & 0_{nu} \end{bmatrix}, \quad \underline{I} \in \mathbb{R}^{(nb-1)nu\times(nb-1)nu}, \ \underline{0} \in \mathbb{R}^{(nb-1)nu\times(na+1)ny},$$

$$B_{\Delta u} = \begin{bmatrix} B_1 \\ 0 \end{bmatrix} = \begin{bmatrix} B_1 \\ 0_{ny\times nu} \\ \vdots \\ 0_{ny\times nu} \end{bmatrix}, \quad B_1 \in \mathbb{R}^{ny\times nu}, \ B_{\Delta u} \in \mathbb{R}^{[(na+1)\cdot ny]\times nu},$$

$$\overline{I} = \begin{bmatrix} I_{nu} \\ 0_{nu} \\ \vdots \\ 0_{nu} \end{bmatrix}, \quad \overline{I} \in \mathbb{R}^{[(nb-1)\cdot nu]\times nu}, \qquad C_y = \Big[\underbrace{I_{ny\times ny} \ 0_{ny\times ny} \ \cdots \ 0_{ny\times ny}}_{na+1} \Big],$$

$$C_{\Delta u} = \Big[\underbrace{0_{ny\times nu} \ 0_{ny\times nu} \ \cdots \ 0_{ny\times nu}}_{nb-1} \Big]. \tag{4}$$

In the system defined by (2), the input and output values read from the plant are used to realign the model, which gives a better disturbance rejection capability to the controller. In González and Odloak [8], it is shown that the model presented above is detectable and stabilizable. More detailed discussions of minimal state space representations and the resulting observability and controllability of time-delayed systems can be found in De La Sen [9].

For control implementation, in order to better locate the uncertainties along the process model, it is adequate to divide the model in two separated parts. The first part of the model is related to the pure integrating outputs, and the second part is related to the stable outputs. Here, it is assumed that the process has no outputs related simultaneously with stable and integrating modes. Then, suppose that the model defined by (2) is written for the integrating outputs:

$$\underbrace{\begin{bmatrix} x_y^i(k+1) \\ x_{\Delta u}^i(k+1) \end{bmatrix}}_{x^i(k+1)} = \underbrace{\begin{bmatrix} A_y^i & A_{\Delta u}^i \\ \underline{0} & \underline{I} \end{bmatrix}}_{A^i} \underbrace{\begin{bmatrix} x_y^i(k) \\ x_{\Delta u}^i(k) \end{bmatrix}}_{x^i(k)} + \underbrace{\begin{bmatrix} B_{\Delta u}^i \\ \overline{I} \end{bmatrix}}_{B^i} \Delta u(k), \tag{5}$$

$$y^i(k) = \begin{bmatrix} C_y^i & C_{\Delta u}^i \end{bmatrix} \begin{bmatrix} x_y^i(k) \\ x_{\Delta u}^i(k) \end{bmatrix}. \tag{6}$$

Analogously, a similar model can be written for the stable outputs:

$$\underbrace{\begin{bmatrix} x_{y,\theta_p}^s(k+1) \\ x_{\Delta u}^s(k+1) \end{bmatrix}}_{x_{\theta_p}^s(k+1)} = \underbrace{\begin{bmatrix} A_y^s(\theta_p) & A_{\Delta u}^s(\theta_p) \\ \underline{0} & \underline{I} \end{bmatrix}}_{A^s(\theta_p)} \underbrace{\begin{bmatrix} x_{y,\theta_p}^s(k) \\ x_{\Delta u}^s(k) \end{bmatrix}}_{x_{\theta_p}^s(k)} + \underbrace{\begin{bmatrix} B_{\Delta u}^s(\theta_p) \\ \overline{I} \end{bmatrix}}_{B^s(\theta_p)} \Delta u(k), \tag{7}$$

$$y_{\theta_p}^s(k) = \begin{bmatrix} C_y^s & C_{\Delta u}^s \end{bmatrix} \begin{bmatrix} x_{y,\theta_p}^s(k) \\ x_{\Delta u}^s(k) \end{bmatrix}. \tag{8}$$

Observe that some of the components of the model defined in (7) and (8) depend on θ_p, that is, the vector of parameters that define the model ($\theta_p = [A_p, B_p]$). These parameters are unknown, which characterizes the uncertainty in the model related to the stable outputs. The simplest form to consider uncertainty is through the multimodel uncertainty where θ_p belongs to a finite set $\theta_p \in \Omega = \{\theta_1, \ldots, \theta_L\}$. This means that the true model of the process is unknown, but it is one of the L models of a finite set of known models.

Within the conventional MPC formulation, the control horizon m is such that $\Delta u(k+j) = 0$, $j = m, m+1, \ldots,$

then, with the model representation defined in (2) and (3), it is easy to show that $x_{\Delta u}(k+n+j) = 0$, $j = 1, 2, \ldots$, where $n = m + nb - 1$ and $x_y(k+n+j) = A_y x_y(k+n+j-1)$, $j = 1, 2, \ldots$.

Then, after time step $k+n$, it is possible to apply the Jordan matrix decomposition of A_y as follows:

$$A_y V = V A_d, \tag{9}$$

where A_d is a block diagonal matrix where the eigenvalues of A_y appear in its main diagonal. If A_y is a full-rank square matrix, then V is invertible and it is possible to make a change of variables such that the integrating states related to the incremental form of the model are separated from the remaining states of the system. For instance, for the integrating system defined in (5) and (6), one can define

$$x_y^i(k) = V_i z^i(k),$$
$$x_y^i(k) = \begin{bmatrix} V_i^{\Delta} & V_i^i \end{bmatrix} \begin{bmatrix} z_i^{\Delta}(k) \\ z_i^i(k) \end{bmatrix}, \tag{10}$$

where z_i^{Δ} corresponds to the integrating states related to the incremental form of the model and z_i^i is the state associated with the real integrating states of the system. V_i is obtained from the Jordan decomposition of matrix A_y^i as follows:

$$A_y^i V_i = V_i A_d^i. \tag{11}$$

It can be shown that $A_d^i = \begin{bmatrix} I_{nyi} & I_{nyi} \\ 0 & I_{nyi} \end{bmatrix}$ where nyi is the number of integrating outputs.

It should be noted that, after time step $k+n$, the transformed state of the integrating system will evolve according to the equation

$$\begin{bmatrix} z_i^{\Delta}(k+n+1) \\ z_i^i(k+n+1) \end{bmatrix} = \begin{bmatrix} I_{ny} & I_{ny} \\ 0 & I_{ny} \end{bmatrix} \begin{bmatrix} z_i^{\Delta}(k+n) \\ z_i^i(k+n) \end{bmatrix}. \tag{12}$$

Analogously, for the system represented through (7) and (8), the following state transformation can be defined:

$$x_{y,\theta_p}^s(k) = V_{s,\theta_p} z_{\theta_p}^s(k) = \begin{bmatrix} V_{s,\theta_p}^{\Delta} & V_{s,\theta_p}^s \end{bmatrix} \begin{bmatrix} z_{s,\theta_p}^{\Delta}(k) \\ z_{s,\theta_p}^s(k) \end{bmatrix}, \tag{13}$$

where z_{s,θ_p}^{Δ} corresponds to the integrating states related to the incremental form of the model and z_{s,θ_p}^s is the state associated with the stable states of the system corresponding to the parameters represented as θ_p. V_{s,θ_p} is obtained from the Jordan decomposition of matrix $A_y^s(\theta_p)$ as follows:

$$A_y^s(\theta_p) V_{s,\theta_p} = V_{s,\theta_p} A_{d,\theta_p}^s. \tag{14}$$

It can be shown that

$$A_{d,\theta_p}^s = \begin{bmatrix} I_{nys} & 0 \\ 0 & F_{\theta_p}^s \end{bmatrix} \tag{15}$$

where nys is the number of stable outputs of the system. If the stable poles of the system are nonrepeated, $F_{\theta_p}^s$ is a diagonal matrix containing the eigenvalues of $A_y^s(\theta_p)$.

After time step $k+n$, the transformed state corresponding to the stable outputs will evolve according to the equation

$$\begin{bmatrix} z_{s,\theta_p}^{\Delta}(k+n+1) \\ z_{s,\theta_p}^s(k+n+1) \end{bmatrix} = \begin{bmatrix} I_{nys} & 0 \\ 0 & F_{\theta_p}^s \end{bmatrix} \begin{bmatrix} z_{s,\theta_p}^{\Delta}(k+n) \\ z_{s,\theta_p}^s(k+n) \end{bmatrix}. \tag{16}$$

3. The Robust MPC with Output Feedback

The system outputs need to be controlled through the manipulation of the nu inputs. Then, the robust MPC proposed here is based on the following objective function:

$$\begin{aligned} J_k(\theta_p) = {} & \sum_{j=1}^{n} \left[y^i(k+j \mid k) - y_i^{sp} - \delta_{\Delta,k}^i - j\delta_{i,k}^i \right]^T \\ & \times Q_i \left[y^i(k+j \mid k) - y_i^{sp} - \delta_{\Delta,k}^i - j\delta_{i,k}^i \right] \\ & + \sum_{j=n+1}^{\infty} \left[y^i(k+j \mid k) - y_i^{sp} - \delta_{\Delta,k}^i - j\delta_{i,k}^i \right]^T \\ & \times Q_i \left[y^i(k+j \mid k) - y_i^{sp} - \delta_{\Delta,k}^i - j\delta_{i,k}^i \right] \\ & + \sum_{j=1}^{n} \left[y_{\theta_p}^s(k+j \mid k) - y_s^{sp} - \delta_{\Delta,k}^s(\theta_p) \right]^T \\ & \times Q_s \left[y_{\theta_p}^s(k+j \mid k) - y_s^{sp} - \delta_{\Delta,k}^s(\theta_p) \right] \\ & + \sum_{j=n+1}^{\infty} \left[y_{\theta_p}^s(k+j \mid k) - y_s^{sp} - \delta_{\Delta,k}^s(\theta_p) \right]^T \\ & \times Q_s \left[y_{\theta_p}^s(k+j \mid k) - y_s^{sp} - \delta_{\Delta,k}^s(\theta_p) \right] \\ & + \sum_{j=0}^{m-1} \Delta u(k+j \mid k)^T R \Delta u(k+j \mid k) \\ & + \left(\delta_{\Delta,k}^i\right)^T S_{\Delta}^i \delta_{\Delta,k}^i + \left(\delta_{\Delta,k}^s(\theta_p)\right)^T S_{\Delta}^s \delta_{\Delta,k}^s(\theta_p) \\ & + \left(\delta_{i,k}^i\right)^T S_i^i \delta_{i,k}^i, \end{aligned} \tag{17}$$

where $y^i(k+j \mid k)$ is prediction of integrating output at time $k+j$ computed at time k and $y_{\theta_p}^s(k+j \mid k)$ is the prediction of the stable output corresponding to model θ_p; $\delta_{\Delta,k}^i$, $\delta_{\Delta,k}^s$, and $\delta_{i,k}^i$ are slack variables that are included in the control problem to guarantee that $J_k(\theta_p)$ is bounded. Q_i, Q_s, R, S_{Δ}^i, S_{Δ}^s and S_i^i are positive weighting matrices. In the second term on the right-hand side of (17), the prediction of the integrating output can be written as follows:

$$\begin{aligned} y^i(k+n+j \mid k) = {} & C_y^i x_y^i(k+n+j \mid k) \\ & + C_{\Delta u}^i x_{\Delta u}^i(k+n+j \mid k). \end{aligned} \tag{18}$$

But, since $x^i_{\Delta u}(k+n+j \mid k) = 0$, (18) becomes

$$y^i(k+n+j \mid k) = C^i_y x^i_y(k+n+j \mid k). \quad (19)$$

Then, using (10) and (12), (19) can be written as follows:

$$y^i(k+n+j \mid k) = C^i_y V^\Delta_i \left[z^\Delta_i(k+n \mid k) + j z^i_i(k+n \mid k) \right] + C^i_y V^i_i z^i_i(k+n \mid k). \quad (20)$$

If (20) is substituted into the infinite sum represented by the second term on the right-hand side of (17), it is easy to see that the objective function defined in (17) will be bounded only if the following constraints are satisfied:

$$C^i_y V^\Delta_i z^\Delta_i(k+n \mid k) + C^i_y V^i_i z^i_i(k+n \mid k) - y^{sp}_i - \delta^i_{\Delta,k} - n\delta^i_{i,k} = 0, \quad (21)$$

$$C^i_y V^\Delta_i z^i_i(k+n \mid k) - \delta^i_{i,k} = 0. \quad (22)$$

Analogously, in the infinite sum represented by the fourth term on the right-hand side of (17), the prediction of the stable output can be written as follows:

$$y^s_{\theta_p}(k+n+j \mid k) = C^s_y x^s_{y,\theta_p}(k+n+j \mid k). \quad (23)$$

Now, using (13) and (16), equation (23) can be written as follows:

$$y^s_{\theta_p}(k+n+j \mid k) = C^s_y V^\Delta_{s,\theta_p} z^\Delta_{y,\theta_p}(k+n \mid k) + C^s_y V^s_{s,\theta_p} \left(F^s_{\theta_p}\right)^j z^s_{y,\theta_p}(k+n \mid k). \quad (24)$$

If the expression for the output prediction obtained in (24) is substituted into the infinite sum represented by the fourth term on the right-hand side of (17), it is easy to see that the cost function will be bounded only if the following constraint is satisfied:

$$C^s_y V^\Delta_{s,\theta_p} z^\Delta_{y,\theta_p}(k+n \mid k) - y^{sp}_s - \delta^s_{\Delta,k}\left(\theta_p\right) = 0. \quad (25)$$

Consequently, the optimization problem that produces the control law proposed here will have to include the constraints defined in (21), (22), and (25).

Equation (21) can be developed in terms of the available states at time k and the vector of the future control moves as follows:

$$\left[C^i_y V^\Delta_i N^i_1 + C^i_y V^i_i N^i_2\right] V^{-1}_i N^i x^i(k+n \mid k) - y^{sp}_i - \delta^i_{\Delta,k} - n\delta^i_{i,k} = 0, \quad (26)$$

where N^i_1 is a matrix of ones and zeros that collects component z^Δ_i from z_i and matrix N^i_2 collects component z^i_i from z_i. The model defined in (5) can be used to represent (26) in terms of the integrating state that at time k, is measured as follows:

$$\left[C^i_y V^\Delta_i N^i_1 + C^i_y V^i_i N^i_2\right] V^{-1}_i N^i \left(\left(A^i\right)^{n-m} x^i(k) + B^i_m \Delta u_k \right) - y^{sp}_i - \delta^i_{\Delta,k} - n\delta^i_{i,k} = 0, \quad (27)$$

where $\Delta u_k = [\Delta u(k \mid k)^T \ \cdots \ \Delta u(k+m-1 \mid k)^T]^T$ and $B^i_m = [(A^i)^{n-1} B^i \ \cdots \ (A^i)^{n-m} B^i]$.

The constraint (22) can also be represented in terms of the present state and the vector of future control moves:

$$C^i_y V^\Delta_i N^i_1 V^{-1}_i N^i \left(\left(A^i\right)^{n-m} x^i(k) + B^i_m \Delta u_k \right) - \delta^i_{i,k} = 0. \quad (28)$$

Analogously, (25) can be written as follows:

$$C^s_y V^\Delta_{s,\theta_p} N^s_1 V^{-1}_{s,\theta_p} N^s_2 \left(A^s\left(\theta_p\right)^{n-m} x^s_{\theta_p}(k) + B^s_m\left(\theta_p\right)\Delta u_k \right) - y^{sp}_s - \delta^s_{\Delta,k}\left(\theta_p\right) = 0, \quad (29)$$

where N^s_1 captures the state component z^Δ_{s,θ_p} of state $z^s_{\theta_p}$ and N^s_2 captures the component x^s_{y,θ_p} of state $x^s_{\theta_p}$. With the constraints discussed above, the control cost function defined in (17) can be written as follows:

$$\begin{aligned} J_k\left(\theta_p\right) = {} & \sum_{j=1}^{n} \left[y^i(k+j \mid k) - y^{sp}_i - \delta^i_{\Delta,k} - j\delta^i_{i,k} \right]^T \\ & \times Q_i \left[y^i(k+j \mid k) - y^{sp}_i - \delta^i_{\Delta,k} - j\delta^i_{i,k} \right] \\ & + \sum_{j=1}^{n} \left[y^s_{\theta_p}(k+j \mid k) - y^{sp}_s - \delta^s_{\Delta,k}\left(\theta_p\right) \right]^T \\ & \times Q_s \left[y^s_{\theta_p}(k+j \mid k) - y^{sp}_s - \delta^s_{\Delta,k}\left(\theta_p\right) \right] \\ & + x^s_{\theta_p}(k+n \mid k)^T (N^s)^T \left(V^{-1}_{s,\theta_p}\right)^T (N^s_3)^T \\ & \times \overline{Q}_s \, N^s_3 \, V^{-1}_{s,\theta_p} N^s x^s_{\theta_p}(k+n \mid k) \\ & + \sum_{j=0}^{m-1} \Delta u(k+j \mid k)^T R \Delta u(k+j \mid k) \\ & + \left(\delta^i_{\Delta,k}\right)^T S^i_\Delta \delta^i_{\Delta,k} + \left(\delta^s_{\Delta,k}\left(\theta_p\right)\right)^T S^s_\Delta \delta^s_{\Delta,k}\left(\theta_p\right) \\ & + \left(\delta^i_{i,k}\right)^T S^i_i \delta^i_{i,k}, \end{aligned} \quad (30)$$

where $\overline{Q}_s$ is obtained from the solution to the following equation:

$$\overline{Q}_s - \left(F^s_{\theta_p}\right)^T \overline{Q}_s F^s_{\theta_p} = \left(F^s_{\theta_p}\right)^T \left(V^s_{s,\theta_p}\right)^T \left(C^s_y\right)^T Q_s C^s_y V^s_{s,\theta_p} F^s_{\theta_p}. \quad (31)$$

Observe that (30) is a finite horizon cost function with a terminal weight computed through (31).

One can now define the robust MPC with output feedback for systems with integrating and stable outputs. The controller is robust in a sense that it maintains stability even in the presence of uncertainty in the model related to the stable outputs. Assuming that θ_N corresponds to the nominal or most probable model, the proposed controller is based on the solution to the following problem.

Problem 1. Consider the following objective function:

$$\min_{\substack{\Delta u_k, \delta^i_{i,k}, \delta^i_{\Delta,k}, \delta^s_{\Delta,k}(\theta_p)\\ p=1,\ldots,L}} J_k(\theta_N) \tag{32}$$

subject to (26), (28) and

$$C^s_y V^{\Delta}_{s,\theta_p} N^s_1 V^{-1}_{s,\theta_p} N^s_2 \left(A^s\left(\theta_p\right)^{n-m} x^s_{\theta_p}(k) + B^s_m\left(\theta_p\right)\Delta u_k\right)$$
$$- y^{sp}_s - \delta^s_{\Delta,k}\left(\theta_p\right) = 0, \quad p = 1,\ldots,L$$

$$-\Delta u_{\max} \le \Delta u(k+j \mid k) \le \Delta u_{\max}, \quad j = 0,\ldots,m-1$$
$$u_{\min} \le u(k+j \mid k) \le u_{\max}, \quad j = 0,\ldots,m-1$$
$$J_k\left(\theta_p\right) \le \widetilde{J}_k\left(\theta_p\right), \quad p = 1,\ldots,L, \tag{33}$$

$$\left\|\delta^i_{i,k}\right\| \le \left\|\widetilde{\delta}^i_{i,k}\right\|, \tag{34}$$

where $\widetilde{J}_k(\theta_p)$ is computed with

$$\Delta\widetilde{u}_k = [\Delta u^*(k \mid k-1) \quad \Delta u^*(k+1 \mid k-1) \quad \cdots \quad \Delta u^*(k+m-2 \mid k-1) \quad 0], \tag{35}$$

where $\Delta u^*(k+j \mid k-1)$ corresponds to the optimal solution to Problem 1 at time step $k-1$ and $\widetilde{\delta}^i_{i,k}$, $\widetilde{\delta}^i_{\Delta,k}$, $\widetilde{\delta}^s_{\Delta,k}(\theta_p)$, $p = 1,\ldots,L$, which are obtained from the solution to the following equations:

$$\begin{aligned}
&C^s_y V^{\Delta}_{s,\theta_p} N^s_1 V^{-1}_{s,\theta_p} N^s_2 \left(A^s\left(\theta_p\right)^{n-m} x^s_{\theta_p}(k) + B^s_m\left(\theta_p\right)\Delta\widetilde{u}_k\right)\\
&\quad - y^{sp}_s - \widetilde{\delta}^s_{\Delta,k}\left(\theta_p\right) = 0, \quad p = 1,\ldots,L,\\
&\left[C^i_y V^{\Delta}_i N^i_1 + C^i_y V^i_i N^i_2\right] V^{-1}_i N^i\left(\left(A^i\right)^{n-m} x^i(k) + B^i_m \Delta\widetilde{u}_k\right)\\
&\quad - y^{sp}_i - \widetilde{\delta}^i_{\Delta,k} - n\widetilde{\delta}^i_{i,k} = 0,\\
&C^i_y V^{\Delta}_i N^i_1 V^{-1}_i N^i\left(\left(A^i\right)^{n-m} x^i(k) + B^i_m \Delta\widetilde{u}_k\right) - \widetilde{\delta}^i_{i,k} = 0.
\end{aligned} \tag{36}$$

Remark 2. The constraint defined in (33) corresponds to the conventional cost contracting constraint first proposed in Badgwell [10] in the development of a robust MPC for the regulator problem and extended in Odloak [11] for the output tracking problem. In these works, it is shown that, for open-loop stable systems with multiplant uncertainty, the inclusion of constraint (33) forces the cost function of the true plant to be a Lyapunov function for the closed-loop system.

Remark 3. The constraint defined in (34) is intended to force slack $\delta^i_{i,k}$ to converge to zero. It will be shown latter that if Problem 1 is feasible with this slack kept equal to zero, the control cost function of the true plant becomes a Lyapunov function for the uncertain system and the controller resulting from the solution to Problem 1. However, if there is uncertainty on the model related to the integrating part of the system, which means that there is uncertainty in matrices A^i and/or B^i, then, it is easy to show that it is not possible to find a solution to Problem 1 that satisfies constraint (28) for the different matrices A^i and B^i with $\delta^i_{i,k}(\theta_p) = 0$, $p = 1,\ldots,L$. This means that the approach followed here cannot be applied for systems with uncertainty in the gains of the integrating outputs. In the next theorems we prove the robust stability of the controller obtained through the solution to Problem 1.

The proof of the stability of an MPC usually involves two ingredients: recursive feasibility of the control problem and the existence of a Lyapnvov function for the closed-loop system. The theorem below shows the recursive feasibility of Problem 1, and the other two theorems show that the cost function can be interpreted as a Lyapunov function for the system.

Theorem 4. *If Problem 1 has a feasible solution at time step k, it will remain feasible at any subsequent time step k + j.*

Proof. Suppose that at time k the optimum solution is represented as follows:

$$\Delta u^*_k, \delta^{*i}_{i,k}, \delta^{*i}_{\Delta,k}, \delta^{*s}_{\Delta,k}\left(\theta_p\right), \quad p = 1,\ldots,L. \tag{37}$$

Then, it is easy to show that, at time step $k+1$, the following solution, inherited from the optimum solution at time k, will be feasible:

$$\begin{aligned}
&\Delta u_{k+1} = [\Delta u^*(k+1 \mid k) \quad \Delta u^*(k+2 \mid k1) \quad \cdots \quad \Delta u^*(k+m-1 \mid k) \quad 0], \quad \delta^i_{i,k+1} = \widetilde{\delta}^i_{i,k+1},\\
&\delta^i_{\Delta,k+1} = \widetilde{\delta}^i_{\Delta,k}, \quad \delta^s_{\Delta,k+1}\left(\theta_p\right) = \widetilde{\delta}^s_{\Delta,k}\left(\theta_p\right), \quad p = 1,\ldots,L,
\end{aligned} \tag{38}$$

where $\widetilde{\delta}^i_{i,k+1}$, $\widetilde{\delta}^i_{\Delta,k+1}$, $\widetilde{\delta}^s_{\Delta,k+1}(\theta_p)$, $p = 1,\ldots,L$ are computed as defined in the enunciate of the theorem and consequently, one can always find a solution to Problem 1. □

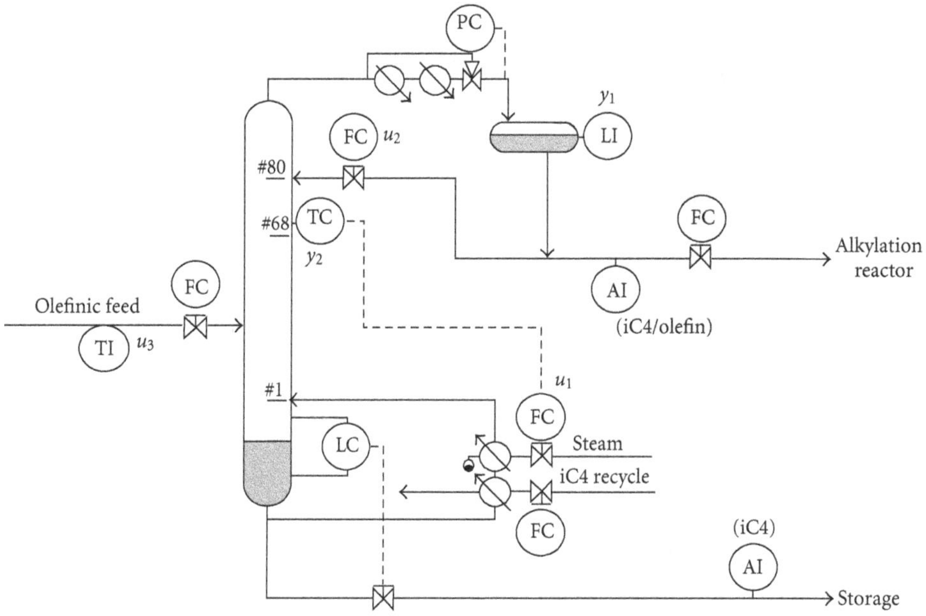

FIGURE 1: Process diagram for alkylation system.

Since Problem 1 has recursive feasibility, the next theorem shows that the consideration of constraint (34) in the control problem guarantees that, if the system is not disturbed, $\delta_{i,k}^{i}$ that is the slack related to the integrating outputs will converge to zero.

Theorem 5. *If the system defined in* (5), (7), (6), *and* (8) *is stabilizable, the solution to Problem 1 at successive time steps drives the slacks $\delta_{i,k}^{i}$ related to the integrating outputs to zero.*

Proof. Suppose that, at time step k, Problem 1 is solved and the solution defined in (37) is obtained. Then, since it is assumed that there is no uncertainty in the model related to the integrating outputs, it is easy to see that for the undisturbed system $\widetilde{\delta}_{\Delta,k+1}^{i} = \delta_{\Delta,k}^{*i}$, and consequently, $\|\delta_{\Delta,k+1}^{*i}\| \leq \|\delta_{\Delta,k}^{*i}\|$. Then, if the system is stabilizable, $\delta_{\Delta,k}^{i}$ will be driven to zero. □

Theorem 6. *Consider the system represented in* (5), (6), (7), *and* (8). *If the conditions defined in Theorems 4 and 5 are satisfied and the desired steady state is reachable, then, the solution of Problem 1 at successive time steps will produce a sequence of control moves that will drive the system outputs to the desired targets.*

Proof. Suppose that the convergence of $\delta_{i,k}^{i}$ to zero has already been achieved and Problem 1 is solved at time step k. Let us designate the resulting optimum solution as

$$\Delta u_k^*, \delta_{i,k}^{*i} = 0, \delta_{\Delta,k}^{*i}, \delta_{\Delta,k}^{*s}\left(\theta_p\right), \quad p = 1, \ldots, L. \tag{39}$$

Let us now define the set of parameters corresponding to the true model as θ_T. The resulting optimum control move $\Delta u^*(k \mid k)$ is implemented in the true system and we move to time instant $k+1$ where Problem 1 is solved again. At this time step, as shown in Theorem 4, $\Delta\widetilde{u}_{k+1}, \widetilde{\delta}_{i,k+1}^{i}, \widetilde{\delta}_{\Delta,k+1}^{i}, \widetilde{\delta}_{\Delta,k+1}^{s}(\theta_p)$, $p = 1, \ldots, L$ is a feasible solution to Problem 1. Also, since $\widetilde{\delta}_{i,k+1}^{i} = \delta_{i,k}^{*i} = 0$, it is easy to show that $\widetilde{\delta}_{\Delta,k+1}^{i} = \delta_{\Delta,k}^{*i}$, $\widetilde{\delta}_{\Delta,k+1}^{s}(\theta_T) = \delta_{\Delta,k}^{*s}(\theta_T)$. Thus, it is easy to see that $\widetilde{J}_{k+1}(\theta_T) = J_k^*(\theta_T)$, and consequently, $J_{k+1}^*(\theta_T) \leq J_k^*(\theta_T)$. Now, since $J_k(\theta_T)$ is positive and bounded below by zero, it can be interpreted as a Lyapunov function for the closed-loop system with the proposed controller. This means that $y^i(k+j \mid k) - y_i^{sp} - \delta_{\Delta,k}^{i}$ and $y_{\theta_T}^{s}(k+j \mid k) - y_s^{sp} - \delta_{\Delta,k}^{s}(\theta_T)$ will converge to zero as $j \to \infty$. Following the same steps as in González et al. [12], it can be shown that if weights S_Δ^i and S_Δ^s are large enough in comparison to R and the desired set point is reachable, slacks $\delta_{\Delta,k}^{i}$ and $\delta_{\Delta,k}^{s}(\theta_T)$ will converge to zero and the outputs will converge to the set points. If the targets are not reachable, the distance to the targets will be minimized according to the norm $(\delta_{\Delta,k}^{i})^T S_\Delta^i \delta_{\Delta,k}^{i} + (\delta_{\Delta,k}^{s}(\theta_T))^T S_\Delta^s \delta_{\Delta,k}^{s}(\theta_T)$. □

4. Simulation Example: The Distillation Column System

The proposed control strategy was tested with a small dimension system of the process industry. The system is part of a distillation column where isobutene is separated from n-butane in the alkylation unit of an oil refinery. This system was studied in Carrapiço et al. [13] that implemented an infinite horizon MPC to control the industrial distillation column. The controlled outputs are the level of liquid in the overhead drum (y_1%) and the temperature at tray number 68 (y_2, °C). The manipulated variables are the steam flow rate to the reboiler (u_1, t/h), the distillate flow rate (u_2, m^3/d), and feed temperature deviation (u_3, °C). Figure 1 shows schematically the structure of the existing regulatory loops of the distillation column.

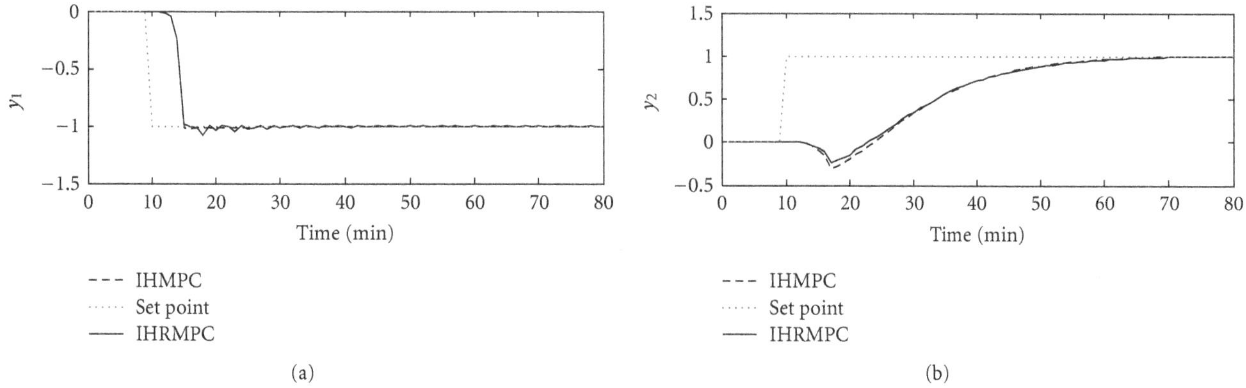

FIGURE 2: Outputs of the distillation system. Nominal case.

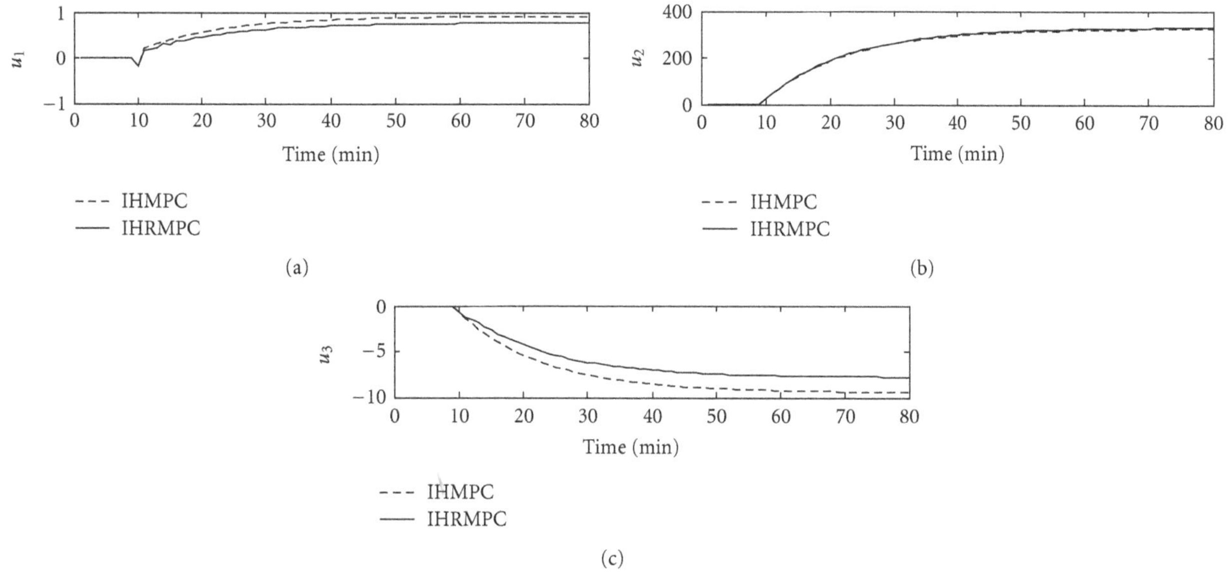

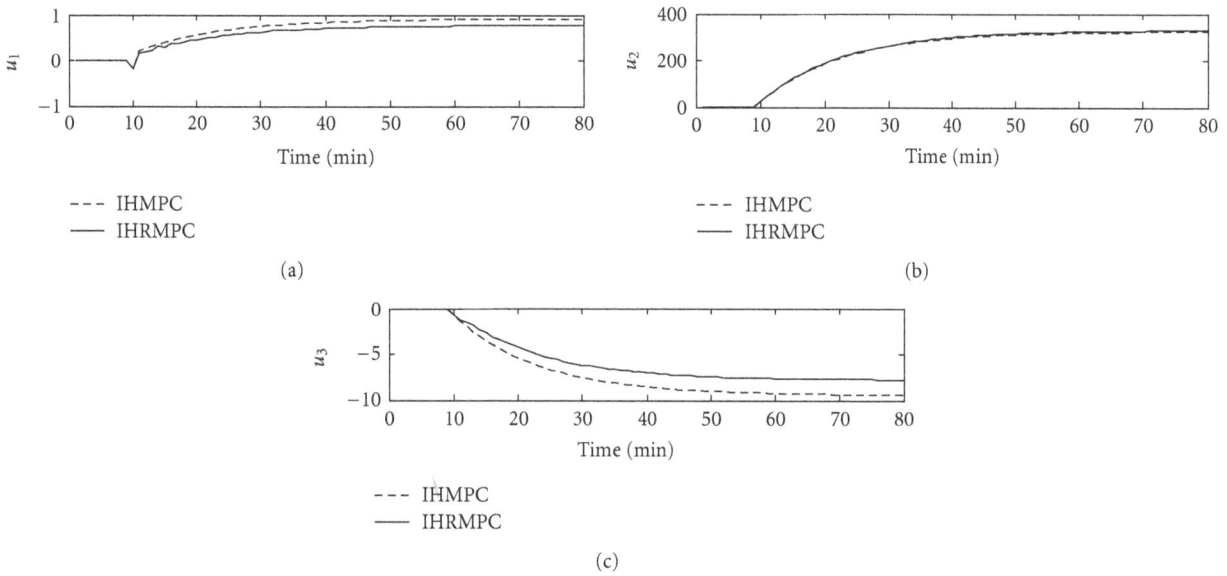

FIGURE 3: Inputs of the distillation system. Nominal case.

From practical tests, for a sampling period $T = 1$ min, the following two models were obtained at different operating conditions:

$$G_1(z) = \begin{bmatrix} \dfrac{2.3z^{-4}}{1-z^{-1}} & \dfrac{-0.7\times 10^{-3}z^{-3}}{1-z^{-1}} & \dfrac{0.2z^{-5}}{1-z^{-1}} \\ \dfrac{0.4604z^{-8}}{1-0.902z^{-1}} & \dfrac{0.1915\times 10^{-3}z^{-3}}{1-0.8632z^{-1}} & \dfrac{0.03304z^{-4}}{1-0.9174z^{-1}} \end{bmatrix} \quad (40)$$

$$G_2(z) = \begin{bmatrix} \dfrac{2.3z^{-4}}{1-z^{-1}} & \dfrac{-0.7\times 10^{-3}z^{-3}}{1-z^{-1}} & \dfrac{0.2z^{-5}}{1-z^{-1}} \\ \dfrac{0.1965z^{-4}}{1-0.9146z^{-1}} & \dfrac{0.3215\times 10^{-3}z^{-2}}{1-0.8852z^{-1}} & \dfrac{0.03332z^{-6}}{1-0.9306z^{-1}} \end{bmatrix} \quad (41)$$

We observe that y_1 is integrating with respect to all the inputs while y_2 is stable. We should notice that there is uncertainty of about 20% in time constants and uncertainty of about 50% in process gains of the stable part of the model. Uncertainty in the dead times is also present in the model.

In the set point tracking problem simulated here, the desired value of the liquid level in the overhead drum (y_1) was reduced by 1% and the desired value of the temperature at stage #68 (y_2) was increased by 1°C. The tuning parameters of the controller that were used in the simulations included here are the following: $m = 2$, $Q_i = 1$, $Q_s = 1$, $R = \text{diag}([0.1\ \ 0.1\ \ 10])$, $S_\Delta^i = 1 \times 10^3$, $S_\Delta^s = 1 \times 10^3$, and $S_i^i = 1 \times 10^3$. Related to the values of the variables at the initial steady state, the input limits are $u_{\max} = [10\ \ 400\ \ 10]$, $u_{\min} = [-10\ \ -400\ \ -10]$, and $\Delta u_{\max} = [0.1\ \ 50\ \ 0.01]$. The nominal model is represented by model $G_1(s)$ defined in (40), and the true plant can be either $G_1(s)$ or $G_2(s)$ defined in (41). In this work, IHMPC represents the same controller as the one presented in Carrapiço et al. [13], which does not consider the uncertainty in the process model. In the

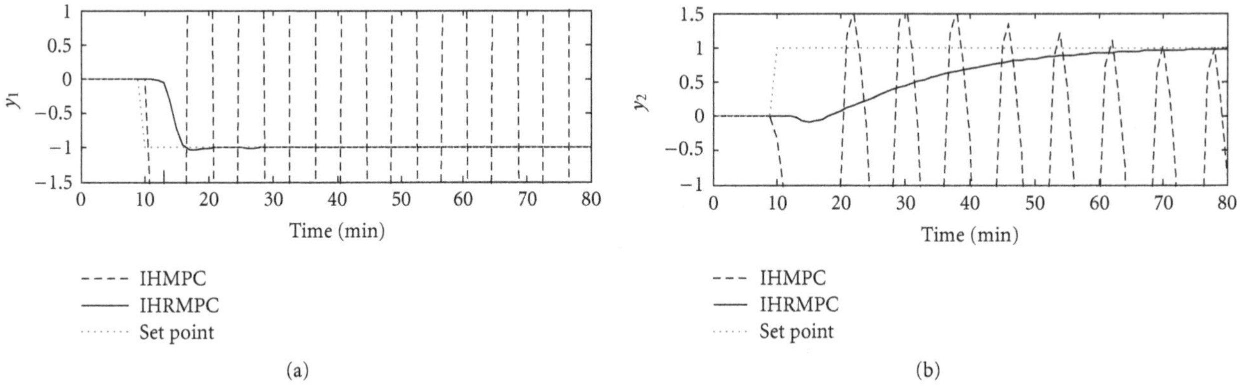

FIGURE 4: Outputs of the distillation system. Uncertain case.

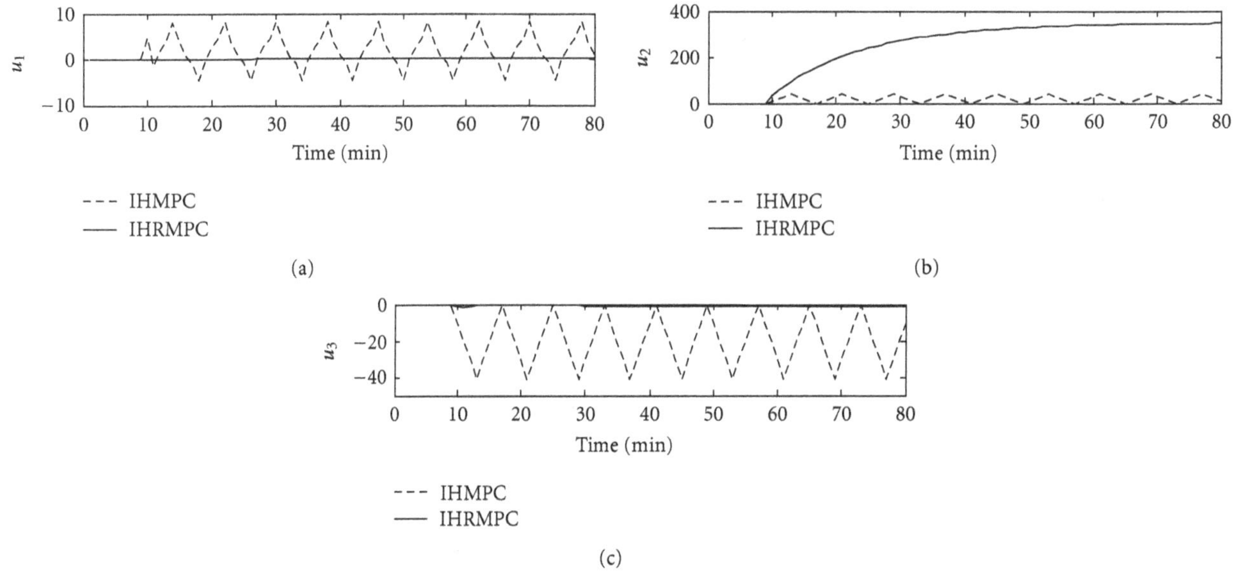

FIGURE 5: Inputs of the distillation system. Uncertain case.

first simulation, it is considered the case in which the true plant has the same model as the nominal plant. Figures 2 and 3 show the system responses for the nominally stable IHMPC proposed in Carrapiço et al. [13] and for the robust controller resulting from the solution to Problem 1 and designated here as IHRMPC. The two controllers have the same tuning parameters. One can see from Figures 2 and 3 that the nominal IHMPC and the robust IHRMPC perform similarly. This is quite surprising because it is widespread in the literature that the robust MPC should have a more conservative behavior than the controller based on the true model as this controller also takes into account the output predictions of model $G_2(s)$ that is quite different from the true model of the plant. This conservative behavior would produce a slower response, which is not observed in the simulation results obtained here. In the second simulation case, model $G_2(s)$ represents the true plant. The IHMPC is still based on model $G_1(s)$ while the proposed IHRMPC contains models $G_1(s)$ and $G_2(s)$. Figures 4 and 5 show the responses of the distillation system with each controller. We can see that the robust controller stabilizes the true model and the performance of the controller is acceptable. However, the IHMPC based only on the nominal model becomes unstable, which indicates that the controller based only on model $G_1(s)$ cannot be used to control the distillation column at the operating point where model $G_2(s)$ represents the true process.

5. Conclusion

In this paper, it was presented a new version of the robust infinitive horizon MPC with output feedback for systems with stable and integrating outputs. The adopted model formulation precludes the need to include a state observer, and the computer burden to run the controller may be reduced. To accommodate uncertainty in the process model, the state space model was built in two separate parts: one part is related to the integrating outputs and the other is related to the stable outputs. With this approach, it was possible to include the multiplant uncertainty in the model of stable outputs. The resulting optimization problem is a convex nonlinear program that can be easily solved with the available NLP packages. A simulation example shows that the implementation of the developed approach at real industrial systems may be achieved at least for systems of small to medium dimension.

References

[1] S. J. Qin and T. A. Badgwell, "A survey of industrial model predictive control technology," *Control Engineering Practice*, vol. 11, no. 7, pp. 733–764, 2003.

[2] J. H. Lee and B. L. Cooley, "Min-max predictive control techniques for a linear state-space system with a bounded set of input matrices," *Automatica*, vol. 36, no. 3, pp. 463–473, 2000.

[3] M. A. Rodrigues and D. Odloak, "Output feedback MPC with guaranteed robust stability," *Journal of Process Control*, vol. 10, no. 6, pp. 557–572, 2000.

[4] M. A. Rodrigues and D. Odloak, "Robust MPC for systems with output feedback and input saturation," *Journal of Process Control*, vol. 15, no. 7, pp. 837–846, 2005.

[5] J. M. Perez, *Robust predictive control with output feedback [M.S. Dissertation]*, University of São Paulo, São Paulo, Brazil, 2006.

[6] J. M. Maciejowski, *Predictive Control with Constraints*, Prentice Hall, London, UK, 2002.

[7] A. H. González, J. M. Perez, and D. Odloak, "Infinite horizon MPC with non-minimal state space feedback," *Journal of Process Control*, vol. 19, no. 3, pp. 473–481, 2009.

[8] A. H. González and D. Odloak, "Robust model predictive controller with output feedback and target tracking," *IET Control Theory and Applications*, vol. 4, no. 8, pp. 1377–1390, 2010.

[9] M. De La Sen, "On minimal realizations and minimal partial realizations of linear time-invariant systems subject to point incommensurate delays," *Mathematical Problems in Engineering*, vol. 2008, Article ID 790530, 19 pages, 2008.

[10] T. A. Badgwell, "Robust model predictive control of stable linear systems," *International Journal of Control*, vol. 68, no. 4, pp. 797–818, 1997.

[11] D. Odloak, "Extended robust model predictive control," *AIChE Journal*, vol. 50, no. 8, pp. 1824–1836, 2004.

[12] A. H. González, J. L. Marchetti, and D. Odloak, "Extended robust model predictive control of integrating systems," *AIChE Journal*, vol. 53, no. 7, pp. 1758–1769, 2007.

[13] O. L. Carrapiço, M. M. Santos, A. C. Zanin, and D. Odloak, "Application of the IHMPC to an industrial process system," in *Proceedings of the7th IFAC International Symposium on Advanced Control of Chemical Processes (ADCHEM '09)*, pp. 851–856, July 2009.

9

Continuous Fuel Level Sensor Based on Spiral Side-Emitting Optical Fiber

Chengrui Zhao, Lin Ye, Xun Yu, and Junfeng Ge

Department of Control Science and Engineering, Huazhong University of Science and Technology, 1037 Luoyu Road, Hubei, Wuhan 430074, China

Correspondence should be addressed to Junfeng Ge, gejf@mail.hust.edu.cn

Academic Editor: Derong Liu

A continuous fuel level sensor using a side-emitting optical fiber is introduced in this paper. This sensor operates on the modulation of the light intensity in fiber, which is caused by the cladding's acceptance angle change when it is immersed in fuel. The fiber is bent as a spiral shape to increase the sensor's sensitivity by increasing the attenuation coefficient and fiber's submerged length compared to liquid level. The attenuation coefficients of fiber with different bent radiuses in the air and water are acquired through experiments. The fiber is designed as a spiral shape with a steadily changing slope, and its response to water level is simulated. The experimental results taken in water and aviation kerosene demonstrate a performance of 0.9 m range and 10 mm resolution.

1. Introduction

Fuel level measurement is a great challenge in aircraft fuel systems [1]. The most frequently used level sensor for aircraft is the capacitive level sensor for its good sensitivity and maintainability. But when the plane takes off from an airport with hot and wet environment, the moisture in the fuel tank will be congealed and mixed in the fuel, then error and short circuit will happen to the capacitive sensor. For this reason, ultrasonic level sensor was used to replace the capacitive sensor in the planes like B777, F22, and so forth. Now the fuel level sensors on the newest aircrafts like B787, A380, and F35 are changed back to the capacitive type, for the ultrasonic level sensor's bad performance in reliability and maintainability during the flight.

Optical fiber liquid level sensors (OFLLS) have been reported to be safe and reliable and have many advantages for aircraft fuel measurement [2]. Above all, the moisture or water mixed in the fuel will have little influence to the OFLLS because the optical fiber is nonconducting. Many OFLLS have been developed, such as float type [3], pressure type [4], optical radar type [5], and TIR type [6]. Side-leaking OFLLSs can make the conducting light leaking from the side of the fiber and have different attenuation coefficients in air and liquid [7]. Compared to other kinds of OFLLS, the main advantages of the side-leaking OFLLSs are immune to air pressure and insensitive to liquid sloshing and have no moving parts. There are many implementations for optical fiber side leakage: reducing the thickness of fiber cladding [8]; removing several zones of the cladding [9]; using fluorescent impurity-doped fiber [10]; side-polishing the cladding and a portion of core on a curved fiber [11]; using an etched FBG [12], or using a tilted FBG [13]. Until now, these sensors have not been commercialized due to low sensitivity [9, 11], limited range [12, 13], expensive cost [8], and complicated manufacturing [8, 12, 13].

Side-emitting optical fiber (SEOF) is a kind of polymer optical fiber of which the conducting light can leak from the side of fiber, and it is usually used for illumination and framing of buildings [14]. Compared to the side-leaking methods mentioned above, the light attenuation of SEOF is uniform, and its manufacturing cost is very low. There are many kinds of SEOF, and the one produced by cladding material crystallization is chosen for liquid level sensing, because it has a larger side emitting intensity compared with other SEOFs [15]. In this kind of fiber, there are many tiny crystals in the cladding; when the conducting light is reflected on the core-cladding interface, it will be scattered by these tiny crystals, and then the conducting light will be leaked. However, the attenuation coefficient changes little from in air to in liquid in a straight SEOF, and fiber bending must be used to increase the attenuation change.

In this paper, a liquid level sensor with a spiral bent SEOF is demonstrated. To investigate the sensor's response with different bent radius, a steadily changing spiral slope is applied to the sensor. The principle of the sensor is explained, while the design, simulation, and a prototype of the sensor are also presented. An InGaAlP Yellow LED (TLYH180P from TOSHIBA Inc.) with a typical peak wavelength of 590 nm is used as light source in all the experiments presented in this paper.

2. Sensor Principle

The basic schematic of a side-leaking OFLLS is shown in Figure 1. In this schematic, the height of the fiber is H, and the liquid level is h; the incident light power is P_i, while the emergent light power is P_R. In the side-leaking OFLLS, light propagating in fiber is attenuated exponentially to the submerged length with the coefficient related to the refractive index of the surrounding medium. In liquid level sensing, there are mainly two mediums: air and liquid, and the attenuation coefficient in liquid is larger than that in air. Define α_a and α_l as the attenuation coefficients in air and liquid, respectively, then the transmitted light attenuates with the coefficient α_a in the fiber exposed in air for a height of $(H - h)$, and the output power in this section is [16]

$$P_a = P_i \exp[-\alpha_a(H - h)]. \quad (1)$$

Then, the transmitted light attenuates with the coefficient α_l in the section of fiber in liquid for a height of h, and the output power in this section is

$$P_R = P_a \exp(-\alpha_l h). \quad (2)$$

Replace P_a in (2) with (1), and make a deformation to the exponential part of the formula; the output power can be simplified to be

$$P_R = P_i \exp(-\alpha_a L) \exp[-(\alpha_l - \alpha_a)h]. \quad (3)$$

In this equation, $\exp(-\alpha_a L)$ decides the ratio of receiving power to the incident power, and $\exp[-(\alpha_l - \alpha_a)h]$ decides the ratio of output change to the maximum output. If α_l is greater than α_a (this will be demonstrated later), the emergent power P_R decreases as the level increases. If $\alpha_l - \alpha_a$ is small enough, P_R responds linearly to liquid level change.

2.1. Light Attenuation of SEOF in Air and Liquids. In SEOF, the attenuation coefficient is larger in liquid than that in the air due to the different acceptance angle of the fiber. As shown in Figure 2, the propagating light is scattered by the tiny scattering crystals in the cladding. According to Rayleigh's equation [17], the scattered light intensity I_s for an unpolarized source is given below:

$$I_s = A(1 + \cos^2\theta) I_i,$$
$$A = \frac{8\pi^4 N r^6}{\lambda^4 R^2} \left(\frac{n_1^2 - n_0^2}{n_1^2 + 2n_0^2} \right)^2, \quad (4)$$

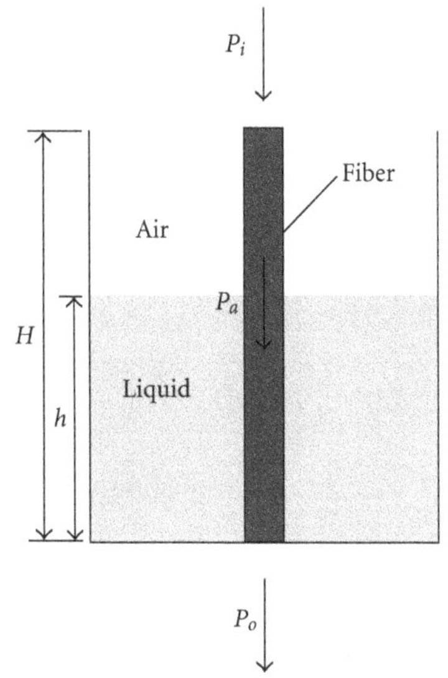

FIGURE 1: Schematic of a basic side-leaking OFLLS.

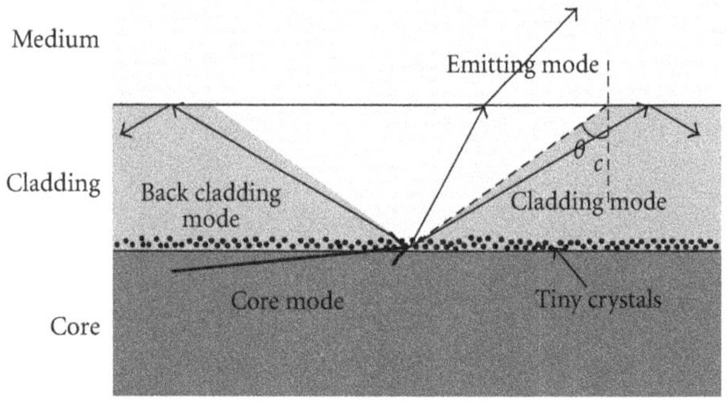

FIGURE 2: Light scattering in the cladding of SEOF.

where I_i (Watts/meter squared) is the intensity of incident light, θ is the included angle between the scattered light and incident light, A is a constant, in which N is the number of scattering particles, r is the radius of particle, λ is the wavelength of incident light, R is the distance between the particle and the observing point, n_1 is the index of particle, and n_0 is the index of the material around the particle. In the SEOF, the angle between transmitted light and fiber axis is very small, so it can be considered that θ approximately equals the angle between the scattering light and fiber axis.

The scattered light can be divided into three parts: part 1 is the backward cladding mode which transmits to the incident end; part 2 is the emitting mode which is refracted into the surrounding medium; part 3 is cladding mode which transmits along the fiber to the emergent end. Their boundaries are decided by the critical angle on the surface between the cladding and medium [18].

$$\theta_c = \arcsin\left(\frac{n_m}{n_{\text{cl}}}\right), \quad (5)$$

where n_{cl} is the index of cladding and n_m is the index of the medium outside of cladding. The power of cladding mode can be integrated as

$$P_r = \int_0^{\arccos(n_m/n_{cl})} 2\pi R^2 I_S(\theta) \sin\theta d\theta, \quad (6)$$

where R is the same as the R in (4), and it will be eliminated in the calculation. The total power of scattering light is

$$P_s = \int_0^{\pi} 2\pi R^2 I_S(\theta) \sin\theta d\theta. \quad (7)$$

Then, P_r and P_s can be calculated, and the ratio of P_r to P_s is

$$\frac{P_r}{P_s} = \frac{1}{8}\left(4 - \frac{3n_m n_{cl}^2 + n_m^3}{n_{cl}^3}\right). \quad (8)$$

The response of (8) to medium's index is shown in Figure 3, which indicates that the cladding mode power decreases while the index of the ambient medium of fiber increases. The value of (8) will reach the minimum value of 0 when $n_m \geq n_{cl}$. In liquid level sensing, the index of the liquid is always larger than that of air, so there will be more power loss in fiber when the fiber is immersed in liquid than that in air. According to the definition of the fiber's attenuation coefficient [19], α_l is larger than α_a.

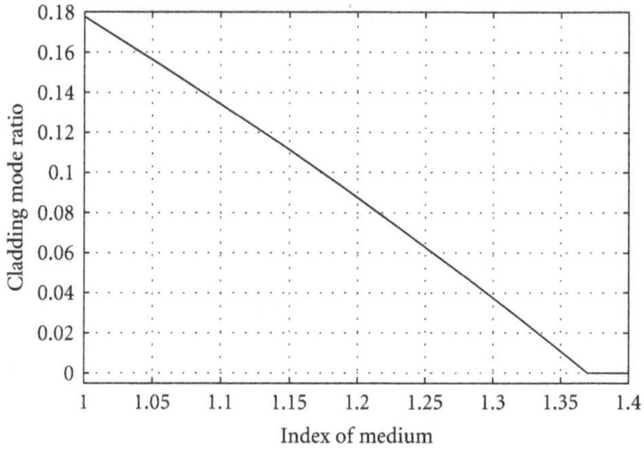

FIGURE 3: Cladding mode ratio versus medium index.

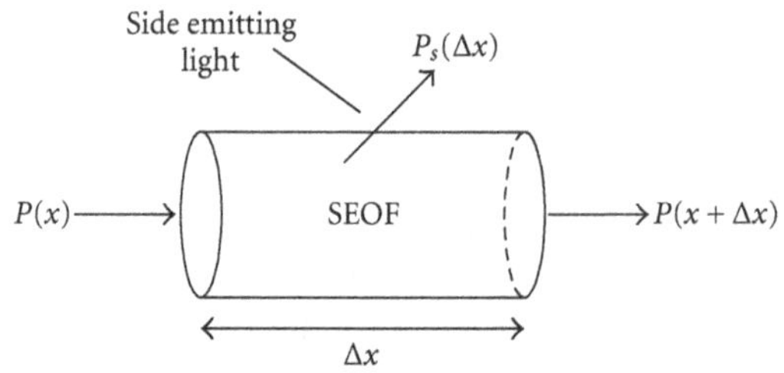

FIGURE 4: Side-emitting in a fragment of fiber.

2.2. Test of Attenuation Coefficients in Straight SEOF. The attenuation coefficients can be achieved by measuring the luminous exitance from the side of SEOF. Define x as the distance from a point in the fiber from the incident end, and in a fragment of fiber at the location x with a length of Δx as it is shown in Figure 4, the incident light power is $P(x)$, the emergent light power is $P(x + \Delta x)$, and the power losses in this section can be calculated according to the definition of fiber's attenuation coefficient:

$$P_s(x, \Delta x) = P_i[\exp(-\alpha x) - \exp(-\alpha x - \alpha\Delta x)]. \quad (9)$$

Luminous exitance can be defined as the side-emitting power in per unit area, and it can be calculated as

$$M(x) = \lim_{\Delta x \to 0} \frac{P_s(x, \Delta x)}{2\pi r_f \Delta x}, \quad (10)$$

where r_f is the radius of SEOF. Equation (10) can be simplified by series expansion, and the result is

$$M(x) = \frac{\alpha P_i}{2\pi r_f} \exp(-\alpha x). \quad (11)$$

Since α is a very small value, (11) approximately equals

$$M(x) = \frac{\alpha P_i}{2\pi r_f}. \quad (12)$$

Then, the attenuation coefficient, α, is proportionate to the luminous exitance.

An experimental setup is established as it is shown in Figure 5 to test the fiber's attenuation coefficients in air and in water. In this setup, an LED is used as the light source; a POF with 1 mm diameter is placed with its end facing the cladding of SEOF at location x, and it is used to guide the side-emitting light to the power meter. In the experiments, the 1 mm POF is moving from the input to the output end of SEOF, and its output power is recorded as $P_p(x)$ at location x. Then, the luminous exitance can be calculated as

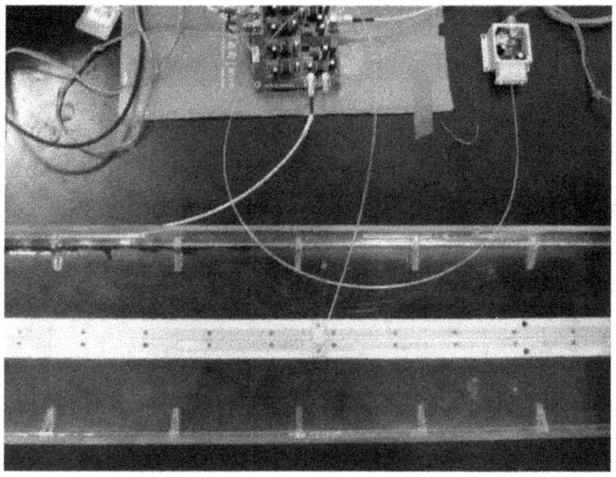

FIGURE 5: Experimental setup of attenuation coefficient measurement for straight SEOF.

$$M(x) = \frac{P_p(x)}{\pi r_p^2}, \quad (13)$$

where r_p is the radius of the 1 mm POF. Then, the attenuation coefficient can be calculated according to (12) and (13) as

$$\alpha = \frac{2r_f P_p(x)}{r_p^2 P_i}. \quad (14)$$

The measured result of the attenuation coefficients are 0.0045/0.0057 (/m) for air/water, respectively, in the middle of the fiber, and $\alpha_l - \alpha_a = 0.0012$ (/m). This is in accordance with the theoretical analysis.

3. Enhancement of Attenuation Coefficient

Because α_l is very close to α_a, the output change is very small compared to the maximum output. In a 1-meter-long SEOF, the maximum change of output is only 0.12% compared to the maximum output. Such a small change of power is difficult to measure, because it will be overwhelmed by the temperature drift of the LED and photodiode. There are two methods to enhance the output change: increasing the difference between α_l and α_a and enlarging the submerged length of fiber, and both of them can be achieved by bending the SEOF as a spiral shape.

3.1. Attenuation Coefficient of Bent SEOF. It is demonstrated that the attenuation coefficient increases exponentially with the decreasing of bent radius [20]. In SEOF, there are too many modes of light, and the boundary conditions are very complex, so it is easier to get its attenuation coefficient by experiments than by calculation. In the experimental setup shown in Figure 6, the fiber is bent as a semicircular with radius varying from 1 cm to 8 cm by a step of 1 cm. In this setup, light emitted by the LED is injected into the fiber directly, and the input power P_i and output power P_o are measured by a power meter. The experiments are taken with the fiber exposed in air (shown in Figure 6(a)) and submerged in water (shown in Figure 6(b)), respectively. Then, the attenuation coefficient per unit length can be calculated as

$$\alpha_L = -\frac{\ln P_o - \ln P_i}{\pi R}, \quad (15)$$

where R is the bent radius of the fiber. The SEOF's attenuation coefficient versus bent radius in the air is shown in Figure 7, and it indicates that the fiber's bend loss increases exponentially while the bent radius decreases. The difference between attenuation coefficients in air and in water versus bent radius is shown in Figure 8, and it shows the value $(\alpha_l - \alpha_a)$ increases when the bent radius decreases. With the tested value of α_a and α_l, the sensor's response to liquid level can be simulated.

3.2. Light Attenuation in a Spiral SEOF. With a two-dimensional bent fiber, the length of fiber immersed in liquid is not linear to the liquid level. Using a spiral shape can solve

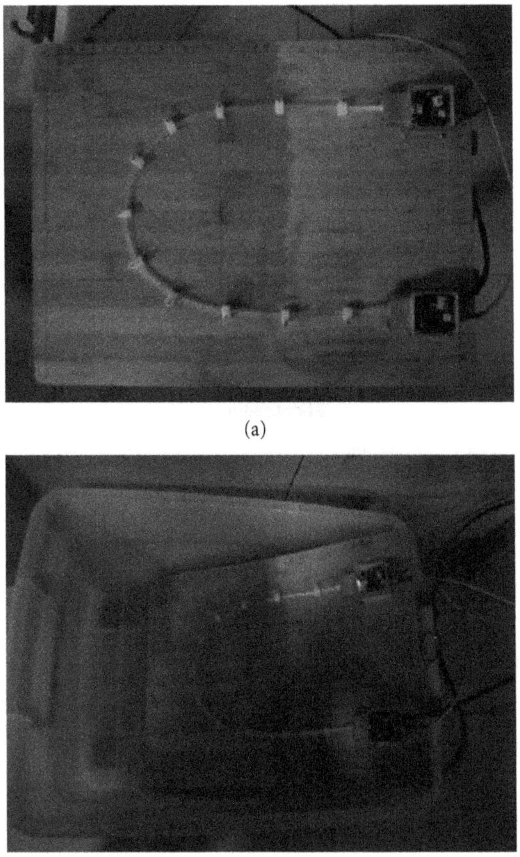

(a)

(b)

FIGURE 6: (a) and (b) are the experimental setup of attenuation coefficient measurement for bended SEOF in air and water.

this problem, and the immersed length can be enlarged in the same fiber. In a spiral line defined as

$$x = a\cos\theta,$$
$$y = a\sin\theta, \quad (16)$$
$$z = b\theta,$$

a is the radius of the core and b is the slope of the spiral line. The arc length in the spiral line is

$$S = \sqrt{z^2 + (a\theta)^2}. \quad (17)$$

Then, the ratio between arc length and height can be calculated by dividing S by z, and the result will be

$$K = \frac{\sqrt{a^2 + b^2}}{b}. \quad (18)$$

Define α_H as the power loss per unit height and α_L as the power loss per unit length, and α_H will be K times as large as α_L according to the definition of the fiber's attenuation coefficient. This means α_H is always larger than α_L.

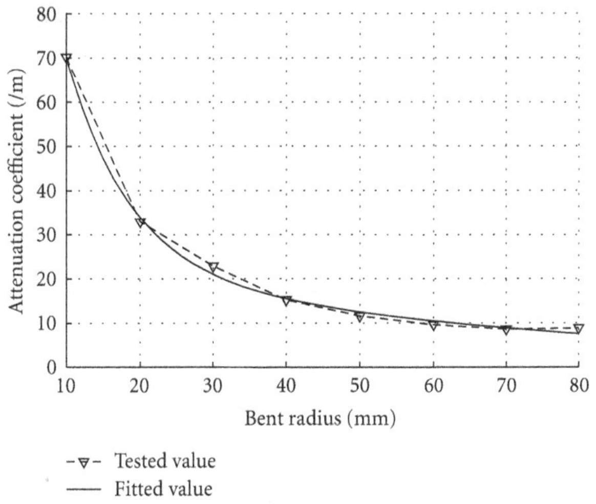

Figure 7: Attenuation coefficient of SEOF in air.

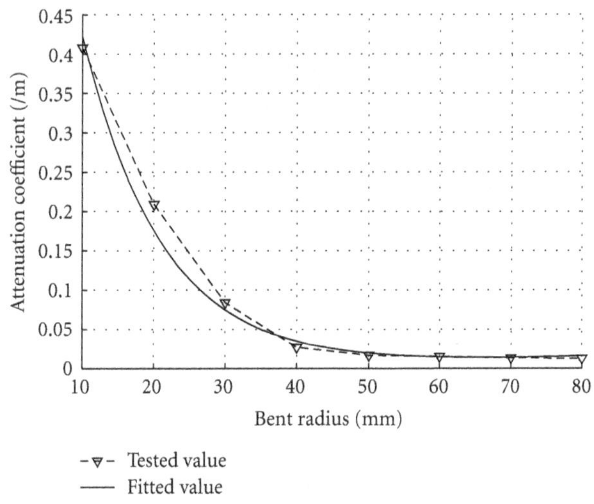

Figure 8: Difference between attenuation coefficient in air and liquid.

4. Experimental Setup

The schematic of the experimental system is shown in Figure 9. This system consists of a SEOF, two crabsticks, an LED, a photodiode, a signal processing circuit, and a data acquisition module DAM3056 to monitor the continuous level change. A PMMA SEOF of core/cladding diameter 2.6/3 mm and refractive indices 1.49/1.37 (at 580 nm) is used. The fiber is coiled downward on a crabstick as a spiral shape and turns upward along the other crabstick as it is shown in Figure 10. The track of fiber is fixed by the nails, and the function of the spiral is

$$x = 12.5 \cos \theta,$$
$$y = 12.5 \sin \theta, \quad (19)$$
$$z = -0.23\theta^2 + 29.05\theta.$$

The spiral has a core radius of 12.5 mm, a height of 900 mm, a minimum bent radius of 15 mm at the top (shown in Figure 10(b)), and a maximum bent radius of 80 mm at the bottom (shown in Figure 10(d)). The slope of the spiral is steadily decreased along the height; then, the sensor's response to liquid level with different bent radius can be tested. The simulated value of the sensor's response to level is shown in Figure 11. In this picture, the total variation of power is 7.9% of the maximum output, and this variation is 66 times as large as the output variation in a straight fiber; the sensitivity is stronger in the section with a smaller slope.

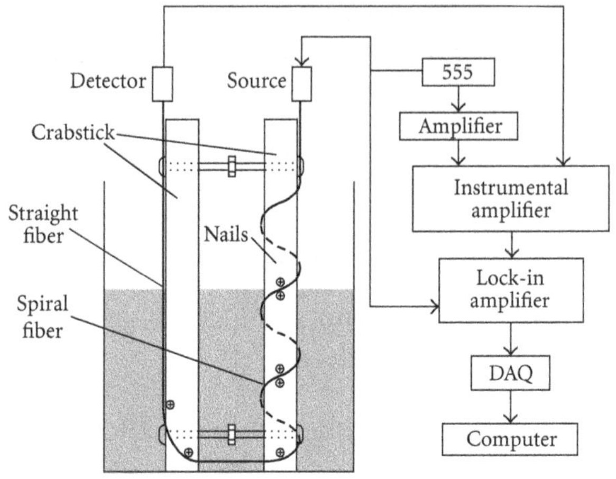

Figure 9: Schematic of liquid level sensor.

In the experimental system, the LED is driven by a 2.2 KHz square wave generated by a 555 timer. The output power is detected by photodiode VTB8440 (PerkinElmer Optoelectronics, Inc.), of which the optical response extends from 330 to 720 nm and has peak sensitivity at 580 nm. The output current of the photodiode is then transformed into voltage by a preamplifier, and this voltage is subtracted by a reference signal in an instrumentation amplifier, and then the limited change of power can be amplified. The reference signal is generated by amplifying the square wave to get a same amplitude and phase with the detector's signal at zero liquid level. The differential signal is then amplified and detected by a lock-in amplifier with a high signal-to-noise ratio, and the noise of the signal is reduced from 3 V to 0.02 V. The output voltage of the lock-in amplifier is monitored by the data acquisition module and then sent to the computer.

5. Results and Discussion

Experiments are taken in water and aviation kerosene. In the water level experiment, a pump is used to let the level rise; after every 10 mm level changing, the flow controller is closed to hold the level and record the output voltage on the computer; when the level reaches 900 mm, the pump is shut down, and the level will decrease under the pull of gravity, and the output voltage is also recorded after every 10 mm level changing. The experimental procedure is almost the same in the aviation kerosene level sensing, except using a hand pump to pump in the fuel.

The experimental results are shown in Figure 12. The solid and dashed curves are the experimental results in water, while the dash-dotted and dotted curves are the experimental

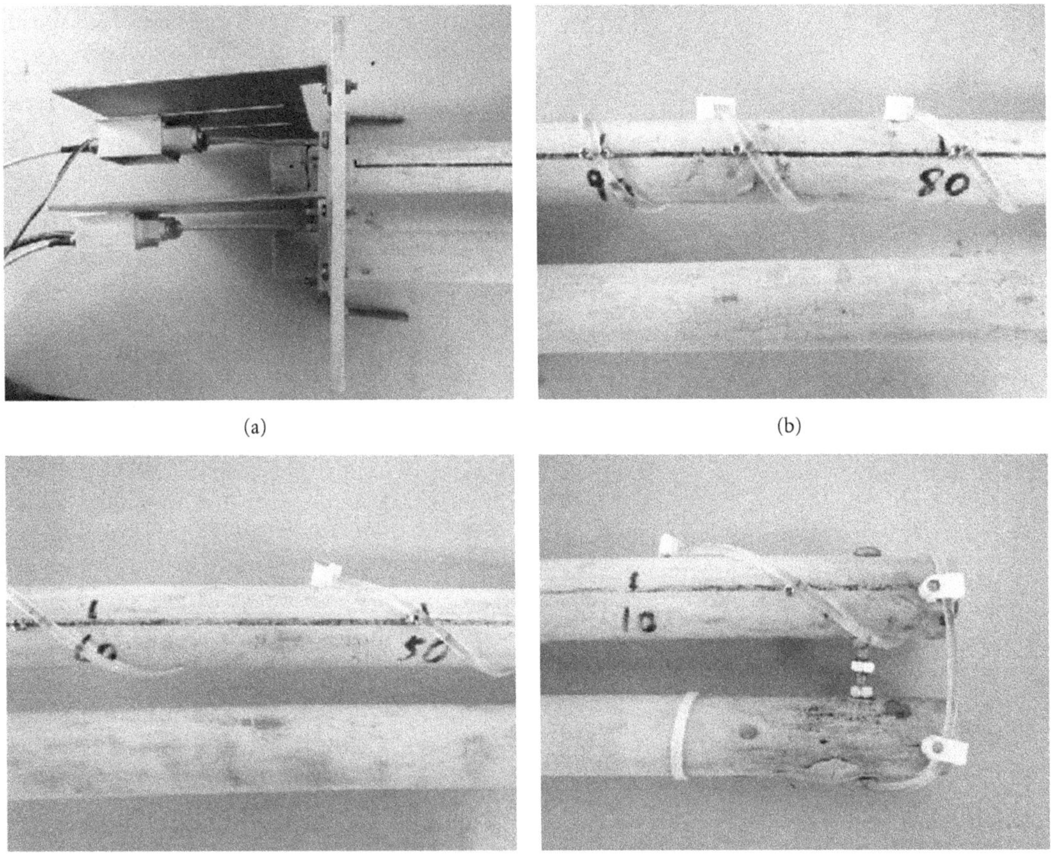

FIGURE 10: (a) source and detector; (b) top section of sensor; (c) middle section of sensor; (d) end section of sensor.

TABLE 1: Performance of several side-leaking optical fiber level sensors.

Side-leaking method	Range (m)	Resolution (mm)
Reduce cladding thickness	1	62.5
Remove zones of cladding	1	25
Fluorescent impurity-doped fiber	1.5	40
Fiber side-polish	0.04	2.2
Etched FBG	0.02	2
Tilted FBG	0.02	4
Side-emitting optical fiber	0.9	10

results in aviation kerosene. The sensor's output responses to the rising and decreasing of levels are shown in separate curves. In all the four curves, a sharp decrease of voltage happened in the beginning, this is caused by the turning section of fiber with a small radius at the bottom of the sensor. As it is shown in Figure 10(d), the fiber has a minimum radius of 12.5 mm in the turning section, which is much smaller than any other positions in the sensor. The average voltage drop is 3.2 V in aviation kerosene, about 1.28 times as large as the voltage drop in the water. The larger voltage change in aviation kerosene is caused by its larger refractive index of 1.43, while the value is 1.33 in water. The drop of voltage at the beginning means the sensor is very sensitive to the existence of the liquid and can be used for low level detection. But for continuous liquid level sensing, this drop limits the amplification factor of the circuit. This disadvantage can be improved by covering the fiber's turning section with some opaque material to prevent the fiber from contacting with the liquid.

The curves are flatter in the following section. Coinciding with the simulated output in Figure 11, the sensor has a larger sensitivity in the section with a smaller bent radius. In the water level sensing the sensitivity of the sensor is -1.14 mV/mm in the section of (0~600) and -3.65 mV/mm in the section of (600~900). In the aviation kerosene level sensing, the sensitivity is -1.31 mV/mm in the section of (0~600) and -3.73 mV/mm in the section of (600~900). In the liquid level sensing, the measurement signal is easier

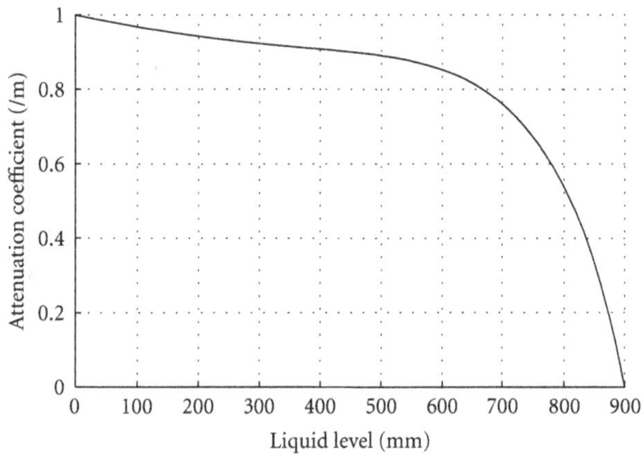

FIGURE 11: Simulation of sensor's output.

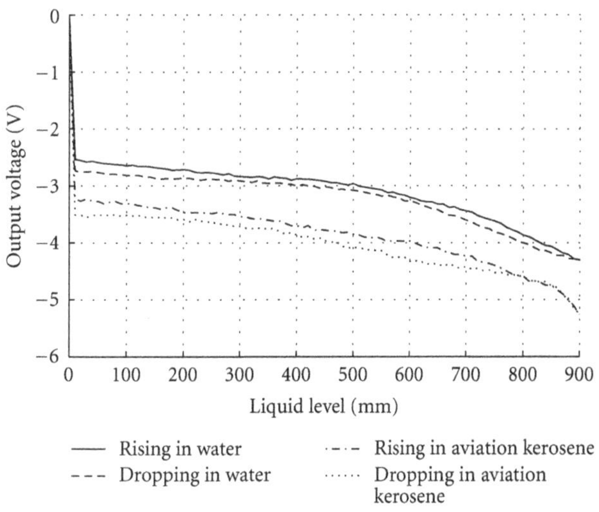

FIGURE 12: Output voltage versus liquid level.

to be amplified with a larger sensitivity. According to the experimental results, using a SEOF with a fixed bent radius of 15 mm will have a linear response to the liquid level, and the output power changing ratio will be enlarged.

The results show a hysteresis between the rising curve and the dropping curve. This hysteresis is caused by the liquid staying in the gap between fiber and crabstick during the level dropping. The difference between rising and dropping curves is larger in aviation kerosene than that in water, because the aviation kerosene has a larger viscosity than water (the viscosity of aviation kerosene/water is 1.25/1.005 mm^2/s at 20°C). The hysteresis of curves causes error in measurement, and half-embedding the fiber in a slot on the sensor's core will eliminate the gap between the fiber and the core; then the hysteresis will be limited.

The performances of this sensor and the OFLLSs mentioned in the introduction are compared in Table 1. In this table, the sensors with a large measurement range have a limited resolution, while the sensors with a high resolution have a limited range. In aircraft fuel measurement, the required measurement range is 1 m, and the required resolution is 5 mm. Among the sensors showed in Table 1, the sensor described in this paper is the one which mostly meets the requirement in aircraft fuel measurement.

6. Conclusions

In this paper, a novel optical fiber level sensor was proposed, based on a spiral bent side-emitting optical fiber, which is produced by cladding material crystallization. The principle of the sensor was interpreted by analyzing the light scattering in the SEOF, and experimental results were presented to explain the reason for using a spiral bent SEOF. A prototype with a steadily changing spiral slope was designed to demonstrate the sensor principle and test the sensor's response with different fiber bent radius. The experimental results taken in water and aviation kerosene showed coincidence to the simulated results. There are also several disadvantages in this sensor, like the long-time instability, the sharp output decrease near the zero level, the hysteresis during level rising and dropping, and the nonlinear response to liquid level. By solving these problems, the sensor's performance can be improved and be suitable for the fuel level measurement in vehicles like a plane, truck, or ship and the level sensing in liquid with low viscosity.

Authors' Contribution

C. Zhao elaborated the basic principle of the sensor, proposed a new method to increase the sensor's sensitivity by fiber bending, realized this sensor, and made experiments with this sensor. L. Ye clarified the theory of side-leaking fiber level sensor and proposed the idea of using a side-emitting optical fiber in this sensor. J. Ge explained the reason of fiber's bend loss and designed experiments to test the fiber's bend loss. X. Yu made experiments to test the fiber's bend loss and took part in the experiments of the sensor.

Acknowledgments

This work is partly supported by the National Nature Science Foundation of China (61104202), and the authors wish to thank Ms. Cheng Yi for the excellent work during this research.

References

[1] R. Langton, C. Clark, M. Hewitt, and L. Richards, *Aircraft Fuel Systems. Chichester*, John Wiley & Sons, United Kingdom, 2009.

[2] J. D. Weiss, "Fluorescent optical liquid-level sensor," *Optical Engineering*, vol. 39, no. 8, pp. 2198–2213, 2000.

[3] K. R. Sohn and J. H. Shim, "Liquid-level monitoring sensor systems using fiber Bragg grating embedded in cantilever," *Sensors and Actuators A*, vol. 152, no. 2, pp. 248–251, 2009.

[4] H. J. Sheng, M. Y. Fu, T. C. Chen, W. F. Liu, and S. S. Bor, "A lateral pressure sensor using a fiber Bragg grating," *IEEE Photonics Technology Letters*, vol. 16, no. 4, pp. 1146–1148, 2004.

[5] H. K. Singh, S. K. Chakroborty, H. Talukdar, N. M. Singh, and T. Bezboruah, "A New Non-Intrusive optical technique to measure transparent liquid level and volume," *IEEE Sensors Journal*, vol. 11, no. 2, pp. 391–398, 2011.

[6] A. A. Kazemi, C. Yang, and S. Chen, "Fiber optic cryogenic liquid level detection system for space applications," in *Photonics in the Transportation Industry: Auto to Aerospace II*, A. A. Kazemi, Ed., p. 73140A, 2009.

[7] F. Pérez-Ocón, M. Rubiño, J. M. Abril, P. Casanova, and J. A. Martínez, "Fiber-optic liquid-level continuous gauge," *Sensors and Actuators A*, vol. 125, no. 2, pp. 124–132, 2006.

[8] G. Betta, L. Ippolito, A. Pietrosanto, and A. Scaglione, "Optical fiber-based technique for continuous-level sensing," *IEEE Transactions on Instrumentation and Measurement*, vol. 44, no. 3, pp. 686–689, 1995.

[9] G. Betta, A. Pietrosanto, and A. Scaglione, "A digital liquid level transducer based on optical fiber," *IEEE Transactions on Instrumentation and Measurement*, vol. 45, no. 2, pp. 551–555, 1996.

[10] J. D. Weiss, "A novel fiber-optic fluid interface sensor. Sensors," 2008, http://www.sensorsmag.com/da-control/a-novel-fiber-optic-fluid-interface-sensor-1476.

[11] M. Lomer, A. Quintela, M. López-Amo, J. Zubia, and J. M. López-Higuera, "A quasi-distributed level sensor based on a bent side-polished plastic optical fibre cable," *Measurement Science and Technology*, vol. 18, no. 7, article no. 061, pp. 2261–2267, 2007.

[12] B. Yun, N. Chen, and Y. Cui, "Highly sensitive liquid-level sensor based on etched fiber bragg grating," *IEEE Photonics Technology Letters*, vol. 19, no. 21, pp. 1747–1749, 2007.

[13] Q. Jiang, D. Hu, and M. Yang, "Simultaneous measurement of liquid level and surrounding refractive index using tilted fiber Bragg grating," *Sensors and Actuators A*, vol. 170, no. 1-2, pp. 62–65, 2011.

[14] J. Spigulis, D. Pfafrods, M. Stafeckis, and W. Jelinska-Platace, "Glowing optical fiber designs and parameters," in *Optical Inorganic Dielectric Materials and Devices*, A. Krumins, D. K. Millers, A. R. Sternberg, and J. Spigulis, Eds., Proceedings of SPIE, p. 231, Riga, Latvia, 1997.

[15] S. Katsuhiko, T. Tsuneo, T. Yasuteru, and I. Kikue, "Illuminating plastic optical fiber and its manufacture," Japan Patent 06-118236, 1992.

[16] J. Hecht and L. Long, *Understanding Fiber Optics*, Publishing House of Electronics Industry, Beijing, China, 5th edition, 2003.

[17] J. C. Stover, *Optical Scattering: Measurement and Analysis*, SPIE-International Society for Optical Engineering, Bellingham, Wash, USA, 2nd edition, 1995.

[18] M. Born, E. Wolf, and A. B. Bhatia, *Principles of Optics: Electromagnetic Theory of Propagation, Interference and Diffraction of Light*, Cambridge University Press, 7th edition, 1999.

[19] J. Hecht and L. Long, *Understanding Fiber Optics*, Prentice Hall Columbus, 4th edition, 2002.

[20] D. Marcuse, "Curvature loss formula for optical fibers," *The Journal of the Optical Society of America*, vol. 66, no. 3, pp. 216–220, 1976.

10

Adaptive Control for Nonlinear Systems with Time-Varying Control Gain

Alejandro Rincon[1] and Fabiola Angulo[2]

[1] *Programa de Ingeniería Ambiental, Facultad de Ingeniería y Arquitectura, Universidad Católica de Manizales, Carrena 23 No. 60-30, Manizales 170002, Colombia*
[2] *Departamento de Ingeniería Eléctrica, Facultad de Ingeniería y Arquitectura, Universidad Nacional de Colombia, Sede Manizales, Electrónica y Computación, Percepción y Control Inteligente, Bloque Q, Campus La Nubia, Manizales 170003, Colombia*

Correspondence should be addressed to Fabiola Angulo, fangulog@unal.edu.co

Academic Editor: Chengyu Cao

We propose a scheme for nonlinear plants with time-varying control gain and time-varying plant coefficients, on the basis of a plant model consisting of a Brunovsky-type model with polynomials as approximators. We develop an adaptive robust control scheme for this plant, under the following assumptions: (i) the plant terms involve time-varying but bounded coefficients, being its upper bound unknown; (ii) the control gain is unknown, not necessarily bounded, and only its signum is known. To achieve robustness, we use a combination of robustifying control inputs and dead zone-type update laws. We apply this methodology to the speed control of a permanent magnet synchronous motor (PMSM), and we achieve proper tracking results.

1. Introduction

Nonlinear behavior is difficult to model accurately, rendering controller design cumbersome. An approach to handle this is to use linear models, which have local validity in the state space, that is, different values of the parameter set would be required for each region in the state space. A second approach is to use a plant model in either the Brunovsky form defined in [1], or the parametric-pure feedback form defined in [2], using the so-called "function approximation techniques", such as locally weighted statistical learning [3], fuzzy sets [4], Fourier series [5], orthonormal functions, [6] or neural networks [7]. In the last case, state and time dependent terms, known as "basis functions," are used to represent the nonlinear and time-varying behaviour. Some identification or learning methods are used to perform the approximation, resulting in a negligible modelling error if there are a sufficient number of basis functions [3]. In some cases, a Brunovsky form model of the plant is used, assuming that there is full-state measurement and that the basis functions are multiplied by unknown constants. Then, a sliding surface model reference adaptive control (SSMRAC) is devised, handling the residual approximation error by means of auxiliary robustifying inputs [7].

Common drawbacks of the above-mentioned schemes are

(i) the convergence of the tracking error to some small residual set is not ensured in a strict sense and depends on the value of the approximation error [7];

(ii) upper or lower bounds of the plant coefficients are required to be known [8];

(iii) discontinuous auxiliary inputs are used, which may lead to chattering [5];

(iv) high enough gains are used, which require excessive values [6].

Due to environment changes, the coefficients of the plant model may experience time-varying but bounded behavior [9, 10]. Both, the control gain and other coefficients may exhibit this behavior. The time-varying behavior of coefficients different from the control gain may be handled by means of robustness techniques. See for instance the robust MRAC scheme in [10], whose drawback is that projection- or σ-type update laws are used, and hence upper bounds of the plant coefficients are required to be known, in order to achieve the convergence of the tracking error to some residual set of user-defined size. The time-varying control gain may be handled by means of robustness techniques [11] or Nussbaum gain technique [12], a projection-type update law is used, such that a lower bound of the control gain is required to be known. In [11], a σ type update law is used, such that bounds of the plant coefficients are required to be known. In summary, the main drawback of using robustness techniques to handle varying control gain is that upper or lower bounds of some plant coefficients are required to be known. On the other hand, the Nussbaum gain technique relaxes this requirement. The main drawback is that the upper bound of the transient behavior of the state S is significantly altered with respect to that of the disturbance-free case: the value of this bound depends on time integrals of the terms that comprise the Nussbaum function. Another drawback is that the control gain is usually required to be upper-bounded by some constant.

The scheme of [13] achieves adequate robustness properties. Mainly, upper bounds of the plant coefficients are not required to be known. The essential element of the technique is to define the quadratic forms related to the sliding surface $\mathcal{S}$ in terms of a dead zone function of $\mathcal{S}$ rather than in terms of $\mathcal{S}$. This scheme has the following benefits.

(Ri) The tracking error converges to a residual set whose size is of the user's choice;

(Rii) known upper bounds of the plant coefficients are not required, such that high enough gains are not used;

(Riii) all the closed-loop signals are bounded (parameter drifting is avoided);

(Riv) auxiliary control signals are not discontinuous in terms of both the tracking error and the sliding surface, hence input chattering is avoided;

(Rv) upper bounds of time-varying but bounded coefficients are not required to be known.

The disadvantages of [13] are (i) the design is only developed for systems with hysteresis nonlinearities in the actuator; (ii) the time-varying control gain is tackled by means of the Nussbaum gain method, which entails higher complexity of the Lyapunov analysis; (iii) the control gain is assumed to be time-varying but bounded.

In contrast to the approaches of [3, 4, 14–16], we propose that an adequate regression model for highly nonlinear systems may be obtained from a Brunovsky type model, inserting polynomials to approximate the nonlinear behavior, and considering also: (i) time-varying but bounded behavior of some plant coefficients; (ii) unknown control gain, time-varying, and not necessarily bounded. We develop a robust adaptive scheme for this plant, achieving benefits (Ri) to (Rv). If we compare the robust technique developed here with the technique developed in [13], the main differences, which are also contributions, are

(Rvi) we consider unknown time-varying control gain, not necessarily bounded, not restricted to actuator nonlinearities;

(Rvii) we consider time-varying but bounded behavior of some plant terms;

(Rviii) we tackle the control gain by means of robustness techniques, which gives a simpler design in comparison with the Nussbaum technique.

This paper is organized as follows. The outline of the scheme is given in Section 2. The function approximation is discussed in Section 3. The plant regression model and the statement of the problem are established in Section 4. The control and update laws are derived in Section 5. The control algorithm and its stability are presented in Section 6. An application of the scheme to a PMSM is given in Section 7. Finally, the conclusions are presented in Section 8.

2. Outline of the Scheme

We propose the use of polynomials to approximate the nonlinear behavior, taking into account the fact that polynomials are universal approximators for continuous functions, according to [17]. We consider a Brunovsky-form plant model, into which we introduce the polynomial terms.

We devise a robust adaptive controller for this plant, achieving benefits (Ri) to (Rix) mentioned in the introduction. We use the SSMRAC method stated in [18], as the basic framework for designing the control and update laws. To handle the effect of modelling error and time-varying behavior of plant terms, we use a robust continuous control input, whose magnitude is adjusted, and a dead zone-type update law. The resulting controller has the following features: (i) the magnitude of the control input is adjusted to cope with the unknown upper bounds of the time-varying coefficients; (ii) the control input is proportional to some continuous function of the sliding surface $\mathcal{S}$, so that chattering is avoided; (iii) an inactivation is introduced in all the update laws, which occurs when the sliding surface $\mathcal{S}$ reaches a target region. If we compare this with projection-type update laws, it has the advantage of not requiring the upper bounds of the plant parameters.

For the stability analysis, we use a truncated version of the quadratic form related to the sliding surface, denoted by $\overline{V_s}$. The validity of this technique, including the conditions of the Lyapunov function, is stated in [19]. It is worth noticing that the standard conditions of the Lyapunov theory for time-variable systems, shown in [18] or [20], are not satisfied because of the truncation. We prove the asymptotic convergence of the tracking error in a rigorous manner by means of Barbalat's Lemma. To that end, we redefine the expression of $\dot{V}$ as an inequality in terms of $\overline{V_s}$.

3. Function Approximation Based on Polynomials

According to [17, page 116], a continuous real-valued function $\overline{f}(x)$, where $x \in D$, $D \subset R^n$, being D a compact set, may be approximated by a polynomial function $\overline{g}$ in the interval $x \in D$, with some finite error, being $\overline{g}$ defined as

$$\overline{g}(x,\theta_o) = \sum_{k_1,k_2,\ldots,k_n} \theta_{o[j]} x_1^{k_1} x_2^{k_2} \cdots x_n^{k_n} : \quad \theta_o \in \Omega,\ \Omega \subset R^p, \tag{1}$$

where Ω is a convex set and $\theta_{o[j]}$ means the jth element of the vector θ_o. Thus, $\overline{f}(x)$ can be expressed in terms of $\overline{g}$ as follows (cf. [17] page 122):

$$\overline{f}(x) = \overline{g}(x,\theta_a) + \epsilon_o(x), \qquad |\epsilon_o(x)| < \epsilon,$$
$$\theta_a = \left\{\theta_o \in \Omega : \theta_o = \arg\min_{\theta_o}\left(\sup_{x\in D}\left|\overline{f}(x) - \overline{g}(x,\theta_o)\right|\right)\right\}, \tag{2}$$

where ϵ is a positive constant, and it is known as approximation error, θ_a is an ideal parameter set and $\overline{g}(x,\theta_a)$ is an ideal representation of $\overline{f}(x)$. The polynomial can be linearly parameterized with respect to its coefficients:

$$\overline{f}(x) = \phi^\top\theta_a + \epsilon_o(x), \tag{3}$$

where ϕ contains polynomial terms, whereas θ_a contains constants. In this work, $\overline{f}$ represents the nonlinear part of the dynamical equation. We will handle the unknown constant vector θ_a by means of adjustment rules and ϵ_o by means of robust techniques.

4. Plant Model and Problem Statement

We assume that the dynamical nonlinear system can be represented by a Brunovsky type model, as defined in [1], with polynomial functions:

$$y^{(n)} = -c_{n-1}y^{(n-1)} - \cdots - c_1\dot{y} - c_o y + \overline{f}(x) + bu, \tag{4}$$

$$x = \left[y, \dot{y}, \ldots, y^{(n-1)}\right]^\top, \tag{5}$$

where $y(t) \in \mathbb{R}^1$ is the system output $u(t) \in \mathbb{R}^1$ is the input and the coefficients $c_{n-1},\ldots,$ and c_o are unknown and time-varying but bounded. The relative degree n may be established by means of previous step response analysis. Inserting the function (3) into (4) gives

$$y^{(n)} = bu + \phi^\top\theta_a + d, \tag{6}$$

$$d = \epsilon_o(x) - c_{n-1}y^{(n-1)} - \cdots - c_1\dot{y} - c_o y,$$
$$|c_{n-1}| \le \mu_{n-1},\ldots,|c_1| \le \mu_1, \tag{7}$$
$$|c_o| \le \mu_n, \qquad |\epsilon_o(x)| \le \mu_{n+1} = \epsilon.$$

We make the following assumptions:

(Ai) the entries of θ_a and the terms $\mu_1,\ldots,$ and μ_{n+1} are constant and unknown;

(Aii) the control gain b varies with respect to the state x, or time, so that it satisfies the following:

$$\text{(i)} \quad 0 < b_m \le |b| \le f_b, \qquad \forall t \ge t_o, \tag{8}$$

$$\text{(ii)} \quad \text{sign}(b)\ \text{constant}, \qquad \forall t \ge t_o, \tag{9}$$

where the value of sign(b) is known, b_m is an unknown positive constant, and f_b is an unknown positive function of time or x. Notice that condition (8) implies that b is always different from zero;

(Aiii) the entries of the vector x are available for measurement;

(Aiv) the entries of the vector ϕ are known functions of x;

(Av) the value of the desired trajectory $y_d(t)$ and its derivatives $y_d^{(n-1)},\ldots,\dot{y}_d$ is bounded.

The desired output y_d is specified in terms of a bounded external command $r(t)$ as follows:

$$y_d^{(n)} + a_{m,n-1}y_d^{(n-1)} + \cdots + a_{m,o}y_d = a_{m,o}r, \tag{10}$$

where $a_{m,n-1},\ldots,a_{m,o}$ are constant coefficients prespecified by the user, such that the polynomial $K(p)$ is Hurwitz with at least one real root, being $K(p)$ defined as $K(p) = p^{(n)} + a_{m,n-1}p^{(n-1)} + \cdots + a_{m,o}$. The external reference signal $r(t)$ must be bounded. The objective of the MRAC design is to formulate a control law $u(t)$ such that the tracking error $e(t) = y(t) - y_d(t)$ asymptotically converges to the residual set D_e, defined as follows:

$$D_e = \{e : |e| \le C_{be}\}, \tag{11}$$

where C_{be} is a user-defined bound.

5. The Control and Update Laws

Let $\mathcal{S}$ the dynamics imposes over the tracking error given by

$$\mathcal{S} = (p+\lambda)^{n-1}e = p^{n-1}e + \lambda_{n-2}p^{n-2}e + \cdots + \lambda_1\dot{e} + \lambda_o e, \tag{12}$$

where λ is a positive constant chosen by the user. Having defined $\mathcal{S}$, we establish the dynamic equation for $\mathcal{S}$ by differentiating with respect to time:

$$\dot{\mathcal{S}} = p^n e + \lambda_{n-2}p^{n-1}e + \cdots + \lambda_1\ddot{e} + \lambda_o\dot{e}, \tag{13}$$

$$\dot{\mathcal{S}} = y^{(n)} - y_d^{(n)} + \lambda_{n-2}p^{n-1}e + \cdots + \lambda_1\ddot{e} + \lambda_o\dot{e}, \tag{14}$$

$$\dot{\mathcal{S}} = y^{(n)} + \varphi_a, \tag{15}$$

$$\varphi_a = -y_d^{(n)} + \lambda_{n-2}p^{n-1}e + \cdots + \lambda_1\ddot{e} + \lambda_o\dot{e}. \tag{16}$$

We insert the expression for $y^{(n)}$ of (6) into (15):

$$\dot{\mathcal{S}} = bu + \theta_a^\top \phi + d + \varphi_a. \quad (17)$$

Define

$$u^* = -\theta_a^\top \phi - \varphi_a - a_m \mathcal{S} = \varphi^\top \theta^*, \quad \theta^* = [\theta_a^\top, 1]^\top, \quad (18)$$

$$\varphi = [-\phi^\top, -\varphi_a - a_m \mathcal{S}]^\top, \quad (19)$$

where φ_a is defined in (16), a_m is a positive constant. We express (17) in terms of u^* as follows:

$$\dot{\mathcal{S}} = -a_m \mathcal{S} + bu + d - u^* = -a_m \mathcal{S} + bu + d - \varphi^\top \theta^*. \quad (20)$$

Let

$$l = n + 1, \quad (21)$$

$$c^* = \frac{2 + l}{2C_{bvs}} \left(\frac{1}{2\sqrt{b_{mn}}} \right)^2 \quad (22)$$

$$C_{bvs} \triangleq \left(\frac{1}{2} \right) (\lambda^{n-1} C_{be})^2, \quad (23)$$

$$f_1 = |\dot{y}|, \ldots, f_{l-2} = \left| y^{(n-1)} \right|, \quad f_{l-1} = |y|, \quad f_l = 1,$$
$$\mu_1 = \max\{|c_1|\}, \ldots, \mu_{l-2} = \max\{|c_{n-1}|\}, \quad \mu_{l-1} = \max\{|c_o|\}, \quad \mu_l = \epsilon. \quad (24)$$

Now, we multiply (20) by $\mathcal{S}$ and add and subtract the term $c^* S^2$:

$$\mathcal{S}\dot{\mathcal{S}} = -a_m \mathcal{S}^2 + b\mathcal{S}u + \mathcal{S}d - \mathcal{S}\varphi^\top \theta^* + c^* S^2 - c^* S^2, \quad (25)$$

where the terms $c^*\mathcal{S}^2 - c^*\mathcal{S}^2$, $-\mathcal{S}\varphi^\top\theta^*$, and $\mathcal{S}d$ can be expressed in terms of adjustment errors:

(i) $$c^* \mathcal{S}^2 - c^* \mathcal{S}^2 = -c^* \mathcal{S}^2 - \tilde{c}\mathcal{S}^2 + \hat{c}\mathcal{S}^2, \quad \tilde{c} = \hat{c} - c^*; \quad (26)$$

(ii) $$-\mathcal{S}\varphi^\top \theta^* = S\varphi^\top \tilde{\theta} - \mathcal{S}\varphi^\top \hat{\theta}, \quad \tilde{\theta} = \hat{\theta} - \theta^*; \quad (27)$$

(iii) $$\mathcal{S}d \le (-1)\tilde{c}_1|\mathcal{S}|f_1 + \hat{c}_1|\mathcal{S}|f_1 + \cdots + (-1)\tilde{c}_l|\mathcal{S}|f_l + \hat{c}_l|\mathcal{S}|f_l, \quad (28)$$

$$\tilde{c}_1 = \hat{c}_1 - \mu_1, \ldots, \tilde{c}_l = \hat{c}_l - \mu_l, \quad (29)$$

where $\hat{c}, \hat{\theta}, \hat{c}_1, \ldots, \hat{c}_l$ are adjusted parameters whose update laws will be defined later. Inserting the above properties into (25) gives

$$\begin{aligned} \mathcal{S}\dot{\mathcal{S}} \le & -a_m \mathcal{S}^2 - \tilde{c}\mathcal{S}^2 + \mathcal{S}\varphi^\top \tilde{\theta} + (-1)\tilde{c}_1|\mathcal{S}|f_1 \\ & + \cdots + (-1)\tilde{c}_l|\mathcal{S}|f_l + \hat{c}\mathcal{S}^2 - \mathcal{S}\varphi^\top \hat{\theta} + \hat{c}_1|\mathcal{S}|f_1 \\ & + \cdots + \hat{c}_l|\mathcal{S}|f_l + b\mathcal{S}u - c^*\mathcal{S}^2. \end{aligned} \quad (30)$$

If b were constant, we would choose the control u so as to cancel the terms involving updated parameters $\hat{c}, \hat{\theta}, \hat{c}_1, \ldots, \hat{c}_l$. Since b is varying, we chose u to attenuate the effect of adjusted parameters, being the remaining error rejected by $-c^*\mathcal{S}^2$:

$$\begin{aligned} u = & (-1)\operatorname{sgn}(b)\hat{c}^2\mathcal{S}^3 + (-1)\operatorname{sgn}(b)\left(\varphi^\top \hat{\theta}\right)^2 \mathcal{S} \\ & + (-1)\operatorname{sgn}(b)\hat{c}_1^2 f_1^2 \mathcal{S} + \cdots + (-1)\operatorname{sgn}(b)\hat{c}_l^2 f_l^2 \mathcal{S}, \end{aligned} \quad (31)$$

where $\hat{c}, \hat{\theta}, \hat{c}_1, \ldots, \hat{c}_l$ are adjusted parameters whose update laws will be defined later. Replacing the above control law into (20) and (30) gives

$$\begin{aligned} \dot{\mathcal{S}} = & -a_m \mathcal{S} - |b|\hat{c}^2\mathcal{S}^3 - |b|\left(\varphi^\top \hat{\theta}\right)^2 \mathcal{S} \\ & - |b|\hat{c}_1^2 f_1^2 \mathcal{S} + \cdots - |b|\hat{c}_l^2 f_l^2 \mathcal{S} + d - \varphi^\top \theta^*, \end{aligned} \quad (32)$$

$$\begin{aligned} \mathcal{S}\dot{\mathcal{S}} \le & -a_m \mathcal{S}^2 - \tilde{c}\mathcal{S}^2 + \mathcal{S}\varphi^\top \tilde{\theta} + (-1)\tilde{c}_1|\mathcal{S}|f_1 \\ & + \cdots + (-1)\tilde{c}_l|\mathcal{S}|f_l + \hat{c}\mathcal{S}^2 - \mathcal{S}\varphi^\top \hat{\theta} + \hat{c}_1|\mathcal{S}|f_1 \\ & + \cdots + \hat{c}_l|\mathcal{S}|f_l - |b|\hat{c}^2\mathcal{S}^4 - |b|\left(\varphi^\top \hat{\theta}\right)^2 \mathcal{S}^2 - |b|\hat{c}_1^2 f_1^2 \mathcal{S}^2 \\ & - \cdots - |b|\hat{c}_l^2 f_l^2 \mathcal{S}^2 - c^*\mathcal{S}^2, \end{aligned} \quad (33)$$

where the last terms of the above equation satisfy the following inequality (see the proof in Appendix A):

$$\begin{aligned} & \hat{c}\mathcal{S}^2 - \mathcal{S}\varphi^\top \hat{\theta} + \hat{c}_1|\mathcal{S}|f_1 + \cdots + \hat{c}_l|\mathcal{S}|f_l - |b|\hat{c}^2\mathcal{S}^4 \\ & - |b|\left(\varphi^\top \hat{\theta}\right)^2 \mathcal{S}^2 - |b|\hat{c}_1^2 f_1^2 \mathcal{S}^2 + \cdots - |b|\hat{c}_l^2 f_l^2 \mathcal{S}^2 \\ & - c^*\mathcal{S}^2 \le 0 \quad \text{if } \mathcal{S}^2 \ge 2C_{bvs}. \end{aligned} \quad (34)$$

Equation (34) expresses the attenuation of the effect of the adjusted parameters. Substituting (34) into (33) gives

$$\begin{aligned} \mathcal{S}\dot{\mathcal{S}} \le & -a_m \mathcal{S}^2 - \tilde{c}\mathcal{S}^2 + \mathcal{S}\varphi^\top \tilde{\theta} - \tilde{c}_1|\mathcal{S}|f_1 \\ & - \cdots - \tilde{c}_l|\mathcal{S}|f_l \quad \text{if } \mathcal{S}^2 \ge 2C_{bvs}, \end{aligned} \quad (35)$$

or equivalently,

$$\mathcal{S}\dot{\mathcal{S}} \le -a_m \mathcal{S}^2 - \tilde{c}\mathcal{S}^2 + \mathcal{S}\varphi^\top \tilde{\theta} - |\mathcal{S}|\tilde{C}_d^\top f \quad \text{if } \mathcal{S}^2 \ge 2C_{bvs}, \quad (36)$$

$$\tilde{C}_d = [\tilde{c}_1, \ldots, \tilde{c}_l]^\top, \quad f = [f_1, \ldots, f_l]^\top. \quad (37)$$

We choose the following update laws:

$$\dot{\hat{c}} = \begin{cases} \gamma_c \mathcal{S}^2 & \text{if } \mathcal{S}^2 \ge 2C_{bvs}, \\ 0 & \text{otherwise,} \end{cases}$$
$$\dot{\hat{C}}_d = \begin{cases} \Gamma_d f|\mathcal{S}| & \text{if } \mathcal{S}^2 \ge 2C_{bvs}, \\ 0 & \text{otherwise,} \end{cases} \quad (38)$$
$$\dot{\hat{\theta}} = \begin{cases} -\Gamma \varphi \mathcal{S} & \text{if } \mathcal{S}^2 \ge 2C_{bvs}, \\ 0 & \text{otherwise,} \end{cases}$$

where $\mathcal{S}$ is defined in (12), C_{bvs} in (23), whereas γ_c, and the diagonal entries of Γ, Γ_d are positive constants chosen by the user, being Γ and Γ_d diagonal matrices.

6. The Control Algorithm

Now we recall the equations corresponding to the controller and establish the tracking convergence theorem. The control law is given by (31); the update laws are given by (38); the terms φ and $\mathcal{S}$ are defined in (19) and (12); φ_a is defined in (16); C_{bvs} is defined in (23) and f is defined in (37). The adjusted parameters $\hat{c}_1, \ldots, \hat{c}_l$, required to compute u, are the entries of the vector $\hat{C}_d$, that is, $\hat{C}_d = [\hat{c}_1, \ldots, \hat{c}_l]^\top$.

6.1. Theorem: Boundedness and Tracking Convergence. If the controller designed in Section 5 is applied to the plant (6), then the tracking error $e(t)$ converges to D_e asymptotically, and the closed-loop signals remain bounded.

Proof. Now we proceed to analyze the stability of the controlled system using the direct Lyapunov method and the Barbalat's Lemma. First, we establish the boundedness of the closed-loop signals on the basis of the time derivative of the Lyapunov function. Then, we establish the convergence of the tracking error to the target region D_e defined in (11), on the basis of the Barbalat's Lemma. The validity of the procedure can be stated using the theory developed and applied in [18, 19, 21, 22].

The closed-loop dynamics is given by (32), and (38). We define the following truncated quadratic form, on the basis of the truncation presented in [19]:

$$\overline{V_s} = \begin{cases} V_s & \text{if } V_s \geq C_{bvs}, \\ C_{bvs} & \text{if } V_s < C_{bvs}, \end{cases} \quad (39)$$

$$V_s \triangleq \left(\frac{1}{2}\right)\mathcal{S}^2, \quad (40)$$

where C_{bvs} is defined in (23). The form $\overline{V_s}$ has the following properties:

$$\begin{aligned} -V_s < -\overline{V_s} + C_{bvs} < 0 \quad & \text{if } V_s > C_{bvs}, \\ -V_s < -\overline{V_s} + C_{bvs} = 0 \quad & \text{if } V_s = C_{bvs}, \\ -V_s \leq 0 = -\overline{V_s} + C_{bvs} = 0 \quad & \text{if } V_s < C_{bvs}. \end{aligned} \quad (41)$$

The states of the closed-loop system may be grouped in the following vector:

$$\overline{x} = \left[\mathcal{S},\ \tilde{c},\ \tilde{C}_d^\top,\ \tilde{\theta}^\top\right]^\top, \quad (42)$$

where $\tilde{c}$, $\tilde{C}_d$, and $\tilde{\theta}$ are defined in (26), (37), (27). We define the following function:

$$\begin{gathered} V(\overline{x}(t)) = \overline{V_s} + V_\theta + V_c + V_d, \\ V_\theta = \left(\frac{1}{2}\right)\tilde{\theta}^\top \Gamma^{-1}\tilde{\theta}, \qquad V_c = \left(\frac{1}{2}\right)\gamma_c^{-1}\tilde{c}^2, \\ V_d = \left(\frac{1}{2}\right)\tilde{C}_d^\top \Gamma_d^{-1}\tilde{C}_d. \end{gathered} \quad (43)$$

Notice that $V(\overline{x})$ does not satisfy standard conditions of Lyapunov theory for time-variable systems, shown in [18]. It satisfies the conditions of [19] instead, which we use to prove the boundedness of the closed-loop signals and the convergence of the tracking error to the residual set D_e (11). The time derivative of V_s along the trajectory (32) is

$$\begin{aligned} \dot{V}_s = \mathcal{S}\dot{\mathcal{S}} &= -a_m\mathcal{S}^2 - |b|\hat{c}^2\mathcal{S}^4 - |b|\left(\varphi^\top\hat{\theta}\right)^2\mathcal{S}^2 \\ &\quad + (-1)|b|\hat{c}_1^2 f_1^2\mathcal{S}^2 + \cdots + (-1)|b|\hat{c}_l^2 f_l^2\mathcal{S}^2 \\ &\quad + \mathcal{S}\left(d - \varphi^\top\theta^*\right) \leq -a_m\mathcal{S}^2 - \tilde{c}\mathcal{S}^2 + \mathcal{S}\varphi^\top\tilde{\theta} \\ &\quad + (-1)|\mathcal{S}|\tilde{C}_d^\top f \quad \text{if } V_s \geq C_{bvs}, \end{aligned} \quad (44)$$

where the last inequality is obtained from (36). The time derivative of $\overline{V_s}$ can be derived from (39) and expressed on the basis of the above equation:

$$\dot{\overline{V_s}} = \begin{cases} \dot{V}_s \leq -a_m\mathcal{S}^2 - \tilde{c}\mathcal{S}^2 + \mathcal{S}\varphi^\top\tilde{\theta} + (-1)|\mathcal{S}|\tilde{C}_d^\top f & \text{if } V_s \geq C_{bvs}, \\ 0 & \text{if } V_s < C_{bvs}. \end{cases} \quad (45)$$

The time derivative $\dot{V}_\theta + \dot{V}_c + \dot{V}_d$ along trajectories (38) is

$$\dot{V}_\theta + \dot{V}_c + \dot{V}_d = \begin{cases} (-1)\mathcal{S}\tilde{\theta}^\top\varphi + \tilde{c}\mathcal{S}^2 + \tilde{C}_d f|\mathcal{S}| & \text{if } V_s \geq C_{bvs}, \\ 0 & \text{otherwise}. \end{cases} \quad (46)$$

Therefore, $\dot{V}$ is given by

$$\begin{aligned} \dot{V} &= \dot{\overline{V_s}} + \dot{V}_\theta + \dot{V}_c + \dot{V}_d \\ &= \begin{cases} \dot{V}_s + \dot{V}_\theta + \dot{V}_c + \dot{V}_d \leq -a_m\mathcal{S}^2 & \text{if } V_s \geq C_{bvs}, \\ 0 & \text{if } V_s < C_{bvs}. \end{cases} \end{aligned} \quad (47)$$

The above equation can be rewritten in terms of V_s defined in (40):

$$\begin{aligned} \dot{V} \leq -2a_m V_s \quad & \text{if } V_s \geq C_{bvs}, \\ \dot{V} = 0 \quad & \text{if } V_s < C_{bvs}. \end{aligned} \quad (48)$$

This implies that $\dot{V} \leq 0$ for all $t \geq t_o$. Thus, according to [19], all the closed-loop signals are bounded, that is, $(\mathcal{S}, \tilde{\theta}, \tilde{c}, \tilde{C}_d) \in L_\infty$, or equivalently, $\overline{x} \in L_\infty$, where $\overline{x}$ is defined in (42). As a consequence, $(V, \overline{V}_s, V_s, V_\theta, V_c, V_d) \in L_\infty$, according to the definitions in (39) and (43). The boundedness of $\overline{x}$ implies $(e, \ldots, e^{(n-1)}) \in L_\infty$ according to (12); hence, $\dot{\mathcal{S}} \in L_\infty$ [22], and $\dot{\overline{V}}_s \in L_\infty$. To establish the convergence of the tracking error, we begin by expressing the equation ahead in terms of $\overline{V_s}$, according to the definition of $\overline{V_s}$ in (39) and the properties of (41):

$$\begin{aligned} &\dot{V} \leq -2a_m V_s < -2a_m\overline{V_s} + 2a_m C_{bvs} \leq 0 \quad \text{if } V_s \geq C_{bvs}, \\ &\dot{V} = 0 = 2a_m\left(-\overline{V_s} + C_{bvs}\right) = 0 \quad \text{if } V_s < C_{bvs}, \\ &\quad \Longrightarrow \dot{V} \leq -2a_m\overline{V_s} + 2a_m C_{bvs} \leq 0. \end{aligned} \quad (49)$$

We reorganize the above equation and integrate as follows:

$$2a_m\int_{t_o}^{t}\left(\overline{V_s} - C_{bvs}\right)d\tau + V(t) \leq V(t_o). \quad (50)$$

Thus, $(\overline{V}_s - C_{bvs}) \in L_1$. Recall that $(\overline{V}_s, \dot{\overline{V}}_s) \in L_\infty$. Thus, $(\overline{V}_s - C_{bvs}) \in L_\infty \cap L_1$, $(d/dt)(\overline{V}_s - C_{bvs}) \in L_\infty$. By invoking the Barbalat Lemma [23], we obtain $(\overline{V}_s - C_{bvs}) \to 0$ as $t \to \infty$. In turn, this implies that $V_s \to D_{vs}$, being D_{vs} defined as

$$D_{vs} = \{V_s : V_s \le C_{bvs}\}, \tag{51}$$

where C_{bvs} is defined in (23). $\mathcal{S}$ can be expressed in terms of V_s, on the basis of the definition of V_s in (40):

$$\mathcal{S} = \sqrt{2V_s}. \tag{52}$$

In the following, we analyze the convergence of $\mathcal{S}$. Taking into account the definition of C_{bvs} in (23), the fact that V_s converges to D_{vs} defined in (51), and (52) the convergence of $\mathcal{S}$ is:

$$\lim_{t\to\infty} \mathcal{S} = D_s, \qquad D_s = \{\mathcal{S} : |\mathcal{S}| \le C_{bs}\},$$
$$C_{bs} = \sqrt{2C_{bvs}} = \lambda^{n-1} C_{be}. \tag{53}$$

To analyze the convergence of e, we begin by expressing e in terms of $\mathcal{S}$, according to (12):

$$e = \frac{1}{(p+\lambda)^{n-1}} \mathcal{S}. \tag{54}$$

The convergence of the tracking error may be established on the basis of (53) and (54) [18, 21]:

$$\lim_{t\to\infty} e(t) = D_e, \qquad D_e = \{e : |e| \le C_{be}\}, \tag{55}$$

which means that the tracking error e converges to a residual set whose size is of the user's choice. □

7. Application to a Permanent Magnet Synchronous Motor

A PMSM is a kind of highly efficient and high-powered motor. The benefits of the PMSM are discussed in [24, 25]. The PMSM has the following difficulties [26]: (i) it is highly nonlinear; (ii) the parameters of the physical model experience unknown time-varying behavior, for example, the stator resistance R and the friction coefficient B; (iii) unknown external disturbances appear, that is, the load torque disturbance T_L. Moreover, by varying the permanent magnet flux λ_{af} the state ω exhibits a pitchfork bifurcation [27]. All the mentioned facts lead to response deterioration of many controllers, specially for high-speed and high-precision tasks in real applications [28].

In view of the complex behavior, with parameters varying with respect to time and state plane, the scheme developed in this work is suitable. We apply the developed scheme by simulation, to a PMSM whose model and parameters are presented in [29]. Therein, the possible manipulated inputs are u_d and u_q; and i_d, i_q, and ω are possible outputs. We choose the motor angular frequency ω as the output to be controlled and the d-axis voltage u_d as the control input, that is, $y = \omega$, $u = u_d$, $v_d = 0$. A similar choice is made in [24], where an adaptive controller is derived for a PMSM. After performing a step response analysis, the variable ω exhibits a second-order behavior when a step change is introduced in u_d. This second order behavior was already noticed in [29]. Thus, we use the regression model (6) with $n = 2$. We can summarize the basic features of the plant as

$$\begin{gathered} n = 2, \qquad \operatorname{sgn}(b) = +1, \\ l = 3, \qquad f_1 = |\dot{y}|, \qquad f_2 = |y|, \\ f_3 = 1, \qquad \Longrightarrow f = [|\dot{y}|, |y|, 1]^\top, \\ \phi = [y^2, y^3]^\top, \qquad \varphi_a = -\ddot{y}_d + \lambda_o \dot{e}, \\ \Longrightarrow \varphi = [-\phi^\top, -\varphi_a - a_m \mathcal{S}]^\top. \end{gathered} \tag{56}$$

We used a factorization with terms y^2 and y^3 to take into account the presence of the pitchfork bifurcation. According to [30] the normal form of the pitchfork bifurcation gives a description of the system behavior in a tight neighborhood of the bifurcation point. For the case of the pitchfork bifurcation, the normal form includes the terms y^2 and y^3. Although the system usually works in different regions far from the bifurcation point, we wanted to include the behavior in the neighborhood of the bifurcation point. Thus, we included the terms of the normal form for the pitchfork bifurcation [30], that is, y^2 and y^3. The terms 1 and y are not included in ϕ because they are already present in d. The remaining parameters of the controller are defined on the basis of the parameters in (56):

$$\begin{aligned} &e = y - y_d, \qquad \mathcal{S} = (p+\lambda)e = \dot{e} + \lambda e \Longrightarrow \lambda_o = \lambda, \\ &C_{bvs} = \left(\frac{1}{2}\right)(\lambda C_{be})^2, \\ &u = (-1)\operatorname{sgn}(b)\hat{c}^2 \mathcal{S}^3 + (-1)\operatorname{sgn}(b)\left(\varphi\hat{\theta}\right)^2 \mathcal{S} \\ &\quad + (-1)\operatorname{sgn}(b)\mathcal{S}\left(\hat{c}_1^2 f_1^2 + \hat{c}_2^2 f_2^2 + \hat{c}_3^2 f_3^2\right), \\ &\quad = (-1)\operatorname{sgn}(b)\mathcal{S}\left(\hat{c}^2\mathcal{S}^2 + \left(\varphi\hat{\theta}\right)^2 + \hat{c}_1^2 f_1^2 + \hat{c}_2^2 f_2^2 + \hat{c}_3^2 f_3^2\right). \end{aligned} \tag{57}$$

In addition, we choose:

$$\begin{gathered} \lambda = 3, \qquad a_m = 10, \qquad a_{m,1} = 70, \\ a_{m,o} = 1225 \text{ (for the reference model)}, \qquad C_{be} = 0.1, \\ \gamma_c = 20, \qquad \Gamma_d = \operatorname{diag}[20, 20, 20], \\ \Gamma = \operatorname{diag}[0.4, 0.4, 0.003]. \end{gathered} \tag{58}$$

The results are shown in Figure 1. The simulation shows expected results: all the closed-loop signals remain bounded, and transient values of the tracking error remain in a small interval. It is worthy of note that the transient values of e depend on its initial value and the initial values of the adjustment errors. To show the effect of the time-varying behavior of the model coefficients, we change the stator resistance R from 1.4 to 1.7 Ω at the time instant 0.6 sec. Results are shown in Figure 2.

In addition, we consider the variation of the damping constant B from 0.00038818 to 0.00046582 Nm/(rad/s) at the

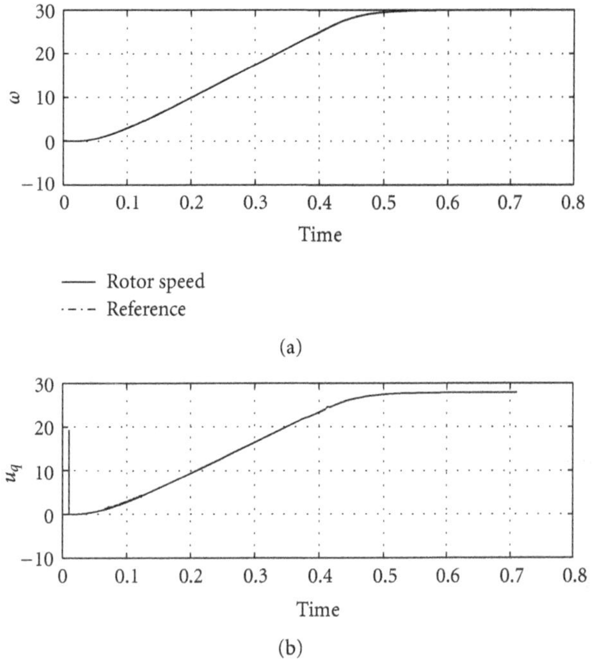

Figure 1: Transient behavior of the output and control input.

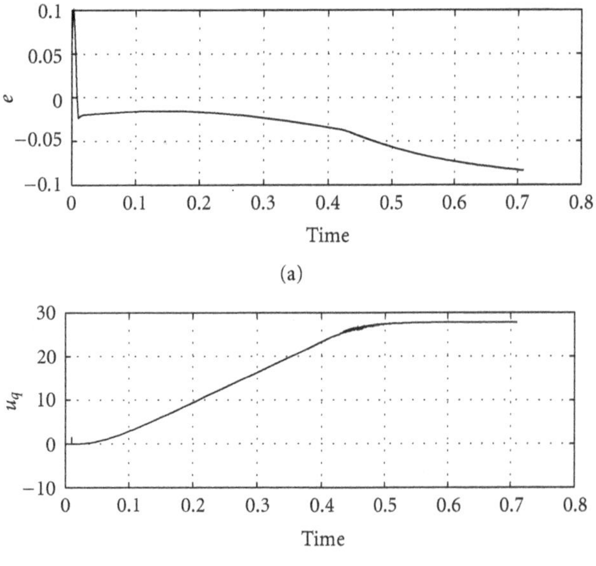

Figure 2: Performance of the tracking error and control input varying the resistance value.

instant 0.6 sec. Results are shown in Figure 3. Notice that in the three cases the final value of $|e|$ is less than $C_{be} = 0.1$. Moreover, the control input u belongs to the interval $[-200\ 200]$ V. Notice in Figures 2 and 3 that the effect of the disturbance on the tracking error is almost negligible. Nevertheless, the control input experiences a large variation, as can be seen in Figure 3.

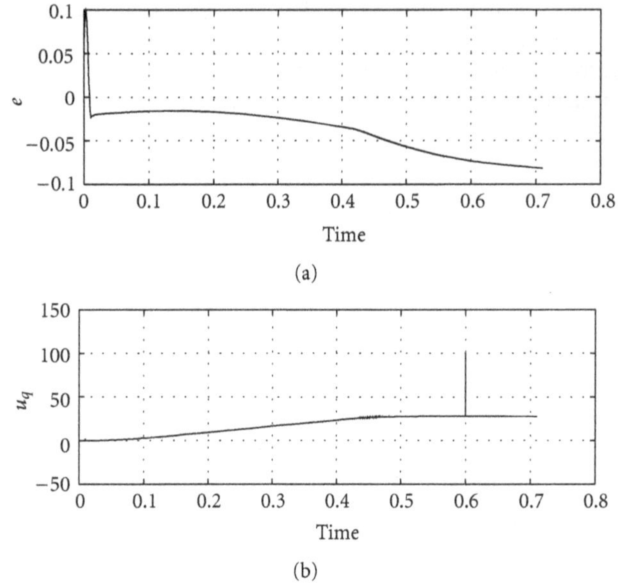

Figure 3: Performance of the tracking error and control input varying the damping constant value.

8. Conclusions

In this work, we have proposed a control scheme for highly nonlinear plants, based on a simple plant model with polynomial approximators, which provides an adequate description of transient behavior. It is worth noticing the fact that benefits Ri to Rviii (Section 2) are achieved at the same time, with minimal requirements on the knowledge of the plant. Indeed, the relative degree and the signum of the control gain can be established by a previous step response analysis. Many techniques are combined at the same time: sliding surface MRAC, dead zone-type update law, robustifying auxiliary control, approximation techniques, and truncation of the quadratic forms.

The disadvantage of polynomials as approximators is that they may be less accurate than other techniques, for example, neural networks or fuzzy sets, leading to higher approximation error. Since the approximation error is bounded, it can be handled by means of robustness techniques without requiring the upper bound to be known. Moreover, we considered the coefficients of the terms $y, \ldots, y^{(n-1)}$ as time-varying but bounded, with constant and unknown bounds. Then, we handled this by means of robust control, without requiring the upper bounds to be known.

We handled the time-varying behavior of the control gain by means of robustness techniques, without using the Nussbaum gain method. The redefinition of the plant terms in terms of adjustment errors and adjustment parameters is a fundamental step. The resulting expression allows a straightforward definition of the control law. The variation of the control gain implies that the terms involving adjusted parameters cannot be cancelled. Rather, we attenuate its effect by means of squared terms and handle the residual error by means of an additional control term that is only a function of the sliding surface. The resulting expression for V, $\dot{V}$, and the design is simpler in comparison with the Nussbaum technique.

Appendix

Proof of (34)

As the first step, we factorize several summands of (34) and apply the property (8):

$$\begin{aligned} &-|b|\hat{c}^2 \mathcal{S}^4 + \hat{c}\mathcal{S}^2 \leq -b_m \hat{c}^2 \mathcal{S}^4 + \hat{c}\mathcal{S}^2 \\ &\quad = -\left[\sqrt{b_m}\hat{c}\mathcal{S}^2 - \frac{1}{2\sqrt{b_m}}\right]^2 + \left(\frac{1}{2\sqrt{b_m}}\right)^2 \leq \left(\frac{1}{2\sqrt{b_m}}\right)^2. \end{aligned} \tag{A.1}$$

Likewise, we obtain:

$$\begin{aligned} &(-1)|b|\left(\varphi^{\top}\hat{\theta}\right)^2 \mathcal{S}^2 - \mathcal{S}\varphi^{\top}\hat{\theta} \leq \left(\frac{1}{2\sqrt{b_m}}\right)^2, \\ &(-1)|b|\hat{c}_1^2 f_1^2 \mathcal{S}^2 + \hat{c}_1|\mathcal{S}|f_1 \leq \left(\frac{1}{2\sqrt{b_m}}\right)^2, \\ &\qquad \vdots \\ &(-1)|b|\hat{c}_l^2 f_l^2 \mathcal{S}^2 + \hat{c}_l|\mathcal{S}|f_l \leq \left(\frac{1}{2\sqrt{b_m}}\right)^2. \end{aligned} \tag{A.2}$$

By adding (A.1) and (A.2) we obtain:

$$\begin{aligned} &-|b|\hat{c}^2 \mathcal{S}^4 + (-1)|b|\left(\varphi^{\top}\hat{\theta}\right)^2 \mathcal{S}^2 + (-1)|b|\hat{c}_1^2 f_1^2 \mathcal{S}^2 \\ &\quad + \cdots + (-1)|b|\hat{c}_l^2 f_l^2 \mathcal{S}^2 + \hat{c}\mathcal{S}^2 - \mathcal{S}\varphi^{\top}\hat{\theta} + \hat{c}_1|\mathcal{S}|f_1 \\ &\quad + \cdots + \hat{c}_l|\mathcal{S}|f_l \leq (2+l)\left(\frac{1}{2\sqrt{b_m}}\right)^2. \end{aligned} \tag{A.3}$$

As a second step, we use the definition of c^* in (22) to rewrite the term $-c^*\mathcal{S}^2$:

$$\begin{aligned} -c^*\mathcal{S}^2 &= -(2+l)\frac{\mathcal{S}^2}{2C_{bvs}}\left(\frac{1}{2\sqrt{b_m}}\right)^2 \\ &\leq -(2+l)\left(\frac{1}{2\sqrt{b_m}}\right)^2 \quad \text{if } \mathcal{S}^2 \geq 2C_{bvs}. \end{aligned} \tag{A.4}$$

As the third step, we add (A.3) and (A.4):

$$\begin{aligned} &-|b|\hat{c}^2 \mathcal{S}^4 + (-1)|b|\left(\varphi^{\top}\hat{\theta}\right)^2 \mathcal{S}^2 + (-1)|b|\hat{c}_1^2 f_1^2 \mathcal{S}^2 \\ &\quad + \cdots + (-1)|b|\hat{c}_l^2 f_l^2 \mathcal{S}^2 + \hat{c}\mathcal{S}^2 - \mathcal{S}\varphi^{\top}\hat{\theta} + \hat{c}_1|\mathcal{S}|f_1 \\ &\quad + \cdots + \hat{c}_l|\mathcal{S}|f_l - c^*\mathcal{S}^2 \leq 0 \quad \text{if } \mathcal{S}^2 \geq 2C_{bvs}, \end{aligned} \tag{A.5}$$

which is (34).

Acknowledgments

This work was partially supported by Universidad Nacional de Colombia-Manizales, project 12475, Vicerrectoría de Investigación, DIMA, resolution number VR-2185.

References

[1] K. S. Narendra and K. George, "Adaptive control of simple nonlinear systems using multiple models," in *Proceedings of the American Control Conference (ACC '02)*, pp. 1779–1784, Anchorage, AK, USA, May 2002.

[2] I. Kanellakopoulos, P. V. Kokotovic, and A. S. Morse, "Systematic design of adaptive controllers for feedback linearizable systems," *IEEE Transactions on Automatic Control*, vol. 36, no. 11, pp. 1241–1253, 1991.

[3] J. Nakanishi, J. A. Farrell, and S. Schaal, "Composite adaptive control with locally weighted statistical learning," *Neural Networks*, vol. 18, no. 1, pp. 71–90, 2005.

[4] H. A. Yousef and M. A. Wahba, "Adaptive fuzzy mimo control of induction motors," *Expert Systems with Applications*, vol. 36, no. 3, pp. 4171–4175, 2009.

[5] A. C. Huang and Y. S. Kuo, "Sliding control of non-linear systems containing time-varying uncertainties with unknown bounds," *International Journal of Control*, vol. 74, no. 3, pp. 252–264, 2001.

[6] A. C. Huang and Y. C. Chen, "Adaptive multiple-surface sliding control for non-autonomous systems with mismatched uncertainties," *Automatica*, vol. 40, no. 11, pp. 1939–1945, 2004.

[7] C. H. Chen, C. M. Lin, and T. Y. Chen, "Intelligent adaptive control for MIMO uncertain nonlinear systems," *Expert Systems with Applications*, vol. 35, no. 3, pp. 865–877, 2008.

[8] Y. Feng, C. Y. Su, H. Hong, and S. S. Ge, "Robust adaptive control for a class of nonlinear systems with generalized Prandtl-Ishlinskii hysteresis," in *Proceedings of the 46th IEEE Conference on Decision and Control 2007 (CDC '07)*, pp. 4833–4838, New Orleans, Calif, USA, December 2007.

[9] K. S. Narendra and J. Balakrishnan, "Adaptive control using multiple models," *IEEE Transactions on Automatic Control*, vol. 42, no. 2, pp. 171–187, 1997.

[10] S. S. Ge and J. Wang, "Robust adaptive tracking for time-varying uncertain nonlinear systems with unknown control coefficients," *IEEE Transactions on Automatic Control*, vol. 48, no. 8, pp. 1463–1469, 2003.

[11] C. P. Bechlioulis and G. A. Rovithakis, "Adaptive control with guaranteed transient and steady state tracking error bounds for strict feedback systems," *Automatica*, vol. 45, no. 2, pp. 532–538, 2009.

[12] C. Y. Su, Y. Feng, H. Hong, and X. Chen, "Adaptive control of system involving complex hysteretic nonlinearities: a generalised Prandtl-Ishlinskii modelling approach," *International Journal of Control*, vol. 82, no. 10, pp. 1786–1793, 2009.

[13] Y. Feng, C. Y. Su, and H. Hong, "Universal construction of robust adaptive control laws for a class of nonlinear systems preceded by generalized Prandtl-Ishlinskii representation," in *Proceedings of the 3rd IEEE Conference on Industrial Electronics and Applications (ICIEA '08)*, pp. 153–158, Singapore, Singapore, June 2008.

[14] C. F. Hsu, C. M. Lin, and T. T. Lee, "Wavelet adaptive backstepping control for a class of nonlinear systems," *IEEE Transactions on Neural Networks*, vol. 17, no. 5, pp. 1175–1183, 2006.

[15] C. Y. Lee, "Adaptive control of a class of nonlinear systems using multiple parameter models," *International Journal of Control, Automation and Systems*, vol. 4, no. 4, pp. 428–437, 2006.

[16] S. Labiod and T. M. Guerra, "Adaptive fuzzy control of a class of SISO nonaffine nonlinear systems," *Fuzzy Sets and Systems*, vol. 158, no. 10, pp. 1126–1137, 2007.

[17] J. Spooner, M. Maggiore, R. Ordóñez, and K. Passino, *Stable Adaptive Control and Estimation for Nonlinear Systems: Neural and Fuzzy Approximator Techniques*, John Wiley & Sons, New York, NY, USA, 2002.

[18] J. J. Slotine and W. Li, *Applied Nonlinear Control*, Prentice Hall, Englewood Cliffs, NJ, USA, 1991.

[19] B. B. Peterson and K. Narendra, "Bounded error adaptive control," *IEEE Transactions on Automatic Control*, vol. AC-27, no. 6, pp. 1161–1168, 1982.

[20] K. Astrom and B. Wittenmark, *Adaptive Control*, Addison-Wesley Publishing Company Inc, Reading, Mass, USA, 1995.

[21] X. S. Wang, C. Y. Su, and H. Hong, "Robust adaptive control of a class of nonlinear systems with unknown dead-zone," *Automatica*, vol. 40, no. 3, pp. 407–413, 2004.

[22] J. Wu, D. Pu, and H. Ding, "Adaptive robust motion control of SISO nonlinear systems with implementation on linear motors," *Mechatronics*, vol. 17, no. 4-5, pp. 263–270, 2007.

[23] P. A. Ioannou and J. Sun, *Robust Adaptive Control*, Prentice-Hall PTR, Upper Saddle River, NJ, USA, 1996.

[24] J. Zhou and Y. Wang, "Real-time nonlinear adaptive backstepping speed control for a PM synchronous motor," *Control Engineering Practice*, vol. 13, no. 10, pp. 1259–1269, 2005.

[25] L. Liu, W. Liu, and D. A. Cartes, "Particle swarm optimization-based parameter identification applied to permanent magnet synchronous motors," *Engineering Applications of Artificial Intelligence*, vol. 21, no. 7, pp. 1092–1100, 2008.

[26] F. Morel, X. Lin-Shi, J. M. Rétif, and B. Allard, "A predictive current control applied to a permanent magnet synchronous machine, comparison with a classical direct torque control," *Electric Power Systems Research*, vol. 78, no. 8, pp. 1437–1447, 2008.

[27] Z. Jing, C. Yu, and G. Chen, "Complex dynamics in a permanent-magnet synchronous motor model," *Chaos, Solitons and Fractals*, vol. 22, no. 4, pp. 831–848, 2004.

[28] A. M. Harb, "Nonlinear chaos control in a permanent magnet reluctance machine," *Chaos, Solitons and Fractals*, vol. 19, no. 5, pp. 1217–1224, 2004.

[29] P. Pillay and R. Krishnan, "Control characteristics and speed controller design for a high performance permanent magnet synchronous motor drive," *IEEE Transactions on Power Electronics*, vol. 5, no. 2, pp. 151–159, 1989.

[30] Y. A. Kuznetsov, *Elements of Applied Bifurcation Theory*, Springer-Verlag, New York, NY, USA, 3rd edition, 2004.

Second-Order Model Reduction Based on Gramians

Cong Teng

School of Mathematics and Quantitative Economics, Shandong University of Finance and Economics, Jinan, Shandong 250014, China

Correspondence should be addressed to Cong Teng, ginger.cong@gmail.com

Academic Editor: S. Skogestad

Some new and simple Gramian-based model order reduction algorithms are presented on second-order linear dynamical systems, namely, SVD methods. Compared to existing Gramian-based algorithms, that is, balanced truncation methods, they are competitive and more favorable for large-scale systems. Numerical examples show the validity of the algorithms. Error bounds on error systems are discussed. Some observations are given on structures of Gramians of second order linear systems.

1. Introduction

Models of linear dynamical systems are often given in second-order form

$$\begin{aligned} M\ddot{q}(t) + G\dot{q}(t) + Kq(t) &= B_0u(t) \\ y(t) &= C_0q(t) + D_0\dot{q}(t), \end{aligned} \tag{1}$$

where $M, G, K \in \mathbb{R}^{n\times n}$, $B_0 \in \mathbb{R}^{n\times m}$, C_0, and $D_0 \in \mathbb{R}^{p\times n}$ are given matrices, $u(t) \in \mathbb{R}^m$ is the given vector of inputs, $y(t) \in \mathbb{R}^p$ is the unknown vector of outputs, $q(t) \in \mathbb{R}^n$ is the unknown vector of internal variables, n is the dimension of the system, and M is assumed to be invertible.

Models of mechanical systems in particular are usually of second-order form (1). For such a system, $M = M^T$, G and $K = K^T$ are respectively the mass, damping, and stiffness matrices, $u(t) \in \mathbb{R}^m$ is the input, $B_0u(t) = f(t) \in \mathbb{R}^n$ is the vector of external forces, and $q(t) \in \mathbb{R}^n$ is the vector of internal variables.

Many applications lead to very large models where n is very large, while $m, p \ll n$. Due to limitations on time and computer storage, it is often difficult or impossible to directly simulate or control these large-scale systems. Therefore, it is often desirable to reduce the system to a smaller-order model:

$$\begin{aligned} \widehat{M}\ddot{q}(t) + \widehat{G}\dot{q}(t) + \widehat{K}q(t) &= \widehat{B}_0u(t) \\ \widehat{y}(t) &= \widehat{C}_0q(t) + \widehat{D}_0\dot{q}(t), \end{aligned} \tag{2}$$

such that

(1) $\|y - \widehat{y}\|/\|u\| <$ tol, that is, outputs corresponding to the same inputs are close.

(2) The reduced system preserves important system properties such as stability and passivity.

(3) The procedure is computationally efficient.

By letting $x = \begin{bmatrix} q \\ \dot{q} \end{bmatrix}$, second-order system (1) is then transformed to a corresponding $2n$-dimensional first-order system

$$\Sigma : \begin{cases} \dot{x}(t) = Ax(t) + Bu(t) \\ y(t) = Cx(t) + Du(t), \end{cases} \tag{3}$$

where

$$\begin{aligned} A &= \begin{bmatrix} 0 & I \\ -M^{-1}K & -M^{-1}G \end{bmatrix} = \begin{bmatrix} 0 & I \\ -K_M & -G_M \end{bmatrix} \\ B &= \begin{bmatrix} 0 \\ M^{-1}B_0 \end{bmatrix} = \begin{bmatrix} 0 \\ B_M \end{bmatrix} \\ C &= \begin{bmatrix} C_0 & D_0 \end{bmatrix} \\ D &= 0. \end{aligned} \tag{4}$$

Since algorithms and theory are well developed for first-order model reduction, it is natural to use these techniques to develop algorithms for second-order systems.

Closely related to first-order system are the two Lyapunov equations

$$A\mathcal{P} + \mathcal{P}A^* + BB^* = 0,$$
$$A^*\mathcal{Q} + \mathcal{Q}A + C^*C = 0. \tag{5}$$

Under assumption that first-order system is stable, it is well known that the reachability and observability Gramians $\mathcal{P}$ and $\mathcal{Q}$ are the unique solutions to Lyapunov equation (5), where both $\mathcal{P}$ and $\mathcal{Q}$ are symmetric positive definite. This Gramian $\mathcal{P}$ has also the following variational interpretation proposed in [1]. Let $J(u, t_1, t_2) = \int_{t_1}^{t_2} u^*(t)u(t)\,dt = \int_{t_1}^{t_2} \|u(t)\|^2\,dt$. The optimal value to optimization problem

$$\min_u J(u, -\infty, 0)$$
$$\text{s.t.} \quad \dot{x} = Ax + Bu, \quad x(0) = x_0 \tag{6}$$

is $x_0^*\mathcal{P}^{-1}x_0$, that is, the minimal energy required to steer the state of the system from $t = -\infty$ to state x_0 at time $t = 0$ is $x_0^*\mathcal{P}^{-1}x_0$. Similarly, the optimal value to its dual system is $x_0^*\mathcal{Q}^{-1}x_0$.

Partition $\mathcal{P}$ and $\mathcal{Q}$ into four equal blocks:

$$\mathcal{P} = \begin{pmatrix} \mathcal{P}_{11} & \mathcal{P}_{12} \\ \mathcal{P}_{21} & \mathcal{P}_{22} \end{pmatrix}, \qquad \mathcal{Q} = \begin{pmatrix} \mathcal{Q}_{11} & \mathcal{Q}_{12} \\ \mathcal{Q}_{21} & \mathcal{Q}_{22} \end{pmatrix}. \tag{7}$$

In [2–4], it is shown that the optimal value to optimization problem

$$\min_{\dot{q}_0 \in \mathbb{R}^n} \min_u J(u, -\infty, 0)$$
$$\text{s.t.} \quad M\ddot{q}(t) + G\dot{q}(t) + Kq(t) = Bu(t), \quad q(0) = q_0 \tag{8}$$

is $q_0^*\mathcal{P}_{11}^{-1}q_0$, that is, the minimal energy required to reach the given position q_0 over all past inputs and initial velocities. And the optimal value to optimization problem

$$\min_{q_0 \in \mathbb{R}^n} \min_u J(u, -\infty, 0)$$
$$\text{s.t.} \quad M\ddot{q}(t) + G\dot{q}(t) + Kq(t) = Bu(t), \quad \dot{q}(0) = \dot{q}_0 \tag{9}$$

is $\dot{q}_0^*\mathcal{P}_{22}^{-1}\dot{q}_0$, that is, the minimal energy required to reach the given velocity $\dot{q}_0$ over all past inputs and initial positions. In [3, 4], $\mathcal{P}_{11}$ and $\mathcal{P}_{22}$ are defined to be position Gramian and velocity Gramian, respectively. By duality, $\mathcal{Q}_{11}$ and $\mathcal{Q}_{22}$ are position Gramian and velocity Gramian, respectively corresponding to $\mathcal{Q}$.

It is well known that

$$\mathcal{P} = \int_0^\infty e^{A\tau}BB^Te^{A^T\tau}\,d\tau,$$
$$\mathcal{Q} = \int_0^\infty e^{A^T\tau}C^TCe^{A\tau}\,d\tau. \tag{10}$$

Suppose $x(t)$ is the solution of first-order system for impulse input $u(t) = \delta(t)$. Then

$$\mathcal{P} = \int_0^\infty x(t)x^*(t)dt. \tag{11}$$

For second-order system and its corresponding first-order system,

$$\begin{aligned} \mathcal{P} &= \int_0^\infty x(t)x^*(t)dt \\ &= \int_0^\infty \begin{pmatrix} q(t) \\ \dot{q}(t) \end{pmatrix} \begin{pmatrix} q^*(t) & \dot{q}^*(t) \end{pmatrix} dt \\ &= \int_0^\infty \begin{pmatrix} q(t)q^*(t) & q(t)\dot{q}^*(t) \\ \dot{q}(t)q^*(t) & \dot{q}(t)\dot{q}^*(t) \end{pmatrix} dt. \end{aligned} \tag{12}$$

Therefore,

$$\mathcal{P}_{11} = \int_0^\infty q(t)q^*(t)dt,$$
$$\mathcal{P}_{22} = \int_0^\infty \dot{q}(t)\dot{q}^*(t)dt, \tag{13}$$

where $\mathcal{P}_{11}$ and $\mathcal{P}_{22}$ are position and velocity Gramians, respectively. Similar results can be applied to $\mathcal{Q}$, $\mathcal{Q}_{11}$, and $\mathcal{Q}_{22}$ through dual systems.

For Gramian based structure preserving mode order reduction of second-order systems, in [3, 5], some balanced truncation methods were derived. In this paper, we present two new algorithms based on SVD rather than balanced truncation. The algorithms are derived from second-order Gramians and system $\mathcal{H}_2$-norm. Compared to existing techniques, they are simple and more suitable for large-scale settings. In Section 4, we apply these methods to three numerical examples, the computational results show the validity of our algorithms. Error bounds and structures of Gramians on second-order systems are also discussed.

2. Balanced Truncation Method

For first-order system

$$\Sigma : \begin{cases} \dot{x}(t) = Ax(t) + Bu(t), \\ y(t) = Cx(t). \end{cases} \tag{14}$$

balanced truncation method is to transform the system to another coordinate system such that both $\mathcal{P}$ and $\mathcal{Q}$ are equal and diagonal:

$$\mathcal{P} = \mathcal{Q} = \operatorname{diag}(\sigma_1, \sigma_2, \ldots, \sigma_N), \tag{15}$$

where $\sigma_1 \geq \sigma_2 \geq \cdots \geq \sigma_N$. It then truncates the model by keeping the states corresponding to k largest eigenvalues of the product $\mathcal{P}\mathcal{Q}$, that is, k largest Hankel singular values of the system.

The main problem is to find balancing projection matrices. A numerically stable way to get the balancing truncated system is given in [6]: suppose the Cholesky factorization of $\mathcal{P}$ and $\mathcal{Q}$ are is

$$\mathcal{P} = U_CU_C^T, \qquad \mathcal{Q} = L_OL_O^T, \tag{16}$$

where U_C is upper triangular and L_O is lower triangular matrices. Take SVD of $U_C^T L_O$,

$$U_C^T L_O = U \Sigma V^T. \quad (17)$$

Let Σ_k be the first k by k principle submatrix of Σ, U_k, and V_k consist of the first k columns of U and V, respectively. Then the balanced projection matrices are

$$W = L_O V_k \Sigma_k^{-1/2}, \qquad V = U_C U_k \Sigma_k^{-1/2}. \quad (18)$$

The reduced system is obtained in:

$$\hat{A} = W^T A V, \qquad \hat{B} = W^T B, \qquad \hat{C} = CV. \quad (19)$$

Balanced truncation method has a global error bound [1]:

$$\|\Sigma_{\text{err}}\|_{\mathcal{H}_\infty} = \|\Sigma - \Sigma_{\text{red}}\|_{\mathcal{H}_\infty} \le 2(\sigma_{k+1} + \cdots + \sigma_n). \quad (20)$$

For second-order system

$$\begin{aligned} M\ddot{q}(t) + G\dot{q}(t) + Kq(t) &= B_0 u(t) \\ y(t) &= C_0 q(t) + D_0 \dot{q}(t), \end{aligned} \quad (21)$$

there are two balanced truncation methods. One was given in [5] which is to balance both $\mathcal{P}_{11}$ and $\mathcal{Q}_{11}$ to form the projection matrices W_{1r} and V_{1r} and then get the reduced system by keeping r largest eigenvalues of $\mathcal{P}_{11}\mathcal{Q}_{11}$. It is equivalent to performing projections to its corresponding first-order system as following:

$$\begin{aligned} \hat{A} &= W^T A V \\ &= \begin{pmatrix} W_{1_r}^T & \\ & W_{1_r}^T \end{pmatrix} \begin{pmatrix} 0 & I \\ -M^{-1}K & -M^{-1}G \end{pmatrix} \begin{pmatrix} V_{1_r} & \\ & V_{1_r} \end{pmatrix} \\ &= \begin{pmatrix} 0 & I \\ -W_{1_r}^{-1} M^{-1} K V_{1_r} & -W_{1_r}^{-1} M^{-1} G V_{1_r} \end{pmatrix}, \\ \hat{B} &= WB = \begin{pmatrix} W_{1_r}^T & \\ & W_{1_r}^T \end{pmatrix} \begin{bmatrix} 0 \\ M^{-1} B_0 \end{bmatrix} \\ &= \begin{bmatrix} 0 \\ W_{1_r}^T M^{-1} B_0 \end{bmatrix} \\ \hat{C} &= CV = \begin{bmatrix} C_0 & D_0 \end{bmatrix} \begin{pmatrix} V_{1_r} & \\ & V_{1_r} \end{pmatrix} = \begin{bmatrix} C_0 V_{1_r} & D_0 V_{1_r} \end{bmatrix}. \end{aligned} \quad (22)$$

The reduction rules are then

$$\begin{aligned} &\widehat{M} = I_r, \quad \widehat{K} = W_{1_r}^T M^{-1} K V_{1_r}, \\ &\widehat{G} = W_{1_r}^T M^{-1} G V_{1_r}, \quad \widehat{B}_{0 = W_{1_r}^T M^{-1} B_0}, \\ &\widehat{C}_0 = C_0 V_{1_r}, \quad \widehat{D}_0 = D_0 V_{1_r}. \end{aligned} \quad (23)$$

Similar results can be applied to $\mathcal{P}_{22}$ and $\mathcal{Q}_{22}$. This gives the following algorithm.

Algorithm 1 (Balanced truncation method based on $\mathcal{P}_{11}$ and $\mathcal{Q}_{11}$ (or $\mathcal{P}_{22}$ and $\mathcal{Q}_{22}$, resp.) [5]).

(1) Compute $\mathcal{P}_{11}$ and $\mathcal{Q}_{11}$.

(2) Compute the balanced truncation matrices W_{1_r}, $V_{1_r} \in \mathbb{R}^{n\times r}$ such that $W_{1_r}^T V_{1_r} = I_r$ and $W_{1_r}^T \mathcal{P}_{11} W_{1_r} = V_{1_r}^T \mathcal{Q}_{11} V_{1_r} = \text{diag}(\sigma_1, \sigma_2, \ldots, \sigma_r)$, where $\sigma_1^2, \sigma_2^2, \ldots, \sigma_r^2$ are the r largest eigenvalues of $\mathcal{P}_{11}\mathcal{Q}_{11}$.

(3) Perform projection to (M, G, K, B_0, C_0, D_0) and get $(\widehat{M}, \widehat{G}, \widehat{K}, \widehat{B}_0, \widehat{C}_0, \widehat{D}_0)$:

$$\begin{aligned} \widehat{M} = I_r, &\quad \widehat{K} = W_{1_r}^T M^{-1} K V_{1_r}, \\ \widehat{G} &= W_{1_r}^T M^{-1} G V_{1_r}, \\ \widehat{B}_0 = W_{1_r}^T M^{-1} B_0, &\quad \widehat{C}_0 = C_0 V_{1_r}, \\ \widehat{D}_0 &= D_0 V_{1_r}. \end{aligned} \quad (24)$$

In [5], it also gave two other similar methods. One is to balance $\mathcal{P}_{11}$ and $\mathcal{Q}_{22}$ to get the projection matrices, the other is to balance $\mathcal{P}_{22}$ and $\mathcal{Q}_{11}$ to get the projection matrices.

Another second-order balanced truncation method is given in [3] which is to balance both $\mathcal{P}_{11}$ and $\mathcal{Q}_{11}$ to get the projection matrices W_{1r} and V_{1r} by keeping r largest eigenvalues of $\mathcal{P}_{11}\mathcal{Q}_{11}$, balance both $\mathcal{P}_{22}$ and $\mathcal{Q}_{22}$ to get the projection matrices W_{2r} and V_{2r}, then use $W = \begin{pmatrix} W_{1r} & \\ & W_{2r} \end{pmatrix}$ and $V = \begin{pmatrix} V_{1r} & \\ & V_{2r} \end{pmatrix}$ as projection matrices for the corresponding first-order system:

$$\begin{aligned} \hat{A} &= W^T A V \\ &= \begin{pmatrix} W_{1_r}^T & \\ & W_{2_r}^T \end{pmatrix} \begin{pmatrix} 0 & I \\ -M^{-1}K & -M^{-1}G \end{pmatrix} \begin{pmatrix} V_{1_r} & \\ & V_{2_r} \end{pmatrix} \\ &= \begin{pmatrix} 0 & W_{1_r}^T V_{2_r} \\ -W_{2_r}^T M^{-1} K V_{1_r} & -W_{2_r}^T M^{-1} G V_{2_r} \end{pmatrix}, \\ \hat{B} &= W^T B = \begin{pmatrix} W_{1_r}^T & \\ & W_{2_r}^T \end{pmatrix} \begin{bmatrix} 0 \\ M^{-1} B_0 \end{bmatrix} \\ &= \begin{bmatrix} 0 \\ W_{2_r}^T M^{-1} B_0 \end{bmatrix} \\ \hat{C} &= CV = \begin{bmatrix} C_0 & D_0 \end{bmatrix} \begin{pmatrix} V_{1_r} & \\ & V_{2_r} \end{pmatrix} = \begin{bmatrix} C_0 V_{1_r} & D_0 V_{2_r} \end{bmatrix}. \end{aligned} \quad (25)$$

In order to let reduced system have the companion form of second-order system, it then takes the following transformation by letting $H^{-1} = W_{1_r}^T V_{2_r}$,

$$\begin{aligned}\tilde{A} &= \begin{pmatrix} I & \\ & H^{-1}\end{pmatrix}\hat{A}\begin{pmatrix} I & \\ & H\end{pmatrix}\\ &= \begin{pmatrix} I & \\ & H^{-1}\end{pmatrix}\begin{pmatrix} 0 & W_{1_r}^T V_{2_r} \\ -W_{2_r}^T M^{-1} K V_{1_r} & -W_{2_r}^T M^{-1} G V_{2_r}\end{pmatrix}\begin{pmatrix} I & \\ & H\end{pmatrix}\\ &= \begin{pmatrix} 0 & I \\ -H^{-1}W_{2_r}^T M^{-1} K V_{1_r} & -H^{-1}W_{2_r}^T M^{-1} G V_{2_r} H\end{pmatrix},\\ \tilde{B} &= \begin{pmatrix} I & \\ & H^{-1}\end{pmatrix}\hat{B} = \begin{pmatrix} I & \\ & H^{-1}\end{pmatrix}\begin{bmatrix} 0 \\ W_{2_r}^T M^{-1} B_0\end{bmatrix}\\ &= \begin{bmatrix} 0 \\ H^{-1}W_{2_r}^T M^{-1} B_0\end{bmatrix},\\ \tilde{C} &= \hat{C}\begin{pmatrix} I & \\ & H\end{pmatrix} = \begin{bmatrix} C_0 V_{1_r} & D_0 V_{2_r}\end{bmatrix}\begin{pmatrix} I & \\ & H\end{pmatrix}\\ &= \begin{bmatrix} C_0 V_{1_r} & D_0 V_{2_r} H\end{bmatrix}.\end{aligned} \tag{26}$$

The corresponding algorithm is as follows:

Algorithm 2 (Balanced truncation method based on $\begin{pmatrix} \mathcal{P}_{11} & \\ & \mathcal{P}_{22}\end{pmatrix}$ and $\begin{pmatrix} \mathcal{Q}_{11} & \\ & \mathcal{Q}_{22}\end{pmatrix}$ [3]).

(1) Compute $\mathcal{P}_{11}$, $\mathcal{P}_{22}$, $\mathcal{Q}_{11}$, and $\mathcal{Q}_{22}$.

(2) Compute the balanced truncation matrices for $\mathcal{P}_{11}$ and $\mathcal{Q}_{11}$ to get $W_{1_r}, V_{1_r} \in \mathbb{R}^{n\times r}$, and balanced transformation matrices for $\mathcal{P}_{22}$ and $\mathcal{Q}_{22}$ to get $W_{2_r}, V_{2_r} \in \mathbb{R}^{n\times r}$.

(3) Perform projection to get reduced system $(\widehat{M}, \widehat{G}, \widehat{K}, \widehat{B}_0, \widehat{C}_0, \widehat{D}_0)$:

$$\begin{gathered}\widehat{M} = I_r, \quad \widehat{K} = H^{-1}\left(W_{2_r}^T M^{-1} K V_{1_r}\right),\\ \widehat{G} = H^{-1}\left(W_{2_r}^T M^{-1} G V_{2_r}\right)H,\\ \widehat{B}_0 = H^{-1}\left(W_{2_r}^T M^{-1} B_0\right), \quad \widehat{C}_0 = C_0 V_{1_r},\\ \widehat{D}_0 = (D_0 V_{2_r})H,\end{gathered} \tag{27}$$

where $H^{-1} = W_{1_r}^T V_{2_r}$.

3. Some New Algorithms Based on SVD

In this section, we propose some new algorithms based on SVD method directly.

First, $\mathcal{H}_2$ norm is an important quantity for bounding linear systems.

Lemma 3 (see [7]). *Given a first-order system*

$$\Sigma = \left(\begin{array}{c|c} A & B \\ \hline C & \end{array}\right), \tag{28}$$

suppose $\mathcal{P}$ and $\mathcal{Q}$ are reachability and observability Gramians. Then, $\mathcal{H}_2$ norm of the system can be expressed as

$$\|\Sigma\|_{\mathcal{H}_2}^2 = \operatorname{trace}\left\{C\mathcal{P}C^T\right\} = \operatorname{trace}\left\{B^T\mathcal{Q}B\right\}. \tag{29}$$

$\mathcal{H}_2$ norm for second-order system (1) with $D_0 = 0$ can then be stated as in the following proposition.

Proposition 4. *Given a second-order system* (1) *with $D_0 = 0$, suppose*

$$\Sigma = \left(\begin{array}{c|c} A & B \\ \hline C & \end{array}\right) \tag{30}$$

is the transformed first-order system, where

$$\begin{aligned} A &= \begin{bmatrix} 0 & I \\ -M^{-1}K & -M^{-1}G\end{bmatrix},\\ B &= \begin{bmatrix} 0 \\ M^{-1}B_0\end{bmatrix},\\ C &= \begin{bmatrix} C_0 & 0\end{bmatrix}.\end{aligned} \tag{31}$$

Assume reachability and observability Gramians $\mathcal{P}$ and $\mathcal{Q}$ are divided into four equal blocks, and $\mathcal{P} = \begin{bmatrix} \mathcal{P}_{11} & \mathcal{P}_{12} \\ \mathcal{P}_{12}^T & \mathcal{P}_{22}\end{bmatrix}$, $\mathcal{Q} = \begin{bmatrix} \mathcal{Q}_{11} & \mathcal{Q}_{12} \\ \mathcal{Q}_{12}^T & \mathcal{Q}_{22}\end{bmatrix}$. Then the following holds:

$$\begin{aligned}\|\Sigma\|_{\mathcal{H}_2}^2 &= \operatorname{trace}\left\{B_0^T M^{-T}\mathcal{Q}_{22}M^{-1}B_0\right\}\\ &= \operatorname{trace}\left\{C_0\mathcal{P}_{11}C_0^T\right\}.\end{aligned} \tag{32}$$

Proof. From Lemma 3,

$$\begin{aligned}\|\Sigma\|_{\mathcal{H}_2}^2 &= \operatorname{trace}\left\{B^T\mathcal{Q}B\right\}\\ &= \operatorname{trace}\left\{\begin{bmatrix} 0 & (M^{-1}B_0)^T\end{bmatrix}\begin{bmatrix} \mathcal{Q}_{11} & \mathcal{Q}_{12} \\ \mathcal{Q}_{12}^T & \mathcal{Q}_{22}\end{bmatrix}\begin{bmatrix} 0 \\ M^{-1}B_0\end{bmatrix}\right\}\\ &= \operatorname{trace}\left\{B_0^T M^{-T}\mathcal{Q}_{22}M^{-1}B_0\right\},\end{aligned} \tag{33}$$

$$\begin{aligned}\|\Sigma\|_{\mathcal{H}_2}^2 &= \operatorname{trace}\left\{\begin{bmatrix} C_0 & 0\end{bmatrix}\begin{bmatrix} \mathcal{P}_{11} & \mathcal{P}_{12} \\ \mathcal{P}_{12}^T & \mathcal{P}_{22}\end{bmatrix}\begin{bmatrix} C_0^T \\ 0\end{bmatrix}\right\}\\ &= \operatorname{trace}\left\{C_0\mathcal{P}_{11}C_0^T\right\}.\end{aligned} \tag{34}$$

□

From (8), we know the positions which are difficult to reach are spanned by the eigenvectors corresponding to small eigenvalues of $\mathcal{P}_{11}$. We would like to eliminate those positions in reduced systems. Proposition 4 shows that $\mathcal{P}_{11}$ can independently decide the system $\mathcal{H}_2$ norm. Therefore, we may get the reduced system by keeping the positions corresponding to r largest eigenvalues of $\mathcal{P}_{11}$. This motivates the following procedure for model order reduction. Suppose SVD of $\mathcal{P}_{11}$ is $\mathcal{P}_{11} = W_1 S V_1^T$, where $S = \text{diag}(\sigma_1, \sigma_2, \ldots, \sigma_n)$, and $\sigma_1 \geq \sigma_2 \geq \cdots \geq \sigma_n > 0$. So W_1, V_1 are orthogonal matrices and $W_1 = V_1$ since $\mathcal{P}_{11}$ is symmetric positive definite. Let the transformation matrices for the corresponding first-order system be

$$W = \begin{pmatrix} W_1 & \\ & W_1 \end{pmatrix}, \qquad V = W, \tag{35}$$

and the projection matrices be

$$W_k = \begin{pmatrix} W_{1_r} & \\ & W_{1_r} \end{pmatrix}, \qquad V_k = W_k, \tag{36}$$

where W_{1_r} consists of the first r columns of W_1. For simplicity, we denote W_k and V_k by W and V, respectively. Then use W and V as projection matrices for the corresponding first-order system, that is, perform projections as in (22) and (23). Same results can be applied to $\mathcal{Q}_{22}$. From the dual system, symmetrically $\mathcal{Q}_{11}$ and $\mathcal{P}_{22}$ are also crucial in weighting the system $\mathcal{H}_2$ norms independently. This results in the following algorithms.

Algorithm 5 (Second-order model reduction—SVD on $\mathcal{P}_{11}$ (or, $\mathcal{P}_{22}$, $\mathcal{Q}_{11}$, and $\mathcal{Q}_{22}$, resp.)).

(1) Compute $\mathcal{P}_{11}$ (or $\mathcal{P}_{22}$, $\mathcal{Q}_{11}$, and $\mathcal{Q}_{22}$, resp.), and denote it as P for simplicity.

(2) Take SVD on P and get $P = W_1 S_1 W_1^T$. Form the orthogonal projection matrices W_{1_r} which consists of the first r columns of W_1.

(3) Perform projection to get the reduced system $(\widehat{M}, \widehat{G}, \widehat{K}, \widehat{B}_0, \widehat{C}_0, \widehat{D}_0)$:

$$\begin{gathered} \widehat{M} = I_r, \qquad \widehat{K} = W_{1_r}^T M^{-1} K W_{1_r}, \\ \widehat{G} = W_{1_r}^T M^{-1} G W_{1_r}, \\ \widehat{B}_0 = W_{1_r}^T M^{-1} B_0, \qquad \widehat{C}_0 = C_0 W_{1_r}, \\ \widehat{D}_0 = D_0 W_{1_r}. \end{gathered} \tag{37}$$

Besides eliminating the positions which are difficult to reach in reduced system, it is also desirable to delete the velocities that are difficult to reach. From (9), these velocities are spanned by the eigenvectors of $\mathcal{P}_{22}$ corresponding to small eigenvalues. This motivates the following procedure for model order reduction. Suppose SVD of $\mathcal{P}_{11}$ is $\mathcal{P}_{11} = W_1 S_1 V_1^T$, where $S_1 = \text{diag}(\sigma_1, \sigma_2, \ldots, \sigma_n)$, $\sigma_1 \geq \sigma_2 \geq \cdots \geq \sigma_n > 0$. And SVD of $\mathcal{P}_{22}$ is $\mathcal{P}_{22} = W_2 S_2 V_2^T$, where $S_2 = \text{diag}(\delta_1, \delta_2, \ldots, \delta_n)$, $\delta_1 \geq \delta_2 \geq \cdots \geq \delta_n > 0$. So W_1, V_1, W_2, and V_2 are all orthogonal matrices, and $W_1 = V_1$, $W_2 = V_2$. Let the transformation matrices for the corresponding first-order system be

$$W = \begin{pmatrix} W_1 & \\ & W_2 \end{pmatrix}, \qquad V = W, \tag{38}$$

and the projection matrices be

$$W_k = \begin{pmatrix} W_{1_r} & \\ & W_{2_r} \end{pmatrix}, \qquad V_k = W_k, \tag{39}$$

where W_{1_r} and W_{2_r} consist of the first r columns of W_1 and W_2, respectively. By using similar idea to Algorithm 2, in order to let the reduced system $\widehat{A}$ have the companion form of second-order model, we adopt a matrix H such that $H^{-1} = W_{1_r}^T V_{2_r}$, and then perform the projections as in (26) and (27). Similar results can be applied to $\mathcal{Q}_{11}$ and $\mathcal{Q}_{22}$. This results in the following algorithm.

Algorithm 6 (Second-order model reduction—SVD on $\mathcal{P}_{11}$ and $\mathcal{P}_{22}$ (or $\mathcal{Q}_{11}$ and $\mathcal{Q}_{22}$, resp.)).

(1) Compute $\mathcal{P}_{11}$ and $\mathcal{P}_{22}$.

(2) Take SVD on $\mathcal{P}_{11}$ and $\mathcal{P}_{22}$ to get $\mathcal{P}_{11} = W_1 S_1 W_1^T$ and $\mathcal{P}_{22} = W_2 S_2 W_2^T$, respectively.

(3) Perform projection to get reduced system $(\widehat{M}, \widehat{G}, \widehat{K}, \widehat{B}_0, \widehat{C}_0, \widehat{D}_0)$:

$$\begin{gathered} \widehat{M} = I_r, \qquad \widehat{K} = H^{-1}\left(W_{2_r}^T M^{-1} K W_{1_r}\right), \\ \widehat{G} = H^{-1}\left(W_{2_r}^T M^{-1} G W_{2_r}\right) H, \\ \widehat{B}_0 = H^{-1}\left(W_{2_r}^T M^{-1} B_0\right), \qquad \widehat{C}_0 = C_0 W_{1_r}, \\ \widehat{D}_0 = \left(D_0 W_{2_r}\right) H, \end{gathered} \tag{40}$$

where $H^{-1} = W_{1_r}^T W_{2_r}$.

Now, consider Algorithm 5 and SISO second-order systems. Note that by taking SVD, $\mathcal{P}_{11} = W_1 S_1 W_1^T$. Denote $S_{11} = \text{diag}(\sigma_1, \sigma_2, \ldots, \sigma_r)$ and $S_{12} = \text{diag}(\sigma_{r+1}, \sigma_{r+2}, \ldots, \sigma_n)$. So $S_1 = \text{diag}(S_{11}, S_{12})$. Suppose $z := W_1^T C_0^T$, that is,

$$\begin{pmatrix} z_1 \\ z_2 \\ \vdots \\ z_n \end{pmatrix} = W_1^T \begin{pmatrix} c_1 \\ c_2 \\ \vdots \\ c_n \end{pmatrix}. \tag{41}$$

Then from Proposition 4,

$$\begin{aligned} \|\Sigma\|_{\mathcal{H}_2}^2 &= C_0 \mathcal{P}_{11} C_0^T \\ &= C_0 W_1 S W_1^T C_0^T = z^T S z \\ &= \sigma_1 z_1^2 + \sigma_2 z_2^2 + \cdots + \sigma_n z_n^2. \end{aligned} \tag{42}$$

Table 1: Errors of reduced models produced by four different methods.

Model	n	r	m	p	$\|\mathcal{H}\|_2$	$\|\varepsilon_{\text{SVD1}}\|_2/\|\mathcal{H}\|_2$	$\|\varepsilon_{\text{SVD12}}\|_2/\|\mathcal{H}\|_2$	$\|\varepsilon_{\text{BT1}}\|_2/\|\mathcal{H}\|_2$	$\|\varepsilon_{\text{BT12}}\|_2/\|\mathcal{H}\|_2$
Building model	24	4	1	1	4.5e−03	4.1e−01	5.1e−01	3.2e−01	3.2e−01
Clamped beam	174	29	1	1	3.3e+02	3.3e−02	1.3e−01	1.3e−03	5.7e−04
International state space	135	27	3	3	1.0e−04	6.7e−04	1.0e−04	3.2e−06	3.2e−06

In reduced system, $\widehat{C}_0 = C_0 W_{1_r}$, $\widehat{\mathcal{P}}_{11} = S_1$. Therefore,

$$\begin{aligned}\|\Sigma_{\text{red}}\|_{\mathcal{H}_2}^2 &= \widehat{C}_0 \widehat{\mathcal{P}}_{11} \widehat{C}_0^T \\ &= (C_0 W_{1_r}) S_1 \left(W_{1_r}^T C_0^T\right) \\ &= \sigma_1 z_1^2 + \sigma_2 z_2^2 + \cdots + \sigma_r z_r^2.\end{aligned} \tag{43}$$

In many cases, the eigenvalues of $\mathcal{P}_{11}$ decay rapidly. Therefore, the reduced system produced by Algorithm 5 takes the main part of the system $\mathcal{H}_2$ norm. Actually we also have,

$$\begin{aligned}\|\Sigma\|_{\mathcal{H}_2}^2 - \|\Sigma_{\text{red}}\|_{\mathcal{H}_2}^2 &= \sigma_{r+1} z_{r+1}^2 + \sigma_{r+2} z_{r+2}^2 + \cdots + \sigma_n z_n^2 \\ &\le \sigma_{r+1}\|z\|_2^2 = \sigma_{r+1}\|C_0\|_2^2.\end{aligned} \tag{44}$$

This is not the $\mathcal{H}_2$-norm of error system $\|\Sigma - \Sigma_{\text{red}}\|_{\mathcal{H}_2}$, but it gives some information to support Algorithm 5. This result for SISO second-order systems can easily extends to MIMO systems. Similar results can be applied to methods in Algorithm 6.

Now, consider complexity of the algorithms. In our SVD method on $\mathcal{P}_{11}$ in Algorithm 5, after computing $\mathcal{P}_{11}$ which takes $O(n^3)$ time, it then uses one-time $O(n^3)$ operation which is SVD on $\mathcal{P}_{11}$ to get the projection matrices. In our Algorithm 6, after computing $\mathcal{P}_{11}$ and $\mathcal{P}_{22}$ each of which takes $O(n^3)$ time, it then uses twice $O(n^3)$ operations which are SVD on $\mathcal{P}_{11}$ and SVD on $\mathcal{P}_{22}$ to get the projection matrices. In balanced truncation method in Algorithm 1, after computing $\mathcal{P}_{11}$ and $\mathcal{Q}_{11}$, it also has one-time SVD. Besides this, it uses two Cholesky factorizations $\mathcal{P}_{11} = U_C U_C^T$ and $\mathcal{Q}_{11} = L_O L_O^T$ which all take time $O(n^3)$ in order to get the projection matrices. Note that Algorithm 2 has more flops than Algorithm 1. Therefore, our SVD algorithms are simpler in time cost than existing balanced truncation methods. There are some other techniques in getting the balanced truncation systems, for example, balance-free truncation [8]. But it is not hard to see that they are also more complex than our SVD methods. This makes our SVD methods more suitable to solve large-scale problems.

4. Computational Results

In this section, we apply the model reduction methods to three numerical examples: the building model, the clamped beam model, and international space station model, see [9] for detailed descriptions. In Table 1, four model reduction methods are compared: SVD method on $\mathcal{P}_{11}$ in Algorithm 5 ("SVD1"), SVD method on $\mathcal{P}_{11}$ and $\mathcal{P}_{22}$ in Algorithm 6 ("SVD12"), balanced truncation on $\mathcal{P}_{11}$ and $\mathcal{Q}_{11}$ in Algorithm 1 ("BT1"), and balanced truncation on $\mathcal{P}_{11}$ and $\mathcal{Q}_{11}$, $\mathcal{P}_{22}$ and $\mathcal{Q}_{22}$ in Algorithm 5 ("BT12"). The comparison is based on the relative error of $\mathcal{H}_2$ norm, where

$$\frac{\|\varepsilon\|_2}{\|\mathcal{H}\|_2} =: \frac{\left\|\Sigma - \widehat{\Sigma}\right\|_{\mathcal{H}_2}}{\|\Sigma\|_{\mathcal{H}_2}}, \tag{45}$$

n is the size of M, r is the size of $\widehat{M}$ in reduced model, and m is the size of input vector $u(t)$, p is the size of output vector $y(t)$. Figures 1 and 2 show the amplitude of frequency response of original and reduced systems for clamped beam model for $r = 20$ and $r = 29$, respectively.

5. Error Bounds and Discussions

Error bounds for first-order model reduction do not apply for second-order systems. Consider second-order system (1) with $D_0 = 0$:

$$\Sigma : \begin{cases} M\ddot{q}(t) + G\dot{q}(t) + Kq(t) = B_0 u(t) \\ y(t) = C_0 q(t). \end{cases} \tag{46}$$

Suppose the reduced system $\widehat{\Sigma}$ is obtained by keeping r largest eigenvalues of the orthogonal eigenspace of $\mathcal{P}_{11}$. Sorensen and Antoulas in [10] provided a priori error bound showing that the $\mathcal{H}_2$ norm of error system is bounded by a constant times the summation of neglected singular values of $\mathcal{P}_{11}$.

It considers structured systems in [10]. Let $Q(s)$ and $P(s)$ be polynomial matrices in s:

$$\begin{aligned} Q(s) &= \sum_{j=1}^{r} Q_j s^j, \qquad Q_j \in \mathbb{R}^{n\times n}, \\ P(s) &= \sum_{j=1}^{r-1} P_j s^j, \qquad sP_j \in \mathbb{R}^{n\times m}, \end{aligned} \tag{47}$$

where Q is invertible, and $Q^{-1}P$ is a strictly proper rational matrix. Denote by $Q(d/dt)$, $P(d/dt)$ the differential operators

$$Q\left(\frac{d}{dt}\right) = \sum_{j=1}^{r} Q_j \frac{d^j}{dt^j}, \qquad P\left(\frac{d}{dt}\right) = \sum_{j=1}^{r-1} P_j \frac{d^j}{dt^j}. \tag{48}$$

The structured systems are defined by the following equations:

$$\Sigma : \begin{cases} Q\left(\frac{d}{dt}\right)x(t) = P\left(\frac{d}{dt}\right)u(t) \\ y(t) = Cx(t), \end{cases} \tag{49}$$

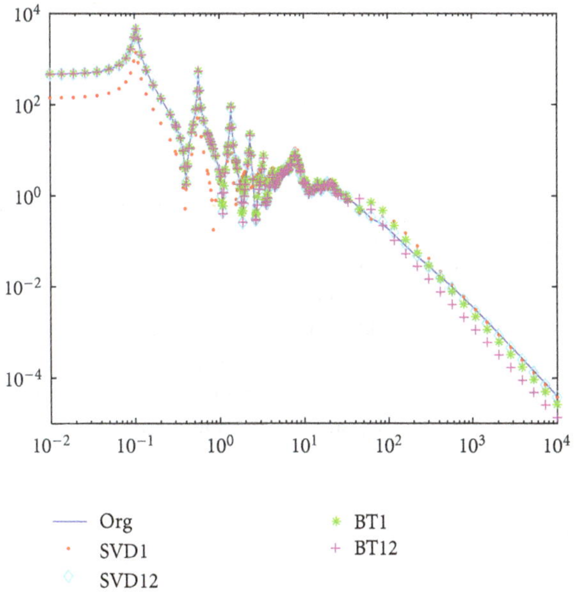

FIGURE 1: Frequency response on clamped beam model with size $n = 174$ and the reduced model of size $r = 20$ by four different methods.

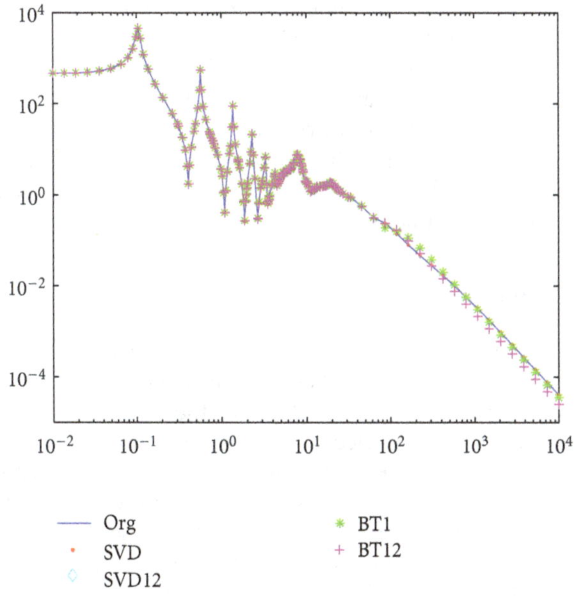

FIGURE 2: Frequency response on clamped beam model with size $n = 174$ and the reduced model of size $r = 29$ by four different methods.

where $C \in \mathbb{R}^{p\times n}$. The Gramian is defined as the Gramian of $x(t)$ when the input is an impulse:

$$\mathcal{P} := \int_0^\infty x(t)x(t)^* dt. \quad (50)$$

Let

$$\mathcal{P} = VSV^* \quad (51)$$

be the eigensystem of $\mathcal{P}$, where V is orthogonal, the diagonal elements of S are in decreasing order:

$$S = \begin{pmatrix} \sigma_1 & & & \\ & \sigma_2 & & \\ & & \ddots & \\ & & & \sigma_n \end{pmatrix}, \quad (52)$$

where $\sigma_1 \ge \sigma_2 \ge \cdots \ge \sigma_n$. Partition $V = [V_1,\ V_2]$ where V_1 consists of the first r columns of V. The reduced model is derived from

$$\hat{Q}_j = V_1^* Q_j V_1, \qquad \hat{P}_j = V_1^* P_j, \qquad \hat{C} = CV_1, \quad (53)$$

and the reduced system is then

$$\hat{\Sigma} : \begin{cases} \hat{Q}\left(\dfrac{d}{dt}\right)\hat{x}(t) = \hat{P}\left(\dfrac{d}{dt}\right)u(t) \\ \hat{y}(t) = \hat{C}\hat{x}(t). \end{cases} \quad (54)$$

Partition accordingly,

$$[C_1, C_2] = CV, \qquad S = \begin{pmatrix} S_1 & \\ & S_2 \end{pmatrix}. \quad (55)$$

Theorem 7 (see[10]). *The reduced model $\hat{\Sigma}$ derived from the dominant eigenspace of the Gramian $\mathcal{P}$ for Σ as described above satisfies*

$$\left\|\Sigma - \hat{\Sigma}\right\|_{\mathcal{H}_2}^2 \le \operatorname{trace}\{C_2 S_2 C_2^*\} + \mathcal{K}\operatorname{trace}\{S_2\}, \quad (56)$$

where $\mathcal{K}$ is a modest constant depending on Σ and $\hat{\Sigma}$, and the diagonal elements of S_2 are the neglected smallest eigenvalues of $\mathcal{P}$.

Specially for second-order system (1) with $D_0 = 0$:

$$\Sigma : \begin{cases} M\ddot{q}(t) + G\dot{q}(t) + Kq(t) = B_0 u(t) \\ y(t) = C_0 q(t), \end{cases} \quad (57)$$

it is easy to see that it can be described by (49) with $Q(s) = Ms^2 + Gs + K$ and $P(s) = B_0$. So there is the following corollary.

Corollary 8. *For second-order system (57), suppose $\mathcal{P}_{11} = \begin{pmatrix} S_1 & \\ & S_2 \end{pmatrix}$ is diagonal, where $S_1 = \operatorname{diag}(\sigma_1, \sigma_2, \ldots, \sigma_r)$, $S_2 = \operatorname{diag}(\sigma_{r+1}, \sigma_{r+2}, \ldots, \sigma_n)$, and $\sigma_1 \ge \sigma_2 \ge \cdots \ge \sigma_n$. Suppose the reduced system $\hat{\Sigma}$ is obtained by keeping r largest eigenvalues of $\mathcal{P}_{11}$ as described in (53) and (54). Then*

$$\begin{aligned} \|\Sigma_{err}\|_{\mathcal{H}_2}^2 &= \left\|\Sigma - \hat{\Sigma}\right\|_{\mathcal{H}_2}^2 \le c_0 \operatorname{trace}\{S_2\} \\ &= c_0(\sigma_{r+1} + \sigma_{r+2} + \ldots + \sigma_n), \end{aligned} \quad (58)$$

where c_0 is a modest constant depending on Σ and $\hat{\Sigma}$.

Now, we explain the corollary and what the constant c_0 is. Note $Q(s) = Ms^2 + Gs + K \in \mathbb{C}^{n\times n}$ and $Q(s) = \begin{bmatrix} Q_{11}(s) & Q_{12}(s) \\ Q_{21}(s) & Q_{22}(s) \end{bmatrix}$ with $Q_{11}(s) \in \mathbb{C}^{r\times r}$. Let $W(s) = Q_{11}(s)^{-1}Q_{12}(s)$. Partition accordingly $C_0 = [C_{0_1}, \ C_{0_2}]$. Then,

$$c_0 = \sup_{\omega\in\mathbb{R}} \left\| (C_{0_1} W(i\omega))^* (C_{0_1} W(i\omega) - 2C_{0_2}) \right\|_2 + \|C_{0_2}\|_2^2. \tag{59}$$

If the reduced system has no poles on the imaginary axis, $\sup_{\omega\in\mathbb{R}} \|W(i\omega)\|_2$ is finite, and so $\sup_{\omega\in\mathbb{R}} \|(C_{0_1} W(i\omega))^*(C_{0_1} W(i\omega) - 2C_{0_2})\|_2$ is finite. Therefore, c_0 is a constant depending on Σ and $\hat{\Sigma}$. There is a similar result for $\mathcal{Q}_{11}$ by duality.

Consider Algorithm 5, SVD method on $\mathcal{P}_{11}$. Note that

$$\mathcal{P}_{11} = \int_0^\infty q(t)q^*(t)\,dt, \tag{60}$$

and $\mathcal{P}_{11} = VSV^*$, $[V_1, V_2] = V$. So the reduction rules in Algorithm 5 are exactly the same as (53). Similar result applies for $\mathcal{Q}_{11}$. Therefore, the error systems produced by our Algorithm 5, SVD method on $\mathcal{P}_{11}$ or $\mathcal{Q}_{11}$, all have good global error bounds. So far, there is no priori error bound for error systems based on $\mathcal{P}_{22}$ or $\mathcal{Q}_{22}$.

Even though balanced truncation methods in Algorithms 1 and 2, and SVD method on both $\mathcal{P}_{11}$ and $\mathcal{P}_{22}$ (or $\mathcal{Q}_{11}$ and $\mathcal{Q}_{22}$) in Algorithm 6, all have good performance in those three numerical examples, so far no priori error bounds are provided for these methods.

For second-order system (1) with $D_0 = 0$, the following theorem provides some observations on structures of gramians $\mathcal{P}$ and $\mathcal{Q}$ in corresponding first-order system.

Theorem 9. *Given second-order system* (1) *with $D_0 = 0$, then in any coordinate system, there is no state space transformation which gives a block diagonal $\mathcal{Q}$.*

Proof. From the observability Lyapunov equation

$$A^T\mathcal{Q} + \mathcal{Q}A + C^TC = 0, \tag{61}$$

one can get

$$\begin{pmatrix} 0 & -K_M^T \\ I & -G_M^T \end{pmatrix}\begin{pmatrix} \mathcal{Q}_{11} & \mathcal{Q}_{12} \\ \mathcal{Q}_{12}^T & \mathcal{Q}_{22} \end{pmatrix} + \begin{pmatrix} \mathcal{Q}_{11} & \mathcal{Q}_{12} \\ \mathcal{Q}_{12}^T & \mathcal{Q}_{22} \end{pmatrix}\begin{pmatrix} 0 & I \\ -K_M & -G_M \end{pmatrix} + \begin{bmatrix} C_0^T \\ 0 \end{bmatrix}\begin{bmatrix} C_0 & 0 \end{bmatrix} = 0. \tag{62}$$

Equating each block gives

$$\begin{aligned} -K_M^T\mathcal{Q}_{12}^T - \mathcal{Q}_{12}K_M + C_0^TC_0 &= 0, \\ -K_M^T\mathcal{Q}_{22} + \mathcal{Q}_{11} - \mathcal{Q}_{12}G_M &= 0, \\ \mathcal{Q}_{12} - G_M^T\mathcal{Q}_{22} + \mathcal{Q}_{12}^T - \mathcal{Q}_{22}G_M &= 0. \end{aligned} \tag{63}$$

From the first equation in (63), one can get $\mathcal{Q}_{12}$ is not zero, otherwise $C_0 = 0$, and then from (61) the solution for this Lyapunov equation is $\mathcal{Q} = \int_0^\infty e^{A^T\tau}C^TCe^{A\tau}\,d\tau = 0$. This contradicts with $\mathcal{Q}$ being symmetric positive definite. So $\mathcal{Q}$ is not block diagonal. □

Algorithm 2 [3], that is, balanced truncation method on $(\mathcal{P}_{11}, \mathcal{P}_{22})$ and $(\mathcal{Q}_{11}, \mathcal{Q}_{22})$, is to balance both $\mathcal{P}_{11}$ and $\mathcal{Q}_{11}$ to get the projection matrices W_{1r} and V_{1r} by keeping r largest eigenvalues of $\mathcal{P}_{11}\mathcal{Q}_{11}$, balance both $\mathcal{P}_{22}$ and $\mathcal{Q}_{22}$ to get the projection matrices W_{2r} and V_{2r}, then use $W = \begin{pmatrix} W_{1r} & \\ & W_{2r} \end{pmatrix}$ and $V = \begin{pmatrix} V_{1r} & \\ & V_{2r} \end{pmatrix}$ as projection matrices for the corresponding first-order system. This is equivalent to assuming both $\mathcal{P}$ and $\mathcal{Q}$ are block diagonal in corresponding first-order system, that is,

$$\mathcal{P} = \begin{pmatrix} \mathcal{P}_{11} & \\ & \mathcal{P}_{22} \end{pmatrix}, \qquad \mathcal{Q} = \begin{pmatrix} \mathcal{Q}_{11} & \\ & \mathcal{Q}_{22} \end{pmatrix}, \tag{64}$$

in order to keep r largest eigenvalues of $\mathcal{P}_{11}\mathcal{Q}_{11}$ and $\mathcal{P}_{22}\mathcal{Q}_{22}$ in reduced corresponding first-order system. Similar results also apply for methods in Algorithm 6, SVD method on $\mathcal{Q}_{11}$ and $\mathcal{Q}_{22}$. From Theorem 9, $\mathcal{Q}$ can not be block diagonal, therefore the reduced system obtained from Algorithm 2 or Algorithm 6 may lose some information. But Algorithm 1 [5] (balanced truncation based on $(\mathcal{P}_{11}, \mathcal{Q}_{11})$ (or $(\mathcal{P}_{22}, \mathcal{Q}_{22})$ resp.)) and Algorithm 5 (SVD on $\mathcal{P}_{11}$ (or $\mathcal{P}_{22}$, $\mathcal{Q}_{11}$, $\mathcal{Q}_{22}$ resp.)), avoid this drawback in assuming $\mathcal{Q}$ being block diagonal, but they only work for either position Gramians or velocity Gramians, while Algorithms 2 and 6 have the advantage in taking into account both position Gramians and velocity Gramians.

6. Conclusions

Two Gramian based second-order model reduction methods are proposed in this paper based on SVD on diagonal blocks of Gramians. Even though they are the well known SVD methods, to our knowledge, they are the simplest ways in Gramian based second-order model reduction. Two existing balanced truncation methods [3, 5] were presented in 2006 and 2008, respectively. When compared to these two existing techniques, our SVD methods are competitive in numerical examples and more suitable for large-scale setting problems. By using the result given in [10], there is global error bound for one method, SVD on $\mathcal{P}_{11}$ or $\mathcal{Q}_{11}$. Except this, so far no priori error bounds are provided for other methods on second-order model reduction, even through they work good in numerical examples. Interesting future work would be to provide global error bounds for these methods.

Acknowledgment

The paper is supported in part by NSFC research Project 61070230, Shandong Social Science Program of China 11CKJJ16, and Grant no. 20110205 by Jinan Science and Technology Bureau of China for Returned Oversea Scholars.

References

[1] K. Glover, "All optimal Hankel-norm approximations of linear multivariable systems and their L^∞ error bounds," *International Journal of Control*, vol. 39, no. 6, pp. 1115–1193, 1984.

[2] D. G. Meyer and S. Srinivasan, "Balancing and model reduction for second-order form linear systems," *IEEE Transactions on Automatic Control*, vol. 41, no. 11, pp. 1632–1644, 1996.

[3] Y. Chahlaoui, D. Lemonnier, A. Vandendorpe, and P. Van Dooren, "Second-order balanced truncation," *Linear Algebra and Its Applications*, vol. 415, no. 2-3, pp. 373–384, 2006.

[4] Y. Chahlaoui, K. Gallivan, A. Vandendorpe, and P. Van Dooren, "Model reduction of second-order systems," in *Dimension Reduction of Large-Scale Systems*, P. Benner, V. Mehrmann, and D. Sorensen, Eds., vol. 45 of *Lecture Notes in Computational Science and Engineering*, pp. 149–170, Springer, Berlin, Germany, 2005.

[5] T. Reis and T. Stykel, "Balanced truncation model reduction of second-order systems," *Mathematical and Computer Modelling of Dynamical Systems*, vol. 14, no. 5, pp. 391–406, 2008.

[6] A. J. Laub, M. T. Heath, C. C. Paige, and R. C. Ward, "Computation of system balancing transformations and other applications of simultaneous diagonalization algorithms," *IEEE Transactions on Automatic Control*, vol. 32, pp. 115–122, 1987.

[7] A. C. Antoulas, *Approximation of Large-Scale Dynamical Systems, Advances in Design and Control*, SIAM, Philadelphia, PA, USA, 2005.

[8] M. G. Safanov and R. Y. Chiang, "A Schur method for balance-trucation model reduction," *IEEE Transactions on Automatic Control*, vol. 34, no. 7, pp. 729–733, 1989.

[9] Y. Chahlaoui and P. Van Dooren, "Benchmark examples for model reduction of linear time-invariant dynamical systems," in *Dimension Reduction of Large-Scale Systems*, P. Benner, V. Mehrmann, and D. Sorensen, Eds., vol. 45 of *Lecture Notes in Computational Science and Engineering*, pp. 381–395, Springer, Berlin, Germany, 2005.

[10] D. C. Sorensen and A. C. Antoulas, "Gramians of structured systems and an error bound for structure-preserving model reduction," in *Dimension Reduction of Large-Scale Systems*, P. Benner, V. Mehrmann, and D. Sorensen, Eds., vol. 45 of *Lecture Notes in Computational Science and Engineering*, pp. 117–130, Springer, Berlin, Germany, 2005.

Distributed Model Predictive Control of the Multi-Agent Systems with Improving Control Performance

Wei Shanbi, Chai Yi, and Li Penghua

College of Automation, Chongqing University, Chongqing 400044, China

Correspondence should be addressed to Wei Shanbi, wsbmei@yahoo.com.cn

Academic Editor: Yugeng Xi

This paper addresses a distributed model predictive control (DMPC) scheme for multiagent systems with improving control performance. In order to penalize the deviation of the computed state trajectory from the assumed state trajectory, the deviation punishment is involved in the local cost function of each agent. The closed-loop stability is guaranteed with a large weight for deviation punishment. However, this large weight leads to much loss of control performance. Hence, the time-varying compatibility constraints of each agent are designed to balance the closed-loop stability and the control performance, so that the closed-loop stability is achieved with a small weight for the deviation punishment. A numerical example is given to illustrate the effectiveness of the proposed scheme.

1. Introduction

Interests in the cooperative control of multiagent systems have been growing significantly over the last years. The main motivation is the wide range of military and civilian applications, including formation flight of UAV and automated traffic systems. Compared with the traditional approach, model predictive control (MPC), or receding horizon control (RHC) has the ability to redefine cost functions and constraints as needed to reflect changes in the system and/or the environment. Therefore, MPC is extensively applied to the cooperative control of multiagent systems, which makes the agents operate close to the constraint boundaries and obtain better performance than traditional approaches [1–3]. Moreover, due to the computational advantages and the convenience of communication, distributed MPC (DMPC) is recognized as a nature technique to address trajectory optimization problems for multiagent systems.

One of the challenges for distributed control is to ensure that local control actions keep consistent with the actions of others agents [4, 5]. For the coupled systems, the local optimization problem is solved based on the states of its neighbors' at sample time instant using Nash-optimization technique in [6]. As the local controllers lack of communication and cooperation, the local control actions cannot keep consistent [7, 8]. require each local controller exchange information with all other local controllers to improve optimality and consistency based on sufficient communication. For the decoupled systems [9], exploits the estimation of the prediction state trajectories of the neighbors' [10]; treats the prediction state trajectories of the neighbor agents as bounded disturbance where a min-max optimal problem is solved for each agent with respect to the worst-case disturbance. In [11, 12], the optimal variables of the local optimization problem contain the control action of its own and its neighbors' which are coupled in collision avoidance constraints and cost function. Obviously, the deviation between the actions of what the agent is actually doing and of what its neighbor estimates for it affects the control performance. Sometimes the consistency and collision avoidance cannot be achieved, and the feasibility and stability of this scheme cannot be guaranteed [13]. proposes a distributed MPC with a fixed compatibility constraint to restrict the deviation. When the bound of this constraint is sufficiently small, the closed-loop system state enter a neighborhood of the objective state [14, 15] give an improvement over [13] by adding *deviation punishment term* to penalize the deviation of the computed state trajectory from the assumed state trajectory. Closed-loop exponential stability follows if the weight on the deviation function term is large enough. But the large weight leads to the loss of the control performance.

A contribution in this paper is to propose an idea to reduce the adverse effect of the deviation punishment on the control performance. At each sample time, the value of compatibility constraint is set as the maximum value of the deviation of the previous sample time. We give the stability condition to guarantee the exponential stability of the global closed-loop system with a small weight on the deviation punishment term, which is obtained by dividing the centralize stability constraint as the manner of [16, 17]. The effectiveness of the scheme is also demonstrated by a numerical example.

Notations. x_k^i is the value of vector x^i at time k · $x_{k,t}^i$ is the value of vector x^i at a future time $k+t$, predicted at time k · $|x| = [|x_1|, |x_2|, \dots, |x_N|]$ is the absolute value for each component of x. For a vector x and positive-definite matrix Q, $\|x\|_Q^2 = x^T Q x$.

2. Problem Statement

Let us consider a system which is composed of N_a agents. The dynamics of agent i [11] is

$$x_{k+1}^i = f^i\left(x_k^i, u_k^i\right), \quad (1)$$

where $u_k^i \in \mathbb{R}^{m_i}$, $x_k^i \in \mathbb{R}^{n_i}$, and $f^i : \mathbb{R}^{n_i} \times \mathbb{R}^{m_i} \mapsto \mathbb{R}^{n_i}$, are the input, state, and state transition function of agent i, respectively. $u_k^i = [u_k^{i,1}, \dots, u_k^{i,m_i}]^{\mathrm{T}}$, $x_k^i = [x_k^{i,1}, \dots, x_k^{i,n_i}]^{\mathrm{T}}$. The sets of feasible input and state of agent i are denoted as $\mathcal{U}^i \subset \mathbb{R}^{m_i}$ and $\mathcal{X}^i \subset \mathbb{R}^{n_i}$, respectively, that is,

$$u_k^i \in \mathcal{U}^i, x_k^i \in \mathcal{X}^i, \quad k \geq 0. \quad (2)$$

At each time k, the control objective is [18] to minimize

$$J_k = \sum_{t=0}^{\infty} \left[\|x_{k,t}\|_Q^2 + \|u_{k,t}\|_R^2 \right] \quad (3)$$

with respect to $u_{k,t}$, $t \geq 0$, where $x = [(x^1)^{\mathrm{T}}, \dots, (x^{N_a})^{\mathrm{T}}]^{\mathrm{T}}$, $u = [(u^1)^{\mathrm{T}}, \dots, (u^{N_a})^{\mathrm{T}}]^{\mathrm{T}}$; $x_{k,t+1}^i = f^i(x_{k,t}^i, u_{k,t}^i)$, $x_{k,0}^i = x_k^i$; $Q = Q^{\mathrm{T}} > 0$, $R = R^{\mathrm{T}} > 0$. $u \in \mathbb{R}^m$, $m = \sum_i m_i$, and $x \in \mathbb{R}^n$, $n = \sum_i n_i$. Then,

$$x_{k+1} = f(x_k, u_k), \quad (4)$$

where $f = [f^1, f^2, \dots, f^{N_a}]^{\mathrm{T}}$, $f : \mathbb{R}^n \times \mathbb{R}^m \mapsto \mathbb{R}^n$. (x_e^i, u_e^i) is the equilibrium point of agent i, and (x_e, u_e) is the corresponding equilibrium point of all agents. $\mathcal{X} = \mathcal{X}^1 \times \mathcal{X}^2 \times \cdots \times \mathcal{X}^{N_a}$, $\mathcal{U} = \mathcal{U}^1 \times \mathcal{U}^2 \times \cdots \times \mathcal{U}^{N_a}$. The models for all agents are completely decoupled. The coupling between agents arises due to the fact that they operate in the same environment, and that the "cooperative" objective is imposed on each agent by the cost function. Hence, there are the coupling cost function and coupling constraints [19]. The coupling constraints can be transformed to coupling cost function term directly or handled as decoupling constraints using the technique of [15]. In the present paper we will not consider this issue.

The *control objective* for all system is to cooperatively asymptotically stabilize all agents to an equilibrium point (x_e, u_e) of (4). In this paper we assumed that the $(x_e, u_e) = (0,0)$, $f(x_e, u_e) = 0$. The corresponding equilibrium point for each agent is $(x_e^i, u_e^i) = (0,0)$, $f^i(x_e^i, u_e^i) = 0$. Assumption $f^i(0,0) = 0$ is not restrictive, since if $(x_e^i, u_e^i) \neq (0,0)$, one can always shift the origin of the system to it.

The resultant control law for minimization of (3) can be implemented in a centralized way. However, the existing methods for centralized MPC are only computationally tractable for small-scale system. Furthermore, the communication cost of implementing a centralized receding horizon control law may be costly. Hence, by means of decomposition, J_k is divided as J_k^i's such that the minimization of (3) is implemented in distributed manner, with

$$J_k^i = \sum_{t=0}^{\infty} \left[\|z_{k,t}^i\|_{\overline{Q}_i}^2 + \|u_{k,t}^i\|_{\overline{R}_i}^2 \right], \qquad J_k = \sum_{i=1}^{N_a} J_k^i, \quad (5)$$

where $z_{k,t}^i = [(x_{k,t}^i)^{\mathrm{T}} (x_{k,t}^{-i})^{\mathrm{T}}]^{\mathrm{T}}$; $x_{k,t}^{-i}$ includes the states of the neighbors. The set of neighbors' of agent i is denoted as $\mathcal{N}_i$. $x_k^{-i} = \{x_k^j \mid j \in \mathcal{N}_i\}$, $x_k^{-i} \in \mathbb{R}^{n^{-i}}$, $n^{-i} = \sum_{j \in \mathcal{N}_i} n^j$. For each agent i, the control objective is to stabilize it to the equilibrium point (x_e^i, u_e^i) · $\overline{Q}_i = \overline{Q}_i^{\mathrm{T}} > 0$, $\overline{R}_i = \overline{R}_i^{\mathrm{T}} > 0$. $\overline{Q}_i$ is obtained by dividing Q using the technique of [19]. For the agents that have decoupled dynamics, the couplings of control moves for all system are not considered. R is a diagonal matrix and $\overline{R}_i$ is directly obtained.

Under the networked environment, the bandwidth limitation can restrict the amount of information exchange [17]. It is thus appropriate to allow agents to exchange information only once in each sampling interval. We assume that the connectivity of the interagent communication network is sufficient for agents to obtain information regarding all the variables that appear in their local problems.

In the receding horizon control manner, a finite-horizon cost function is exploited to approximate J_k^i. According to the (5), the evolution of the control moves with predictive horizon for agent i is based on the estimation of the state trajectories $x_{k,t}^{-i}, t \leq N$ of the neighbors', which are substituted by the assumed state trajectories $\hat{x}_{k,t}^{-i}, t \leq N$ as [11]. In each control interval, the transmitted information between agents is the assumed state trajectories. As the cooperative consistency and efficiency of distributed control moves is affected for the existence of the deviation of the computed state trajectory from the assumed state trajectory, it is appreciate to penalize it by adding the deviation punishment term into the local cost function.

Define

$$u_{k,t}^i = F_i(k) x_{k,t}^i, \qquad \forall t \geq N. \quad (6)$$

$F_i(k)$ is the gain of distributed state feedback controller.

Consider

$$\begin{aligned} \tilde{J}_k^i = & \sum_{t=0}^{N-1}\left[\|\hat{z}_{k,t}^i\|_{\overline{Q}_i}^2 + \|u_{k,t}^i\|_{\overline{R}_i}^2 + \|x_{k,t}^i - \hat{x}_{k,t}^i\|_{\overline{T}_i}^2\right] \\ & + \sum_{t=N}^{\infty}\left[\|x_{k,t}^i\|_{Q_i}^2 + \|u_{k,t}^i\|_{R_i}^2\right], \end{aligned} \tag{7}$$

where

$$\hat{z}_{k,t}^i = \left[\left(x_{k,t}^i\right)^{\mathrm{T}}\left(\hat{x}_{k,t}^{-i}\right)^{\mathrm{T}}\right]^{\mathrm{T}}, \quad \hat{x}_{k,0}^i = x_k^i, \tag{8}$$

$\hat{x}_{k,t}^{-i}$ includes the assumed states of the neighbors. $Q_i = Q_i^{\mathrm{T}} > 0$ and $R_i = R_i^{\mathrm{T}} = \overline{R}_i$ satisfy

$$\operatorname{diag}\{Q_1, Q_2, \ldots, Q_{N_a}\} \geq Q, \qquad \operatorname{diag}\{R_1, R_2, \ldots, R_{N_a}\} = R. \tag{9}$$

Obviously, Q_i is designed to stabilize the agent i to the local equilibrium point, independently. $\overline{Q}_i$ is designed to stabilize the agent i to the local equilibrium point with neighbor agents, cooperatively. $\overline{T}_i$ is the weight on the deviation punishment term, to penalize the deviation of the computed state trajectory from the assumed state trajectory.

At each time k, the optimization problem for distributed MPC is transformed as:

$$\min_{\overline{U}_k^i, F_i(k)} \tilde{J}_k^i, s.t. (1), (2), (6), (7). \tag{10}$$

$\overline{U}_k^{*i} = [(u_{k,0}^{*i})^{\mathrm{T}}, (u_{k,1}^{*i})^{\mathrm{T}}, \ldots, (u_{k,N-1}^{*i})^{\mathrm{T}}]^{\mathrm{T}}$, only when $u_k^{*i} = u_{k,0}^{*i}$ is implemented, and the problem (9) is solved again at time $k+1$.

Remark 1. The local deviate punishment by each agent effects the control performance, that is, incurs the loss of optimality.

3. Stability of Distributed MPC

The stability of distributed MPC by simply applying the procedure as in the centralized MPC will be discussed. The compact and convex terminal set Ω^i is defined

$$\Omega^i = \left\{x^i \in \mathbb{R}^{n_i} \mid \left(x^i\right)^{\mathrm{T}} P_i x^i \leq \alpha^i\right\}, \tag{11}$$

where $P_i > 0$, $\alpha^i > 0$ are specified such that Ω^i is a control invariant set. So using the idea of [20, 21], one simultaneously determines a linear feedback such that Ω^i is a positively invariant under this feedback.

Define the local linearization at the equilibrium point

$$A_i = \frac{\partial f^i}{\partial x^i}(0,0), \qquad B_i = \frac{\partial f^i}{\partial u^i}(0,0). \tag{12}$$

and assume that (A_i, B_i) is stabilizable. When $x_{k,N+t}^i$, $t \geq 0$ enters into the terminal set Ω^i, the local linear feedback control law is assumed as $u_{k,N+t}^i = F_i(k)x_{k,N+t}^i = K_i x_{k,N+t}^i$. K_i is a constant which is calculated off line as follows.

3.1. Design of the Local Control Law. The following equation follows for achieving closed-loop stability:

$$\begin{aligned} \left\|x_{k,N+t+1}^i\right\|_{P_i}^2 - \left\|x_{k,N+t}^i\right\|_{P_i}^2 \leq & -\left\|x_{k,N+t}^i\right\|_{Q_i}^2 \\ & - \left\|u_{k,N+t}^i\right\|_{R_i}^2, \quad t \geq 0. \end{aligned} \tag{13}$$

Lemma 1. *Suppose that there exist $Q_i > 0$, $R_i > 0$, $P_i > 0$, which satisfy the Lyapunov-equation:*

$$(A_i + B_i K_i)^T P_i (A_i + B_i K_i) - P_i = -\kappa_i P_i - Q_i - K_i^T R_i K_i, \tag{14}$$

for some $\kappa_i > 0$. Then, there exists a constant $\alpha^i > 0$ such that Ω_i defined in (11) satisfies (13).

Remark 2. Lemma 1 is directly obtained by referring to "*Lemma* 1" in [21]. For MPC, the stability margin can be adjusted by turning the value of κ_i according to Lemma 1. With regard to DMPC, [11] adjusts the stability margin by tuning the weight in the local cost function. The control objective is to asymptotically stabilize the closed-loop system, so that $x_{k,\infty}^i = 0$ and $u_{k,\infty}^i = 0$. For $t = 0, \ldots, \infty$, summing (13) obtains

$$\sum_{t=N}^{\infty}\left[\left\|x_{k,t}^i\right\|_{Q_i}^2 + \left\|u_{k,t}^i\right\|_{R_i}^2\right] \leq \left\|x_{k,N}^i\right\|_{P_i}^2. \tag{15}$$

Considering both (7) and (15), yields

$$\begin{aligned} \tilde{J}_k^i \leq \overline{J}_k^i = & \sum_{t=0}^{N-1}\left[\left\|\hat{z}_{k,t}^i\right\|_{\overline{Q}_i}^2 + \left\|u_{k,t}^i\right\|_{\overline{R}_i}^2 + \left\|x_{k,t}^i - \hat{x}_{k,t}^i\right\|_{\overline{T}_i}^2\right] \\ & + \left\|x_{k,N}^i\right\|_{P_i}^2, \end{aligned} \tag{16}$$

where $\overline{J}_k^i$ is a finite-horizon cost function, which consists of a finite horizon standard cost, to specify the desired control performance and a terminal cost, to penalize the states at the end of the finite horizon.

The terminal region Ω^i for agent i is designed, so that it is invariant for nonlinear system controlled by a local linear state feedback. The quadratic terminal cost $\|x_{k,N}^i\|_{P_i}^2$ bounds the infinite horizon cost of the nonlinear system starting from Ω^i and controlled by the local linear state feedback.

3.2. Compatibility Constraint for Stability. As in [18], we define two terms, $\xi^{-i} = x^{-*i} - \hat{x}^{-i}$, $\xi^i = x^{*i} - \hat{x}^i$,

$$\begin{aligned} \overline{Q}_i &= \begin{bmatrix} \overline{Q}_i^1 & \overline{Q}_i^{12} \\ \left(\overline{Q}_i^{12}\right)^{\mathrm{T}} & \overline{Q}_i^3 \end{bmatrix}, \\ C_x^*(k) &= \sum_{i=1}^{N_a}\sum_{t=1}^{N-1}\left\{2\left(x_{k,t}^{*i}\right)^{\mathrm{T}}\overline{Q}_i^{12}\xi_{k,t}^{-i}\right. \\ & \qquad \left. +2\left(\hat{x}_{k,t}^{-i}\right)^{\mathrm{T}}\overline{Q}_i^3\xi_{k,t}^{-i} + \left(\xi_{k,t}^{-i}\right)^{\mathrm{T}}\overline{Q}_i^3\xi_{k,t}^{-i}\right\}, \\ C_\xi^*(k) &= \sum_{i=1}^{N_a}\sum_{t=1}^{N-1}\left(\xi_{k,t}^i\right)^{\mathrm{T}}\overline{T}_i\xi_{k,t}^i, \end{aligned} \tag{17}$$

Lemma 2. *Suppose that (9) holds and there exits $\rho(k)$ such that, for all $k > 0$,*

$$0 \le \rho(k) \le 1,$$
$$-\rho(k)\sum_{i=1}^{N_a}\left\{\left\|\left(x^i(k)\right)^T, \left(\hat{x}_k^{-i}\right)^T\right\|_{\overline{Q}_i}^2 + \left\|u^{*i}(0 \mid k)\right\|_{\overline{R}_i}^2\right\} \quad (18)$$
$$+ C_x^*(k) - C_\xi^*(k) \le 0.$$

Then, by solving the receding-horizon optimization problem

$$\min_{\overline{U}^i(k)} \overline{J}_k^i, s.t. (1), (2), (14), (16), \qquad u_{k,N}^i = K_i x_{k,N}^i, x_{k,N}^i \in \Omega^i, \quad (19)$$

and implementing $u_{k,0}^{*i}$, the stability of the global closed-loop system is guaranteed, once a feasible solution at time $k = 0$ is found.

Proof. Define $\overline{J}(k) = \sum_{i=1}^{N_a} \overline{J}_k^i$. Suppose, at time k, there are optimal solution $\overline{U}_k^{*i}$, $i \in \{1, \ldots, N_a\}$, which yields

$$\overline{J}^*(k) = \sum_{i=1}^{N_a}\left\{\left\|\left(x_k^i\right)^{\mathrm{T}}, \left(\hat{x}_k^{-i}\right)^{\mathrm{T}}\right\|_{\overline{Q}_i}^2 + \left\|u_{k,0}^{*i}\right\|_{\overline{R}_i}^2\right\}$$
$$+ \sum_{i=1}^{N_a}\sum_{t=1}^{N-1}\left\{\left\|\left(x_{k,t}^{*i}\right)^{\mathrm{T}}, \left(\hat{x}_{k,t}^{-i}\right)^{\mathrm{T}}\right\|_{\overline{Q}_i}^2 + \left\|u_{k,t}^{*i}\right\|_{\overline{R}_i}^2\right.$$
$$\left. + \left\|\left(x_{k,t}^{*i}\right)^{\mathrm{T}} - \left(\hat{x}_{k,t}^{i}\right)^{\mathrm{T}}\right\|_{T_i}^2\right\} + \sum_{i=1}^{N_a}\left\|x_{k,N}^{*i}\right\|_{P_i}^2. \quad (20)$$

At time $t + 1$, according to Lemma 2, $\overline{U}_{k+1}^i = \{u_{k,1}^{*i}, \ldots, u_{k,N-1}^{*i}, K_i x_{k,N}^{*i}\}$ is feasible, which yields

$$\overline{J}(k+1) = \sum_{i=1}^{N_a}\sum_{t=1}^{N}\left\{\left\|\left(x_{k,t}^{*i}\right)^{\mathrm{T}}, \left(x_{k,t}^{-*i}\right)^{\mathrm{T}}\right\|_{\overline{Q}_i}^2 + \left\|u_{k,t}^{*i}\right\|_{\overline{R}_i}^2\right\}$$
$$+ \sum_{i=1}^{N_a}\left\|x_{k,N+1}^{*i}\right\|_{P_i}^2 \quad (21)$$
$$= \sum_{i=1}^{N_a}\sum_{t=1}^{N-1}\left\{\left\|\left(x_{k,t}^{*i}\right)^{\mathrm{T}}, \left(x_{k,t}^{-*i}\right)^{\mathrm{T}}\right\|_{\overline{Q}_i}^2 + \left\|u_{k,t}^{*i}\right\|_{\overline{R}_i}^2\right\}$$
$$+ \left\|x_{k,N}^*\right\|_Q^2 + \left\|u_{k,N}^*\right\|_R^2 + \left\|x_{k,N+1}^*\right\|_P^2,$$

where $P = \mathrm{diag}\{P_1, P_2, \ldots, P_{N_a}\}$. By applying (9) and Lemma 2, (11) guarantees that

$$\left\|x_{k,N+1}^*\right\|_P^2 - \left\|x_{k,N}^*\right\|_P^2 \le -\left\|x_{k,N}^*\right\|_Q^2 - \left\|u_{k,N}^*\right\|_R^2. \quad (22)$$

Substituting (22) into $\overline{J}(k+1)$ yields

$$\overline{J}(k+1) \le \sum_{i=1}^{N_a}\sum_{t=1}^{N-1}\left\{\left\|\left(x_{k,t}^{*i}\right)^{\mathrm{T}}, \left(x_{k,t}^{-*i}\right)^{\mathrm{T}}\right\|_{\overline{Q}_i}^2 + \left\|u_{k,t}^{*i}\right\|_{\overline{R}_i}^2\right\}$$
$$+ \sum_{i=1}^{N_a}\left\|x_{k,N}^{*i}\right\|_{P_i}^2. \quad (23)$$

By applying (17)–(19),

$$\overline{J}(k+1) - \overline{J}^*(k)$$
$$\le -(1-\rho(k))\sum_{i=1}^{N_a}\left\{\left\|\left(x_k^i\right)^{\mathrm{T}}, \left(\hat{x}_k^{-i}\right)^{\mathrm{T}}\right\|_{\overline{Q}_i}^2 + \left\|u_{k,0}^{*i}\right\|_{\overline{R}_i}^2\right\}$$
$$\le -(1-\rho(k))\left\|x_k\right\|_Q^2. \quad (24)$$

At time $k + 1$, by reoptimization, $\overline{J}^*(k+1) \le \overline{J}(k+1)$. Hence, it leads to

$$\overline{J}^*(k+1) - \overline{J}^*(k) \le -(1-\rho(k))\left\|x_k\right\|_Q^2$$
$$\le -(1-\rho(k))\lambda_{\min}(Q)\left\|x(k)\right\|_Q^2, \quad (25)$$

where $\lambda_{\min}(Q)$ is the minimum eigenvalue of Q. This indicates that the closed-loop system is exponentially stable.

Satisfaction of (18) indicates that all $x_{k,t}^i$ should not deviate too far from their assumed values $\hat{x}_{k,t}^i$ [13]. Hence, (18) can be taken as a new version of the compatibility condition. This compatibility condition is derived from a single compatibility condition that collects all the states (whether predicted or assumed) with in the switching horizon and is disassembled to each agent in distributed manner, which results in local compatibility constraint for each agent. □

3.3. Synthesis Approach of Distributed MPC. In the synthesis approach, the local optimization problem incorporates the above compatibility condition. Since $x_{k,t}^*$ for all agent i is coupled with other agents through (18), it is necessary to assign the constraint to each agent so as to satisfy (18) along the optimization. The continued discussion on stability depends on handling of (18).

Denote $\xi_k^i = [\xi_k^{i,1}, \ldots, \xi_k^{i,n_i}]^{\mathrm{T}}$, $\xi_k^{-i} = \{\xi_k^j \mid j \in \mathcal{N}_i\}$. At time $k > 0$, by solving the optimization problem, there exits a parameter $\varepsilon_k^{i,l}$, $l = 1, \ldots, n_i$, for each element of $\xi_k^{i,l}$, $l = 1, \ldots, n_i$.

Define

$$\varepsilon_k^{i,l} = \max_t\left|\xi_{k-1,t}^{i,l}\right|, \quad (26)$$

and denote $\varepsilon_k^i = [\varepsilon_k^{i,1}, \ldots, \varepsilon_k^{i,n_i}]^{\mathrm{T}}$, $\varepsilon_k^{-i} = \{\varepsilon_k^j \mid j \in \mathcal{N}_i\}$. At time $k + 1 > 0$, set following constraint for each agent i:

$$\left|\left(x_{k+1,t}^i\right)^{\mathrm{T}} - \left(\hat{x}_{k+1,t}^i\right)^{\mathrm{T}}\right| < \varepsilon_k^i. \quad (27)$$

From (26) and (27), it is shown that $\xi_{k+1,t}^i < \varepsilon_k^i$ and $\xi_{k+1,t}^{-i} < \varepsilon_k^{-i}$.

Denote

$$C_x^{*i}(k) = \sum_{t=1}^{N-1}\left\{2\left(x_{k,t}^{*i}\right)^{\mathrm{T}}\overline{Q}_i^{12}\varepsilon_k^{-i}\right.$$
$$\left. +2\left(\hat{x}_{k,t}^{-i}\right)^{\mathrm{T}}\overline{Q}_i^{3}\varepsilon_k^{-i} + \left(\varepsilon_k^{-i}\right)^{\mathrm{T}}\overline{Q}_i^{3}\left(\varepsilon_k^{-i}\right)^{\mathrm{T}}\right\}, \quad (28)$$

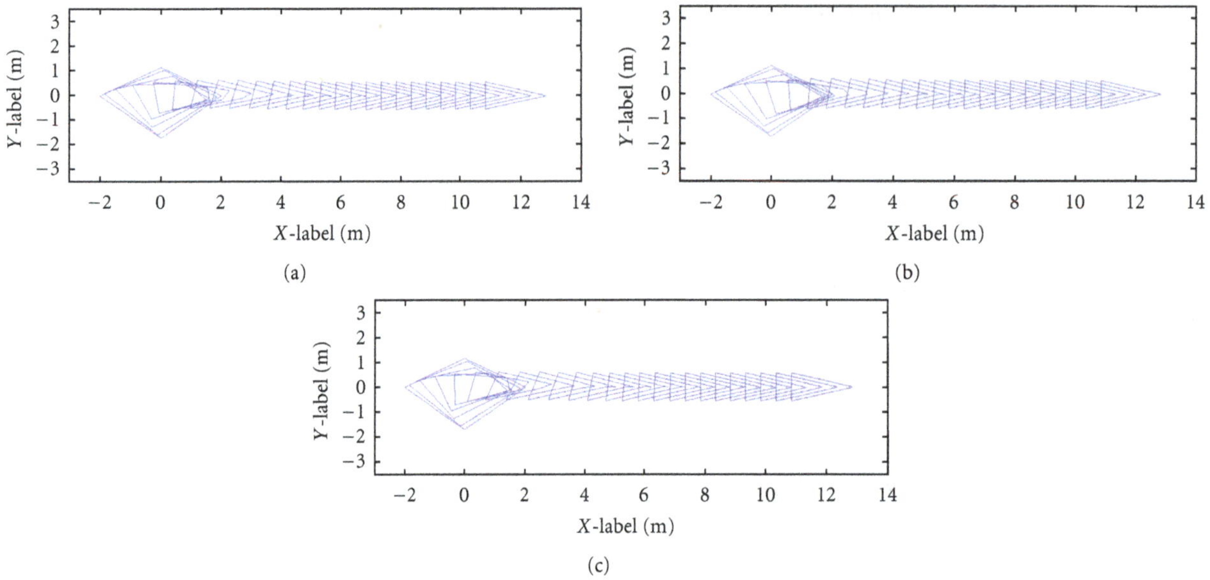

Figure 1: Evolutions of the formation with different control schemes.

$$C_{\xi}^{*i}(k) = \sum_{t=1}^{N-1} \left(\xi_{k,t}^{i}\right)^{\mathrm{T}} \overline{T}_i \xi_{k,t}^{i}. \quad (29)$$

Then $C_x^*(k) \le \sum_{i=1}^{N_a} C_x^{*i}(k)$, $C_\xi^*(k) = \sum_{i=1}^{N_a} C_\xi^{*i}(k)$.

By applying (26)–(28), it is shown that (18) is guaranteed by assigning

$$\begin{aligned} & 0 \le \rho_i(k) \le 1, \\ & \sum_{i=1}^{N_a} - \rho_i(k) \left\{ \left\| \left(x_k^i\right)^{\mathrm{T}}, \left(\hat{x}_k^{-i}\right)^{\mathrm{T}} \right\|_{\overline{Q}_i}^2 + \left\| u_{k,0}^{*i} \right\|_{\overline{R}_i}^2 \right\} \\ & \quad + \sum_{i=1}^{N_a} C_x^{*i}(k) - \sum_{i=1}^{N_a} C_\xi^{*i}(k) \le 0. \end{aligned} \quad (30)$$

is dispensed to agent i:

$$\begin{aligned} & 0 \le \rho_i(k) \le 1, \\ & \sum_{t=1}^{N-1} \left\| \xi_{k,t}^i \right\|_{\overline{T}_i}^2 \ge -\rho_i(k) \left\{ \left\| \left(x_k^i\right)^{\mathrm{T}}, \left(\hat{x}_k^{-i}\right)^{\mathrm{T}} \right\|_{\overline{Q}_i}^2 \right. \\ & \quad \left. + \left\| u_{k,0}^{*i} \right\|_{\overline{R}_i}^2 \right\} + C_x^{*i}(k). \end{aligned} \quad (31)$$

By using (26)–(28), conservativeness is introduced. Hence, (31) is more stringent than (18).

Remark 3. By adding the deviation punishment term in the local cost function, the closed-loop stability follows with a large weight. The larger weight means the more loss of the performance [14, 19]. For a small value of $\overline{T}_i$, we can adjust the value of $\rho_i(k)$ to obtain exponential stability. As the $\rho_i(k)$ is set by optimization, this scheme has more freedom to tuning parameters, to balance the closed-loop stability and control performance.

Remark 4. According to (31), the maximum value and minimum value of $\overline{T}_i$ can be calculated by considering the range of each variable. We choose the middle value for $\overline{T}_i$. Obviously, the $\overline{T}_i$ is time varying and denoted as $\overline{T}_i(k)$.

4. Control Strategy

For practical implementation, distributed MPC is formulated in the following algorithm.

Algorithm. Off-line stage:

(i) Set the value of the prediction horizon N.

(ii) According to (3), (5) and (9), find Q_i, R_i, $\overline{Q}_i$, $\overline{R}_i$, $t = 0, \ldots, N-1$, for all agents.

(iii) Set the value of the compatibility constraint for all agents $\mathcal{E}_i(0) = +\infty$, $j \in \mathcal{N}_i$.

(iv) Calculate the terminal weight P_i, local linear feedback control gain K_i and the terminal set Ω^i.

On-line stage: For agent i, perform the following steps at $k = 0$:

(i) Take the measurement of x_0^i. Set $\overline{T}_i = 0$.

(ii) Send x_0^i to its neighbor j, $j \in \mathcal{N}_i$ of agent i. Receive x_0^j.

(iii) Set $\hat{x}_{t,0}^j = x_{0,0}^j$, $j \in \mathcal{N}_i$, $t = 0, \ldots, N-1$ and $\hat{x}_{0,t}^i = x_0^i$.

(iv) Solve problem (19).

(v) Implement $u_0^i = u_{0,0}^{*i}$.

(vi) Get $\hat{x}_{t,0}^i$ and the value of compatibility constraint $\mathcal{E}_i(1)$.

(vii) Send $\hat{x}^i_{0,t}$ and $\mathcal{E}_i(1)$ to its neighbor j, $j \in N_i$. Receive $\hat{x}^j_{0,t}$ and $\mathcal{E}_j(1)$. Calculate $\overline{T}_i(k)$.

For the agent i, perform the following steps at $k > 0$:

(i) Take the measurement of x^i_k.

(ii) Solve problem (19).

(iii) Implement $u^i_k = u^{*i}_{0,k}$.

(iv) Get $\hat{x}^i_{k,t}$ and the new value of compatibility constraint $\mathcal{E}_i(k+1)$.

(v) Send $\hat{x}^i_{k,t}$ and $\mathcal{E}_i(k+1)$ to its neighbor j, $j \in \mathcal{N}_i$. Receive $\hat{x}^j_{k,t}$ and $\mathcal{E}_j(k+1)$.

(vi) Calculate $\overline{T}_i(k)$.

5. Numerical Example

We consider the model of agent i [22] as

$$x^i_{k+1} = \begin{bmatrix} I_2 & I_2 \\ 0 & I_2 \end{bmatrix} x^i_k + \begin{bmatrix} 0.5I_2 \\ I_2 \end{bmatrix} u^i_k, \tag{32}$$

which is obtained by discretizing the continuous-time model

$$\dot{x}^i = \begin{bmatrix} 0 & I_2 \\ 0 & 0 \end{bmatrix} x^i + \begin{bmatrix} 0 \\ I_2 \end{bmatrix} u^i. \tag{33}$$

($x^i_k = [q^{i,x}_k, q^{i,y}_k, v^{i,x}_k, v^{i,y}_k]^{\mathrm{T}}$, $q^{i,x}_k$ and $q^{i,y}_k$ are positions in the horizontal and vertical directions, resp. $v^{i,x}_k$ and $v^{i,y}_k$ are velocities in the horizontal and vertical directions, resp.) with sampling time interval of 0.5 second. There are four agents. A set of positions of the four agents constitute a formation. The initial positions of the four agents are

$$\left[q^{1,x}_o, q^{1,y}_o\right] = [0, 2], \qquad \left[q^{2,x}_o, q^{2,y}_o\right] = [-2, 0], \tag{34}$$

$$\left[q^{3,x}_o, q^{3,y}_o\right] = [0, -3], \qquad \left[q^{4,x}_o, q^{4,y}_o\right] = [2, 0]. \tag{35}$$

Linear constraints on states and input are

$$\left|x^i\right| \le \begin{bmatrix}100 & 100 & 15 & 15\end{bmatrix}^{\mathrm{T}}, \qquad \left|u^i\right| \le \begin{bmatrix}2 & 2\end{bmatrix}^{\mathrm{T}}. \tag{36}$$

The agent i, $i = 1, 2, 3$ are selected as the core agents of the formation. $\mathcal{A}_0$ is designed as $\mathcal{A}_0 = \{(1, 2); (1, 3); (2, 4)\}$. If all systems achieve the desire formation and the core agents cooperatively cover the virtue leader, then $u^{i,x}_k(k) = 0$, $u^{i,y}_k = 0$. The global cost function is obtained as

$$\begin{aligned} J(k) = \sum_{t=0}^{\infty} \Big[&\left\|q^1_{k,t} - q^2_{k,t} + c_{12}\right\|^2 + \left\|q^1_{k,t} - q^3_{k,t} + c_{13}\right\|^2 \\ &+ \left\|q^2_{k,t} - q^4_{k,t} + c_{24}\right\|^2 + \frac{1}{9}\left\|\left(q^1_{k,t} + q^2_{k,t} + q^3_{k,t}\right) - q_c\right\|^2 \\ &+ \left\|v^1_{k,t}\right\|^2 + \left\|v^2_{k,t}\right\|^2 + \left\|v^3_{k,t}\right\|^2 + \left\|v^4_{k,t}\right\|^2 + \left\|u_{k,t}\right\|^2 \Big]. \end{aligned} \tag{37}$$

They cooperatively track the virtual leader whose reference is $q_c = (0.5 * k, 0)$. The distance between agents is defined as $c_{12} = (-2, 1)$, $c_{13} = (-2, -1)$, $c_{24} = (-2, 1)$. Choose $\mathcal{N}_1 = \{2\}$, $\mathcal{N}_2 = \{1\}$, $\mathcal{N}_3 = \{1\}$, $\mathcal{N}_4 = \{2\}$. Then,

$$\begin{gathered} Q = \begin{bmatrix} 2\frac{1}{9}I_2 & 0 & -\frac{8}{9}I_2 & 0 & -\frac{8}{9}I_2 & 0 & 0 & 0 \\ 0 & I_2 & 0 & 0 & 0 & 0 & 0 & 0 \\ -\frac{8}{9}I_2 & 0 & 2\frac{1}{9}I_2 & 0 & \frac{1}{9}I_2 & 0 & -I_2 & 0 \\ 0 & 0 & 0 & I_2 & 0 & 0 & 0 & 0 \\ -\frac{8}{9}I_2 & 0 & \frac{1}{9}I_2 & 0 & 1\frac{1}{9}I_2 & 0 & 0 & 0 \\ 0 & 0 & 0 & 0 & 0 & I_2 & 0 & 0 \\ 0 & 0 & -I_2 & 0 & 0 & 0 & I_2 & 0 \\ 0 & 0 & 0 & 0 & 0 & 0 & 0 & I_2 \end{bmatrix}, \quad R = I_8. \\ \overline{Q}_1 = \begin{bmatrix} \frac{7}{9}I_2 & 0 & -\frac{4}{9}I_2 & 0 \\ 0 & \frac{1}{3}I_2 & 0 & 0 \\ -\frac{4}{9}I_2 & 0 & I_2 & 0 \\ 0 & 0 & 0 & \frac{1}{3}I_2 \end{bmatrix}, \\ \overline{Q}_2 = \begin{bmatrix} 1\frac{1}{9}I_2 & 0 & -\frac{4}{9}I_2 & 0 \\ 0 & \frac{1}{3}I_2 & 0 & 0 \\ -\frac{4}{9}I_2 & 0 & \frac{4}{9}I_2 & 0 \\ 0 & 0 & 0 & \frac{1}{3}I_2 \end{bmatrix}, \\ \overline{Q}_3 = \begin{bmatrix} 1\frac{1}{9}I_2 & 0 & -\frac{8}{9}I_2 & 0 \\ 0 & \frac{1}{2}I_2 & 0 & 0 \\ -\frac{8}{9}I_2 & 0 & \frac{8}{9}I_2 & 0 \\ 0 & 0 & 0 & \frac{1}{3}I_2 \end{bmatrix}, \\ \overline{Q}_4 = \begin{bmatrix} I_2 & 0 & -I_2 & 0 \\ 0 & I_2 & 0 & 0 \\ -I_2 & 0 & I_2 & 0 \\ 0 & 0 & 0 & \frac{1}{3}I_2 \end{bmatrix}, \end{gathered} \tag{38}$$

and $\overline{R}_i = I_2$, $i \in \{1, 2, 3, 4\}$. Choose $Q_i = 6.85 * I_4$ and $R_i = I_2$, $i \in \{1, 2, 3, 4\}$, $N = 10$. The terminal set is $\alpha_i = 0.22$. The above choice of model, cost, and constraints allow us to rewrite problem (19) as a quadratic programming with quadratic constraint. To solve the optimal control problems numerically, the package NPSOL 5.02 is used. From top to bottom, the first subgraph of Figure 1 is the evolution of the formation with central MPC; the second sub-graph of Figure 1 is the evolution of the formation with distributed MPC with time-varying compatible constraint; the third sub-graph of Figure 1 is the evolution of the formation with distributed MPC with a fixed compatibility constraint.

With the three control schemes, the formation of all agents can be achieved. The obtained $J_{\text{true's}}$ are 2.5779×10^6, 4.8725×10^6, and 5.654×10^6, respectively. Compared with the second sub-graph, the third sub-graph have a

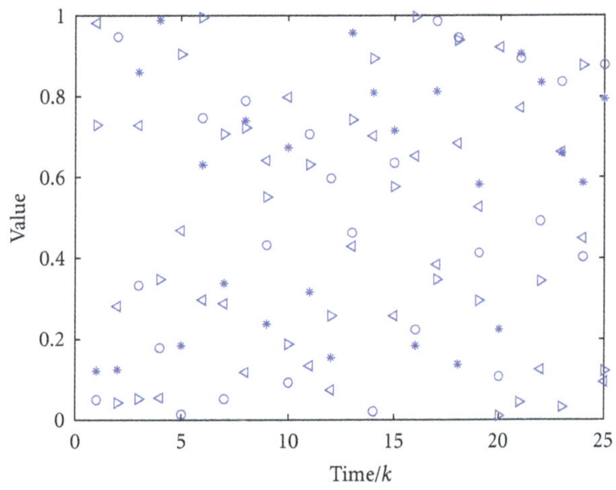

FIGURE 2: The value of $\rho_i(k)$.

large overshoot at the time-instant $k = 9$ (nearby the position $(3, 0)$). The distributed MPC with the time-varying compatible constraint has a better control process comparing to the one with fixed compatible constraint. The value of $\rho_i(k)$ is shown in Figure 2. "*" for agent 1; "O" for agent 2; ">" for agent 3; "<" for agent 4.

Remark 5. For the second simulation, the value of the fixed compatible constraint is 0.2. For the third simulation, the values of the time-varying compatible constraint is calculated according to the states deviation of the previous horizon.

6. Conclusions

In this paper, we have proposed an improved distributed MPC scheme for multiagent systems based on deviation punishment. One of the features of the proposed scheme is that the cost function of each agent penalizes the deviation between the predicted state trajectory and the assumed state trajectory, which improves the consistency and optimal control trajectory. At each sample time, the value of compatibility constraint is set by the deviation of previous sample time-instant. The closed-loop stability is guaranteed with a small value for the weight of the deviation function term. Furthermore, the effectiveness of the scheme has been investigated by a numerical example. One of the future works will focus on feasibility of optimization.

Acknowledgment

This work is supported by a Grant from the Fundamental Research Funds for the Central Universities of China, no. *CDJZR10170006*.

References

[1] J. A. Primbs, "The analysis of optimization based controllers," *Automatica*, vol. 37, no. 6, pp. 933–938, 2001.

[2] J. M. Maciejowski, *Predictive Control with Constraints*, Prentice Hall, Englewood Cliffs, NJ, USA, 2002.

[3] J. A. Rossiter, *Model-Based Predictive Control: A Practical*, CRC, Boca Raton, Fla, USA, 2003.

[4] Y. Kuwata, A. Richards, T. Schouwenaars, and J. P. How, "Distributed robust receding horizon control for multivehicle guidance," *IEEE Transactions on Control Systems Technology*, vol. 15, no. 4, pp. 627–641, 2007.

[5] E. Camponogara, D. Jia, B. H. Krogh, and S. Talukdar, "Distributed model predictive control," *IEEE Control Systems Magazine*, vol. 22, no. 1, pp. 44–52, 2002.

[6] S. Li, Y. Zhang, and Q. Zhu, "Nash-optimization enhanced distributed model predictive control applied to the Shell benchmark problem," *Information Sciences*, vol. 170, no. 2–4, pp. 329–349, 2005.

[7] A. N. Venkat, J. B. Rawlings, and S. J. Wright, "Stability and optimality of distributed model predictive control," in *Proceedings of the 44th IEEE Conference on Decision and Control, and the European Control Conference (CDC-ECC '05)*, vol. 2005, pp. 6680–6685, Seville, Spain, December 2005.

[8] A. N. Venkat, J. B. Rawlings, and S. J. Wright, "Distributed model predictive control of large-scale systems," in *Assessment and Future Directions of Nonlinear Model Predictive Control*, vol. 358 of *Lecture Notes in Control and Information Sciences*, pp. 591–605, 2007.

[9] M. Mercangöz and F. J. Doyle III, "Distributed model predictive control of an experimental four-tank system," *Journal of Process Control*, vol. 17, no. 3, pp. 297–308, 2007.

[10] D. Jia and B. Krogh, "Min-max feedback model predictive control for distributed control with communication," in *Proceedings of the American Control Conference*, pp. 4507–4512, May 2002.

[11] T. Keviczky, F. Borrelli, and G. J. Balas, "Decentralized receding horizon control for large scale dynamically decoupled systems," *Automatica*, vol. 42, no. 12, pp. 2105–2115, 2006.

[12] F. Borrelli, T. Keviczky, G. J. Balas, G. Stewart, K. Fregene, and D. Godbole, "Hybrid decentralized control of large scale systems," in *Hybrid Systems: Computation and Control*, vol. 3414 of *Lecture Notes in Computer Science*, pp. 168–183, Springer, 2005.

[13] W. B. Dunbar and R. M. Murray, "Distributed receding horizon control for multi-vehicle formation stabilization," *Automatica*, vol. 42, no. 4, pp. 549–558, 2006.

[14] W. B. Dunbar, "Distributed receding horizon control of cost coupled systems," in *Proceedings of the 46th IEEE Conference on Decision and Control (CDC '07)*, pp. 2510–2515, December 2007.

[15] S. Wei, Y. Chai, and B. Ding, "Distributed model predictive control for multiagent systems with improved consistency," *Journal of Control Theory and Applications*, vol. 8, no. 1, pp. 117–122, 2010.

[16] B. Ding, "Distributed robust MPC for constrained systems with polytopic description," *Asian Journal of Control*, vol. 13, no. 1, pp. 198–212, 2011.

[17] B. Ding, L. Xie, and W. Cai, "Distributed model predictive control for constrained linear systems," *International Journal of Robust and Nonlinear Control*, vol. 20, no. 11, pp. 1285–1298, 2010.

[18] B. Ding and S. Y. Li, "Design and analysis of constrained nonlinear quadratic regulator," *ISA Transactions*, vol. 42, no. 2, pp. 251–258, 2003.

[19] W. Shanbi, B. Ding, C. Gang, and C. Yi, "Distributed model predictive control for multi-agent systems with coupling constraints," *International Journal of Modelling, Identification and Control*, vol. 10, no. 3-4, pp. 238–245, 2010.

[20] H. Chen and F. Allgöwer, "A quasi-infinite horizon nonlinear model predictive control scheme with guaranteed stability," *Automatica*, vol. 34, no. 10, pp. 1205–1217, 1998.
[21] T. A. Johansen, "Approximate explicit receding horizon control of constrained nonlinear systems," *Automatica*, vol. 40, no. 2, pp. 293–300, 2004.
[22] W. B. Dunbar, *Distributed receding horizon control for multiagent systems*, Ph.D. thesis, California Institute of Technology, Pasadena, Calif, USA, 2004.

13

Performance Improvements of a Permanent Magnet Synchronous Machine via Functional Model Predictive Control

Ahmed M. Kassem[1] and A. A. Hassan[2]

[1] *Control Technology Department, Beni-Suef University, Beni-Suef 62511, Egypt*
[2] *Department of Electrical, Faculty of Engineering, Minia University, El Menia 61519, Egypt*

Correspondence should be addressed to Ahmed M. Kassem, kassem_ahmed53@hotmail.com

Academic Editor: Ricardo Dunia

This paper investigates the application of the model predictive control (MPC) approach to control the speed of a permanent magnet synchronous motor (PMSM) drive system. The MPC is used to calculate the optimal control actions including system constraints. To alleviate computational effort and to reduce numerical problems, particularly in large prediction horizon, an exponentially weighted functional model predictive control (FMPC) is employed. In order to validate the effectiveness of the proposed FMPC scheme, the performance of the proposed controller is compared with a classical PI controller through simulation studies. Obtained results show that accurate tracking performance of the PMSM has been achieved.

1. Introduction

Permanent magnet synchronous motors fed by PWM inverters are widely used for industrial applications, especially servo drive applications, in which constant torque operation is desired. In traction and spindle drives, on the other hand, constant power operation is desired [1]. The inherent advantages of these machines are light weight, small size, simple mechanical construction, easy maintenance, good reliability, and high efficiency. Generally speaking, the applications of the PMSM drive system include two major areas: the adjustable-speed drive system and the position control system. The adjustable-speed drive system has two control-loops: the current loop and the speed loop. To improve the performance of the PMSM drive system, a lot of research has been done. In general, the research has focused on improvement of the performance related to current loop, speed loop, and/or position loop.

The PMSM drive system has been controlled using a PI controller due to its simplicity. The PI controller, however, cannot provide good performance in both transient and load disturbance conditions. Several researchers have investigated the speed controller design of adjustable-speed PMSM systems to improve their transient responses, load disturbance rejection capability, tracking ability, and robustness [2–11].

The MPC controller generally requires a significant computational effort. As the performance of the available computing hardware has rapidly increased and new faster algorithms have been developed, it is now possible to implement MPC to command fast systems with shorter time steps, as electrical drives. Electric drives are of particular interest for the application of MPC for at least two reasons:

(1) they fit in the class of systems for which a quite good linear model can be obtained both by analytical means and by identification techniques;

(2) bounds on drive variables play a key role in the dynamics of the system; indeed, two main approaches are available to deal with system constraints: anti-windup techniques, widely used in the classical PI controllers, and MPC. The presence of the constraint is one of the main reasons why, for example, state space controllers have limited application in electrical drives.

In spite of these advantages, MPC applications to electrical drives are still largely unexplored and only few research laboratories are involved in them. For example, generalized Predictive Control (GPC)—a special case of MPC—has been applied to induction motors for only current regulation [12] and later for speed and current control [13]. In [14], the more general MPC solution has been adopted for the design of the current controller in the same drive.

In this paper, a centralized MPC with large prediction horizon for PMSM speed control is presented. The proposed centralized scheme improves the control performance in a coordinated manner.

Another challenge of centralized MPC for PMSM is its large computational effort needed. To overcome this drawback, a functional MPC with orthonormal basis Laguerre function [15] is presented. The presented functional MPC reduces computational effort significantly which makes it more appropriate for practical implementation. In addition, an exponential data weighting is used to reduce numerical issue in MPC with large prediction horizon [16]. To verify the effectiveness of the proposed scheme, time-based simulations are carried out. The results obtained proved that the functional MPC is able to control successfully the PMSM system in the transient and steady state cases.

2. Dynamic Model of PMSM

The dynamic model of the PMSM can be described in the d-q rotor frame as follows [17]:

$$\begin{aligned}\frac{d}{dt}i_d &= \frac{1}{L}\left(v_d - ri_d + \frac{P}{2}\omega_r L i_q\right),\\ \frac{d}{dt}i_q &= \frac{1}{L}\left(v_q - ri_q - \frac{P}{2}\omega_r (Li_d + \lambda_m)\right),\\ \frac{d}{dt}\omega_r &= \frac{1}{j}(T_e - T_L - B\omega_r),\\ \frac{d}{dt}\theta_r &= \omega_r,\end{aligned} \tag{1}$$

where

$$T_e = \frac{3}{2}\frac{P}{2}\left(\lambda_{ds}\, i_q - \lambda_{qs}\, i_d\right), \qquad \lambda_{ds} = Li_d + \lambda_m, \tag{2}$$

where r is the stator resistance, i_d is the d-axis current, i_q is the q-axis current, v_d is the d-axis stator voltage, v_q is the q-axis stator voltage, L is the stator inductance, where stator inductance in d- and q-axis are assumed to be equal ($L_d = L_q$), d/dt is the differential operator, P is the pole numbers, ω_r is the mechanical rotor speed, λ_m is the flux linkage generated from the permanent magnet material, j is the motor moment of inertia, B is the motor viscous friction coefficient, T_e is the electromagnetic torque, T_L is the load torque and θ_r is the rotor position.

3. Linearised Model

The basic principle in controlling the PMSM is based on field orientation. This is obtained by letting the permanent magnet flux linkage be aligned with the d-axis, and the stator current vector is kept along the q-axis direction. This means that the value of i_d is kept zero in order to achieve the field orientation condition. Since the permanent magnet flux is constant, therefore the electromagnetic torque is linearly proportional to the q-axis current which is determined by closed loop control. As a result, maximum torque per ampere can be obtained from the machine in addition to the achievement of high dynamic performance. So, the electromagnetic torque can be written as follows:

$$T_e = \frac{3}{2}\frac{P}{2}\lambda_m i_q = k_t i_q. \tag{3}$$

Applying the field orientation concept (the electromagnetic torque is linearly proportional with i_q) in (1), the linearised model of the PMSM can be described in a state space form as

$$\begin{aligned}\dot{x} &= Ax + Bu + Ed,\\ y &= Cx + Du,\end{aligned} \tag{4}$$

where

$$x = \begin{bmatrix} i_d & i_q & \omega_r & \theta_r \end{bmatrix}^T,$$

$$u = \begin{bmatrix} v_q & v_d \end{bmatrix}^T, \qquad d = [T_L],$$

$$A = \begin{bmatrix} -\frac{r}{L} & \frac{P}{2}\omega_{ro} & \frac{P}{2}i_{qo} & 0\\ \frac{-P}{2}\omega_{ro} & -\frac{r}{L} & \frac{-P}{2}\left(i_{do} + \frac{\lambda_m}{L}\right) & 0\\ 0 & \frac{k_t}{J} & \frac{-B}{J} & 0\\ 0 & 0 & 1 & 0 \end{bmatrix}, \tag{5}$$

$$B = \begin{bmatrix} \frac{1}{L} & 0 & 0 & 0\\ 0 & \frac{1}{L} & 0 & 0 \end{bmatrix}^T, \qquad E = \begin{bmatrix} 0 & 0 & \frac{-1}{J} & 0 \end{bmatrix}^T.$$

4. Functional Model Predictive Control

4.1. Model Predictive Control. Model predictive control uses an explicit model of system to predict future trajectory of system states, and outputs. This prediction capability allows solving optimal control problem online, where prediction error (i.e., containing difference between the predicted output and reference output) and control input action are minimized over a future horizon, possibly subject to constraints on the manipulated inputs, states, and outputs. The optimization yields an optimal control sequence as input and only the first input from the sequence is used as the input to the system. At the next sampling interval, the horizon is shifted and the whole optimization procedure is repeated. The main reason for using this procedure, which is called receding horizon control (RHC), is that it allows compensating for future disturbance and modeling error.

The basic structure of model predictive control is depicted in Figure 1. An explicit model of the system is used to predict future output response chain $\hat{y}$. Based on the predicted system output and current system output, the error is calculated. The errors, then, are fed to the optimizer. In the optimizer, the future optimal control sequence, Δu, is calculated based on the objective function and system constraints.

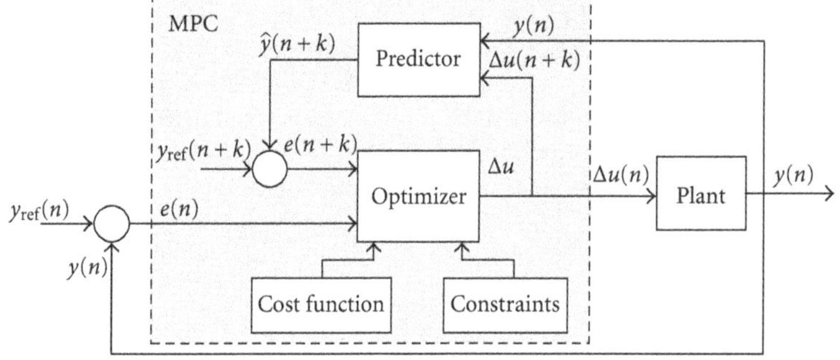

FIGURE 1: Basic structure of model predictive control.

In this paper, the state space model of the system is used in the model predictive control. The general discrete form of the state space model used in model predictive control is of the form:

$$x(k+1) = Ax(k) + Bu(k) + Ed(k) + Fw(k),$$
$$y(k) = Cx(k), \quad (6)$$

where k is the sampling instant, x is state vector, u is input vector, d represents system disturbance, and w represents system noise model. A, B, C, E, and F are coefficients of system state space model and reflect the PMSM model in (4).

The final aim of model predictive control is to provide zero output error with minimal control effort.

Therefore, the cost function J that reflects the control objectives is defined as

$$J(n) = \sum_{k=1}^{N_p} \mu_k (y'(n+k) - y_{\text{ref}}(n+k))^2 + \sum_{k=1}^{N_c} v_k \Delta u(n+k)^2, \quad (7)$$

where μ_k and v_k are respectively, the weighting factors for the prediction error and control energy, $y'(n+k)$: the kth step output prediction, $y_{\text{ref}}(n+k)$: the kth step reference trajectory, and $\Delta u(n+k)$: the kth step control action, where the first term reflects the future output error and second term reflects the consideration given to the control effort. The predicted output vector has dimension of $1 \times N_p$, where N_p is the prediction horizon. Δu is control action vector with dimension of $1 \times N_c$, where N_c is control horizon. In the model predictive control, the control horizon, N_c, is always smaller than or equal to prediction horizon (N_p). μ_k and v_k reflect the weights on the predicted error of predicted outputs and change in the control action.

The constraints of model predictive control include constraints of magnitude and change of input, state, and output variables that can be defined in the following form

$$u_{\min} \le u(n+k) \le u_{\max}, \qquad \Delta u_{\min} \le \Delta u(n+k) \le \Delta u_{\max},$$
$$x_{\min} \le x(n+k) \le x_{\max}, \qquad \Delta x_{\min} \le \Delta x(n+k) \le \Delta x_{\max},$$
$$y_{\min} \le y(n+k) \le y_{\max}, \qquad \Delta y_{\min} \le \Delta y(n+k) \le \Delta y_{\max}. \quad (8)$$

Solving the objective function (7) with system constraint (8) gives the optimal input control sequence.

4.2. Laguerre Based Model Predictive Control. In the classical model predictive control, the future control signal is modeled as a vector of forward shift operator with length of N_c:

$$\Delta U = [\Delta u(n), \ldots, \Delta u(n+k), \ldots, \Delta u(n+N_c-1)]^T, \quad (9)$$

where N_c unknown control variables are achieved in the optimization procedure. However, large prediction horizon is needed to achieve high closed-loop performance. That would require large computational burden. Therefore, MPC may not be fast enough to be used as a real-time optimal control for such case.

A solution to this drawback is the use of functional MPC. In the functional MPC, future input is assumed to be a linear combination of a few simple base functions. In principle, these could be any appropriate functions. However in practice, a polynomial basis is usually used [18–21]. This approximation of input trajectory can be more accurate by proper selection of base function. Using functional MPC, the term used in the optimization procedure can be reduced to a fraction of that required by classical MPC. Therefore, the computational load will be reduced largely.

In this paper, orthonormal basis Laguerre function is used for modeling input trajectory. Laguerre polynomial is one of the most popular orthonormal base functions which has extensive applications in system identification [15]. The z-transform of m'th Laguerre function is given by

$$\Gamma_m = \frac{\sqrt{1-a^2}}{z-a}\left[\frac{1-az}{z-a}\right]^{m-1}, \quad (10)$$

where $0 \le a \le 1$ is the pole of Laguerre polynomial and is called scaling factor in the literature. The control input sequence can be described by the following Laguerre functions:

$$\Delta u(n+k) \approx \sum_{m=1}^{N} c_m l_m(k), \quad (11)$$

where l_m is the inverse z-transform of Γ_m in the discrete domain. The coefficients c_m are unknowns and should be obtained in the optimization procedure. The parameters a and N are tuning parameters and should be adjusted by user. Usually, the value of N is selected smaller than 10 that is enough for most practical applications. Generally, choosing larger value for N increases the accuracy of input sequence estimation.

4.3. Exponentially Weighted Model Predictive Control. Closed-loop performance of MPC depends on the magnitude of prediction horizon N_p. Generally, by increasing the magnitude of prediction horizon, the closed-loop performance will be improved. However practically, selection of large prediction horizon is limited by numerical issue, particularly in the process with high sampling rate. One approach to overcome this drawback is to use exponential data weighting in model predictive control [16].

4.4. Design of the Proposed Functional Model Predictive Control. In this section, the Laguerre-based model predictive control and exponentially weighted model predictive control are combined in order to alleviate computational effort and reduce numerical problems. At first, a discrete model predictive control with exponential data weighting is designed. The input, state, and output vectors are changed in the following way:

$$\begin{aligned}\Delta\hat{U}^T &= \left[\alpha^{-0}\Delta u(n),\ldots,\alpha^{-(N_c-1)}\Delta u(n+N_c-1)\right],\\ \hat{X}^T &= \left[\alpha^{-1}x(n+1),\ldots,\alpha^{-N_p}x\left(n+N_p\right)\right],\\ \hat{Y}^T &= \left[\alpha^{-1}y(n+1),\ldots,\alpha^{-N_p}x\left(n+N_p\right)\right],\end{aligned} \tag{12}$$

where α is tuning parameter in exponential data weighting and is larger than 1. The state space representation of system with transformed variable is

$$\begin{aligned}\hat{x}(n+1) &= \hat{A}\hat{x}(n)+\hat{B}\Delta\hat{u}(n),\\ \hat{y}(n) &= \hat{C}\hat{x}(n),\end{aligned} \tag{13}$$

where $\hat{A}=A/\alpha$, $\hat{B}=B/\alpha$, $\hat{C}=C/\alpha$.

The optimal control trajectory with transformed variables can be achieved by solving the new objective function and constraints:

$$\hat{J}(n) = \sum_{k=1}^{N_p}\mu_k\left(\hat{y}(n+k)-y_{\text{ref}}(n+k)\right)+\sum_{k=1}^{N_c}\nu_k\Delta\hat{u}(n+k)^2, \tag{14}$$

$$\begin{aligned}\alpha^{-k}u_{\min} &\le \hat{u}(n+k)\le\alpha^{-k}u_{\max},\\ \alpha^{-k}\Delta u_{\min} &\le \Delta\hat{u}(n+k)\le\alpha^{-k}\Delta u_{\max}\\ \alpha^{-k}x_{\min} &\le \hat{x}(n+k)\le\alpha^{-k}x_{\max},\\ \alpha^{-k}\Delta x_{\min} &\le \Delta\hat{x}(n+k)\le\alpha^{-k}\Delta x_{\max}\\ \alpha^{-k}y_{\min} &\le \hat{y}(n+k)\le\alpha^{-k}y_{\max},\\ \alpha^{-k}\Delta y_{\min} &\le \Delta\hat{y}(n+k)\le\alpha^{-k}\Delta y_{\max}.\end{aligned} \tag{15}$$

By choosing $a > 1$, the condition number of Hessian matrix will be reduced significantly, especially for large values of prediction horizon (N_p). This leads to a more reliable numerical approach.

After solving new objective function with new variables, the calculated input trajectory should be transformed into standard variable with the following equation:

$$\Delta U^T = \left\lfloor a^0\Delta\hat{u}(k),\ldots,a^{(N_c-1)}\Delta\hat{u}(k+N_c-1)\right\rfloor. \tag{16}$$

The Laguerre-based model predictive control and exponentially weighted model predictive control can be combined using the following systematic procedure:

(i) choosing of the proper tuning parameter α;

(ii) transforming the system parameters (A, B, C) and the system variables (U, X, Y) are transformed using (13) and (14);

(iii) the objective function with its constraints is created based on (15) and (16);

(iv) optimizing objective function based on Laguerre polynomial and then calculating unknown Laguerre coefficients;

(v) calculating input chain from (11);

(vi) the calculated weighted input chain is transformed into unweighted input chain using (16) and applied on the plant.

5. System Configuration

The block diagram of the field-oriented PMSM with the proposed FMPC is shown in Figure 2. All the commanded values are superscripted with asterisk in the diagram. The proposed system control consists of three loops. The first loop for the speed and based on FMPC and the others for the d-q currents and based on PI controllers. Simulations are carried out to compare the performance of designed speed controller by FMPC with conventional PI controller. The input and the output of the FMPC are considered as speed error and reference q-axis current, respectively. The control parameters are assumed as next:

input weight matrix: $\mu = 0.15\times I_{N_c\times N_c}$,

output weight matrix: $\nu = 1\times I_{N_p\times N_p}$.

The constraints are chosen such that the d-and q-axis stator voltages are normalized to be between 0 and 1, where 0 corresponds to zero and 1 corresponds to maximum stator voltage. Thus,

$$u_{\min} = 0 \le \begin{bmatrix} v_d \\ v_q \end{bmatrix} \le 1 = u_{\max}. \tag{17}$$

The constraints imposed on the control signal are hard, whereas the constraints on the states are soft, that is, small violations can be accepted. The constraints on the states are chosen so as to guarantee signals stay at physically reasonable values as follows:

$$x_{\min} = \begin{pmatrix} 0 \\ 0 \end{pmatrix} \le \begin{pmatrix} |i_q| \\ |\omega_r| \end{pmatrix} \le \begin{pmatrix} 22 \\ 430 \end{pmatrix} = x_{\max}. \tag{18}$$

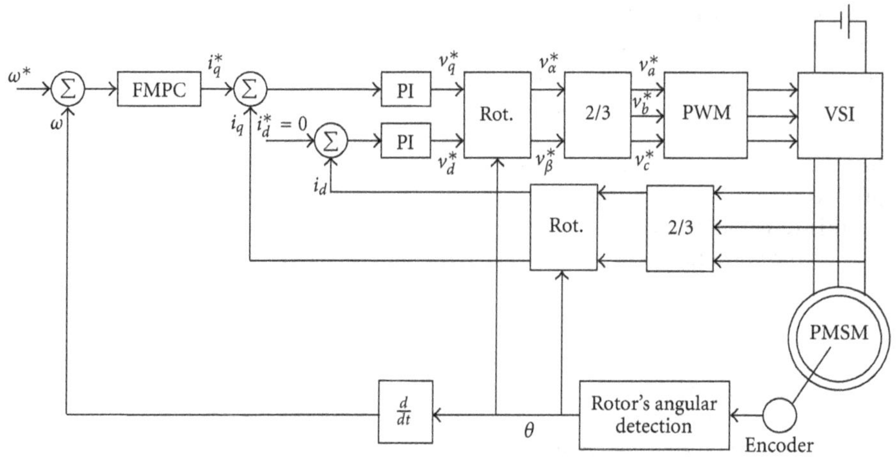

FIGURE 2: Block diagram of the proposed PMSM speed control system.

The speed error is fed to the speed controller (FMPC) in order to generate the torque current command i_q^*. The flux current command i_d^* is set to zero to satisfy the field orientation condition. The reference currents i_q^* and i_d^* are compared with their respective actual currents. The resulted errors are used to generate the voltage commands v_d^* and v_q^* which are converted to three phase reference values v_a^*, v_b^*, and v_c^* in the stator frame. These voltage signals are compared with triangular carrier signal and the output logic is used to control the PWM inverter.

The entire system has been simulated on the digital computer using the Matlab/Simulink/Powerlib software package. The motor used in the simulation procedure has the following specifications [22]:

PMSM: 1.5 kw, 240 V, 2-pole, 4250 rpm,

stator resistance: 1.6 ohm,

stator inductances: $L = 6.37$ m·H,

permanent magnet flux: 0.19 Wb,

moment of inertia: 0.0001854 kg·m^2,

friction coefficient: 5.396e-005 N·m·s/rad.

Computer simulations have been carried out in order to validate the effectiveness of the proposed scheme. The simulation tests are carried out using Matlab/Simulink software package. Wherever, the state space model of the permanent magnet synchronous motor is programmed with the functional model predictive algorithms in MATLAB work space.

The MPC control algorithm depends on the solution of a constrained optimization problem. Most designers choose N_p (prediction horizon) and N_c (control horizon) in a way such that the controller performance is insensitive to small adjustments in these horizons. Here are typical rules of thumb for obtaining a stable process:

(1) choose the control interval such that the plant's open-loop settling time is approximately 20–30 sampling periods (i.e., the sampling period is approximately one-fifth of the dominant time constant),

(2) choose prediction horizon to be the number of sampling periods used in step 1,

(3) use a relatively small control horizon, for example, 3–5.

Selection of suitable values of a and N will increase the system output predicted values accuracy and help to improve the system performance with small control effort. The tuning parameter α is chosen in order to decrease the numerical problems and decrease the simulation time and hence make the system more suitable for implementation. Therefore, the system state space with transformed values $\hat{A}$, $\hat{B}$, $\hat{C}$, and $\hat{D}$ is obtained using the system state space model A, B, C, and D and tuning parameter α, where $\hat{A} = A/\alpha$, $\hat{B} = B/\alpha$, $\hat{C} = C/\alpha$ and $\hat{D} = D/\alpha$. Then, the control objectives are achieved by solving the new cost function $\hat{J}$ and new constraints.

In the proposed system under study, the parameters of the FMPC are adjusted to be $a = 0.38$, $N = 6$, $\alpha = 1.04$, $N_p = 200$ and $N_c = 5$. The system performance with the proposed FMPC controller is compared with the corresponding one using the conventional PI controller. The gains of the PI controller are adjusted as follows: proportional gain $K_p = 6$ and integral gain $K_i = 2.5$. The following simulation tests are carried out to show the validity of the proposed FMPC controller.

5.1. High Speed Case. It is assumed that the machine follows a certain speed trajectory starting from 400 rad/sec., stepped to 300 rad/sec. at time $t = 0.03$ sec., then returned back to 400 rad/sec at $t = 0.05$ sec. The load torque is kept constant at the value 3 N·m. during the simulation period. Figures 3 to 7 show the dynamic responses of the speed, torque, rotor position, and stator currents of the PMSM system based on both FMPC and PI controllers. It has been shown that the proposed system has better transient response. This is clear in Figures 3 and 4 where the system with PI controller oscillates many times before the steady state values are attained. In contrast, the system with the proposed controller has attained the steady state value very quickly. That is can be

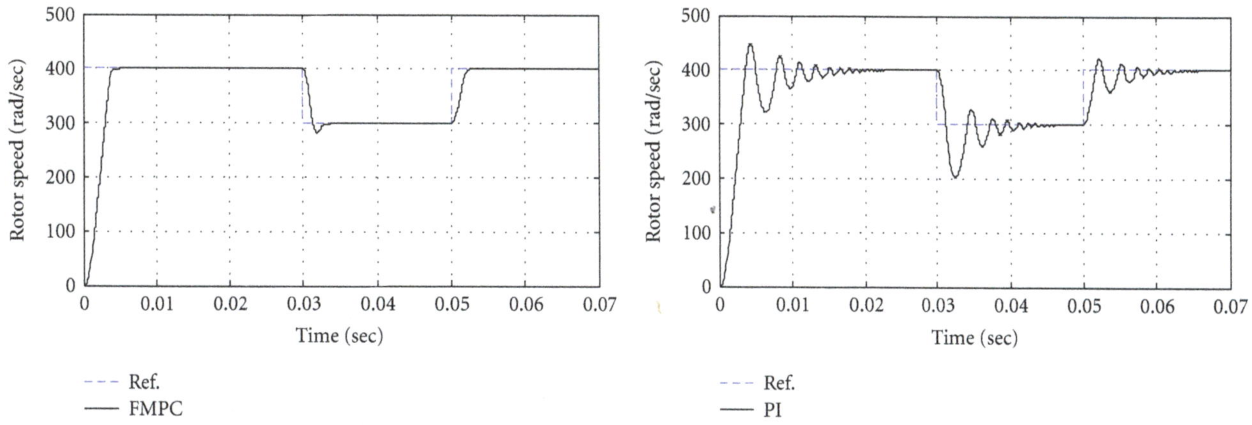

(a) Using FMPC controller (b) Using PI controller

Figure 3: Speed response of the PMSM system based on FMPC and PI controllers.

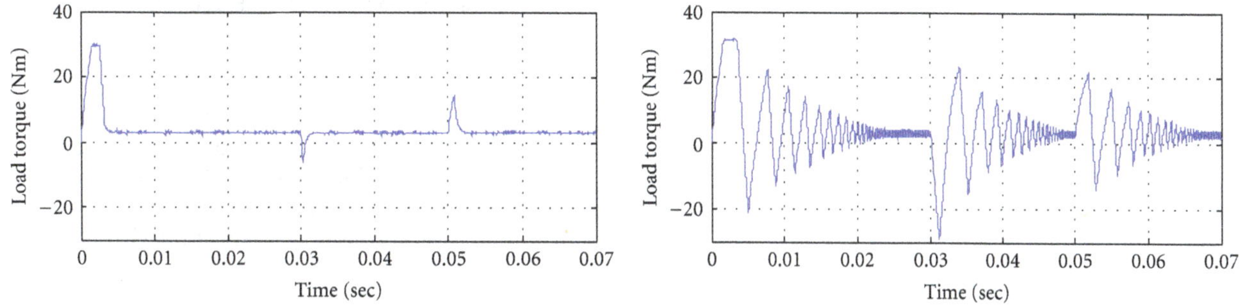

(a) Using FMPC controller (b) Using PI controller

Figure 4: Torque response of the PMSM system based on FMPC and PI controllers.

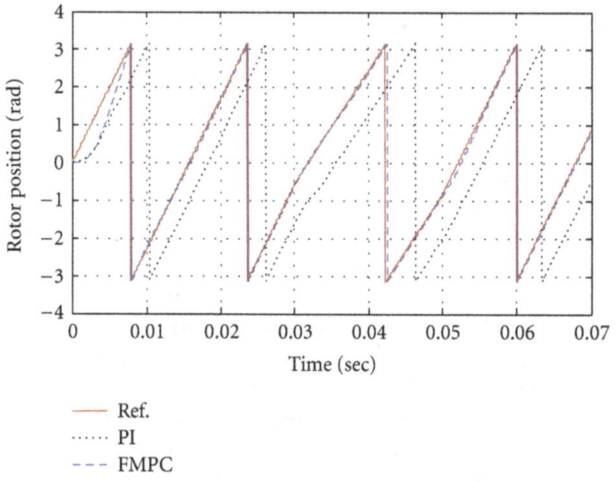

Figure 5: Rotor position response of the PMSM system based on FMPC and PI controllers.

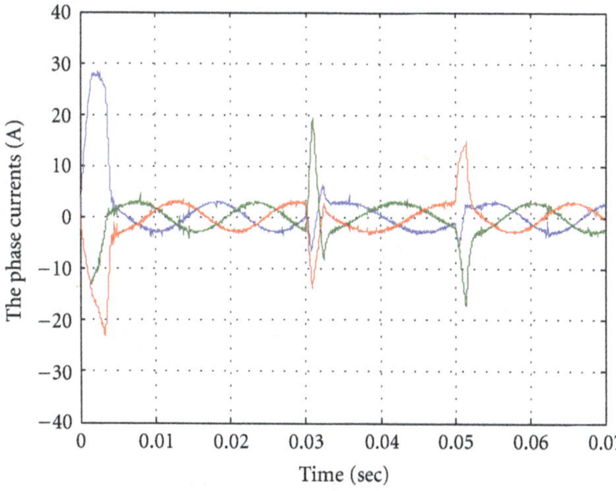

Figure 6: Stator current response of the PMSM system based on the FMPC controller.

shown in Figures 3 to 9, where overshoot and settling time of system are reduced when FMPC controller is used. The settling time of PI controller is 18 ms, where in the case of FMPC the settling time equal 5 ms as shown in Figure 3. Also, Figure 5 illustrates that the PI controller produces large phase difference in the rotor position response which adversely affects the axes transformation and the flux orientation, and thereby reduces the system performance. On the other hand,

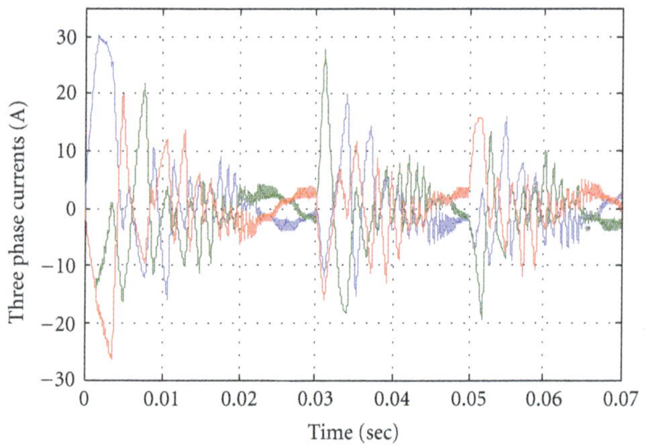

FIGURE 7: Stator current response of the PMSM system based on PI controller.

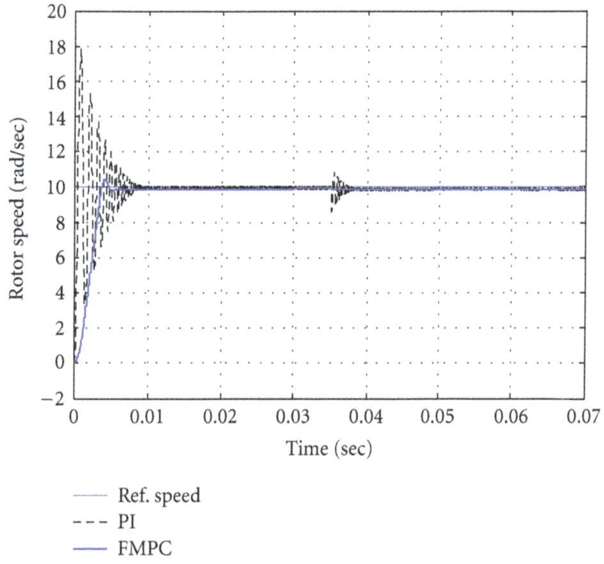

FIGURE 8: Low speed response at variable load based on proposed FMPC and PI controllers.

the FMPC tracks well the rotor position reference and the field orientation condition is satisfied.

Figures 6 and 7 show the stator current response based on FMPC and PI controllers. It is obvious that, with the FMPC controller, the stator current has less ripple content and over shoot than using PI controller.

5.2. Low Speed Case. The performance of the PMSM scheme with the proposed FMPC controller is investigated at low speed (10 rad/sec). The load torque is assumed to be stepped from 2 N·m. to 5 N·m. at time $t = 0.035$ second. Figures 8 and 9 show the system responses using the FMPC and PI controllers. It is clear that the system has poor transient response using PI controller especially at starting and at the instant of load change. Also, more ripples are noticed in the torque response. These drawbacks are nearly eliminated using the FMPC controller.

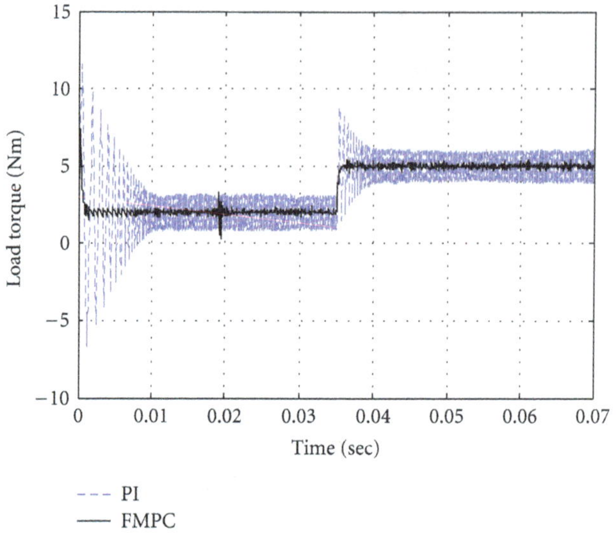

FIGURE 9: Torque response based on FMPC and PI controllers at low speed.

Also in the FMPC, the unknown variables are 16 times less than the classical MPC. In each time interval, the calculation time needed for classical MPC is 4.6 ms, whereas this time is reduced to 0.48 ms in the FMPC. This is a great computational advantage of using functional MPC.

6. Conclusions

In this paper, a centralized functional model predictive controller is proposed to control the speed and torque of the permanent magnet synchronous motor drive system. The proposed predictive controller uses orthonormal Laguerre functions to describe control input trajectory which reduces real-time computation largely. Also, exponential data weighing is used to decrease numerical issue, particularly with large prediction horizon. Constraints are imposed on both the q-axis current and the motor speed.

Computer simulations have been carried out in order to evaluate the effectiveness of the proposed controller. The results proved that the proposed system has accurate tracking performance at low speeds as well as high speeds. Also, small ripple contents are noticed in the torque and stator current waveforms. Moreover, the proposed controller has significantly better performance relative to PI controller especially at starting and load change conditions. The main reasons of this superiority are centralized structure of the proposed controller which reduces negative interaction between local control actions, proper constraints that improve optimal calculation of control trajectory, and, finally, using large prediction horizon which gives a performance close to global.

References

[1] S. Morimoto, M. Sanada, and Y. Takeda, "Wide-speed operation of interior permanent magnet synchronous motors with high-performance current regulator," *IEEE Transactions on Industry Applications*, vol. 30, no. 4, pp. 920–926, 1994.

[2] Y. A. R. I. Mohamed, "Adaptive self-tuning speed control for permanent-magnet synchronous motor drive with dead time," *IEEE Transactions on Energy Conversion*, vol. 21, no. 4, pp. 855–862, 2006.
[3] C. H. Fang, C. M. Huang, and S. K. Lin, "Adaptive sliding-mode torque control of a PM synchronous motor," *IEE Proceedings: Electric Power Applications*, vol. 149, no. 3, pp. 228–236, 2002.
[4] A. M. Howlader, N. Urasaki, T. Senjyu, A. Yona, and A. Y. Saber, "Optimal PAM control for a buck boost DC-DC converter with a wide-speed-range of operation for a PMSM," *Journal of Power Electronics*, vol. 10, no. 5, pp. 477–484, 2010.
[5] Y. S. Kung and M. H. Tsai, "FPGA-based speed control IC for PMSM drive with adaptive fuzzy control," *IEEE Transactions on Power Electronics*, vol. 22, no. 6, pp. 2476–2486, 2007.
[6] N. S. Park, M. H. Jang, J. S. Lee, K. S. Hong, and J. M. Kim, "Performance improvement of a PMSM sensorless control algorithm using a stator resistance error compensator in the low speed region," *Journal of Power Electronics*, vol. 10, no. 5, pp. 485–490, 2010.
[7] Y. S. Kung and M. H. Tsai, "FPGA-based speed control IC for PMSM drive with adaptive fuzzy control," *IEEE Transactions on Power Electronics*, vol. 22, no. 6, pp. 2476–2486, 2007.
[8] D. Yan and Z. Ji, "Backstepping speed control of PMSM based on equivalent input disturbance estimator," in *Proceedings of the 29th Chinese Control Conference (CCC '10)*, pp. 3377–3382, Wuxi, China, July 2010.
[9] R. M. Jan, C. S. Tseng, and R. J. Liu, "Robust PID control design for permanent magnet synchronous motor: a genetic approach," *Electric Power Systems Research*, vol. 78, no. 7, pp. 1161–1168, 2008.
[10] F. M. E. Fayez El-Sousy, "A vector-controlled PMSM drive with a continually on-line learning hybrid neural-network model-following speed controller," *Journal of Power Electronics*, vol. 5, no. 2, pp. 129–141, 2005.
[11] M. Karabacak and H. I. Eskikurt, "Speed and current regulation of a permanent magnet synchronous motor via nonlinear and adaptive backstepping control," *Mathematical and Computer Modelling*, vol. 53, no. 9-10, pp. 2015–2030, 2011.
[12] L. Zhang, R. Norman, and W. Shepherd, "Long-range predictive control of current regulated PWM for induction motor drives using the synchronous reference frame," *IEEE Transactions on Control Systems Technology*, vol. 5, no. 1, pp. 119–126, 1997.
[13] R. Kennel, A. Linder, and M. Linke, "Generalized predictive control (GPC)—ready for use in drive applications?" in *Proceedings of the IEEE 32nd Annual Power Electronics Specialists Conference*, pp. 1839–1844, June 2001.
[14] A. Linder and R. Kennel, "Model predictive control for electrical drives," in *Proceedings of the IEEE Power Electronics Specialists Conference (PESC '05)*, pp. 1793–1799, 2005.
[15] L. Wang, "Discrete model predictive controller design using Laguerre functions," *Journal of Process Control*, vol. 14, no. 2, pp. 131–142, 2004.
[16] L. Wang, "Use of exponential data weighting in model predictive control design," in *Proceedings of the 40th IEEE Conference on Decision and Control (CDC '01)*, pp. 4857–4862, December 2001.
[17] S. Özçira, N. Bekiroğlu, and E. Ayçiçek, "Speed control of permanent magnet synchronous motor based on direct torque control method," in *Proceedings of the International Symposium on Power Electronics, Electrical Drives, Automation and Motion (SPEEDAM '08)*, pp. 268–272, June 2008.
[18] J. M. Maciejowski, "Predictive Control with Constraints," Pearson Education, POD, 2002.
[19] A. Bilbao-Guillerna, M. De la Sen, A. Ibeas, and S. Alonso-Quesada, "Robustly stable multiestimation scheme for adaptive control and identification with model reduction issues," *Discrete Dynamics in Nature and Society*, vol. 2005, no. 1, pp. 31–67, 2005.
[20] T. Murao, H. Kawai, and M. Fujita, "Stabilizing predictive visual feedback control for fixed camera systems," *Electronics and Communications in Japan*, vol. 94, no. 8, pp. 1–11, 2011.
[21] T. Geyer, "Computationally efficient model predictive direct torque control," *IEEE Transactions on Power Electronics*, vol. 26, no. 10, pp. 2804–2816, 2011.
[22] "MATLAB Math Library User's Guide," by the Math Works, 2010.

14

Implementation of Control System for Hydrokinetic Energy Converter

Katarina Yuen, Senad Apelfröjd, and Mats Leijon

Division of Electricity, Department of Engineering Science, Uppsala University, P.O. Box 534, 75121 Uppsala, Sweden

Correspondence should be addressed to Katarina Yuen; katarina.yuen@angstrom.uu.se

Academic Editor: Pierluigi Siano

At Uppsala University, a research group is investigating a system for converting the power in freely flowing water using a vertical-axis turbine directly connected to a permanent magnet generator. An experimental setup comprising a turbine, a generator, and a control system has been constructed and will be deployed in the Dalälven river in the town of Söderfors in Sweden. The design, construction, simulations, and laboratory tests of the control system are presented in this paper. The control system includes a startup sequence for the turbine and load control. These functions have performed satisfactorily in laboratory tests. Simulations of the system show that the power output is not maximized at the same tip-speed ratio as that which maximizes the turbine power capture.

1. Introduction

Unregulated rivers, tides, and other ocean currents comprise a renewable energy resource that may contribute to mankind's energy use. At Uppsala University, a research group is investigating a system for converting the power in freely flowing (undammed) water using a vertical axis turbine directly connected to a permanent magnet generator [1–4].

Several ideas for converting flowing water into electricity are under investigation worldwide, with both companies (http://www.orpc.co/, http://www.marineturbines.com/, http://verdantpower.com/, http://www.hammerfeststrom.com/, http://www.cleancurrent.com/, http://www.openhydro.com/home.html, http://www.minesto.com/) and academic research groups (http://www.eng.ox.ac.uk/tidal, http://oelab.naoe.inha.ac.kr/, http://www.energy.soton.ac.uk/marine/marine.html) as actors. The predominant concept is to use a turbine to transfer power from the flowing water to a shaft, possibly including a gear-box, and then use a generator to convert the mechanical torque on the shaft to electricity, which then can be transferred to an electric grid.

The concept studied at Uppsala Universtity entails a cross-flow, fixed-pitch turbine and a directly driven permanent-magnet generator. By using a directly driven generator and a fixed-pitch turbine, the mechanical complexity of the unit is limited, thus reducing the expected need for maintainance. The generator can be placed outside of the flow through the turbine, and a turbine with a vertical axis can utilize water currents from any direction in the horizontal plane. Challenges pertained to this concept involve the low rotational speed of the generator, and as in all other hydrokinetic energy projects, creating systems that can function well enough in the aquatic environment they are subjected to, withstanding forces, corrosion, biofouling, and floating debris.

The research group at Uppsala University is currently finalizing an experimental station to be deployed in the Dalälven river in the town of Söderfors [5, 6]. This paper describes the control system implemented at the test-site and some simulations of the system.

2. The Söderfors Experimental Setup

The purpose of the Söderfors project is to build and test a system for converting hydrokinetic power to electricity in a natural setting for an extended period of time. The project involves building and deploying a vertical-axis turbine, a directly connected permanent-magnet generator, a control system, and an on-land station to which the power is delivered.

2.1. Site. The site chosen for the experimental turbine and generator is in a river, Dalälven, in Sweden. In the town of Söderfors, a hydropower plant regulates the river. The outlet channel of the hydropower plant is about 1 km long. At the end of the channel, a bridge for the main road through the town crosses the river. Near the bridge, the water depth is between 6 and 7 meters. The width of the river is approximately 100 m (see Figure 1).

Velocity measurements were combined with discharge data for a period of five years. From this, it was found that the velocity on the downstream side of the bridge was mostly around 1–1.4 m/s and that the maximum velocity was about 1.9 m/s. Low velocities, under 2 m/s, are of interest to the research group, since the ability to utilize such sites would increase the number of possible sites for hydro-kinetic energy conversion.

Parameters considered when looking for a suitable site for an experimental station included velocity distribution, depth, deployment possibilities, proximity to Uppsala, and likelyhood to acquire permits. Information about the site and proposed experimental equipment have previously been published [5–7].

2.2. Turbine. The turbine for the Söderfors station is a five-bladed vertical-axis turbine with fixed blade pitch. Blades and struts are made of a carbon fiber reinforced polymer and attached to a central hub made of steel. The profiles are NACA 0021, with a chord length of 0.18 m. The radius is 3 m and the height 3.5 m. Simulations, as described previously [5, 6], predict a maximum power capture C_p of about 0.36 obtained at a tip-speed ratio λ of 3.5 in 1.3 m/s (see Table 1).

2.3. Generator. The turbine and generator share a shaft; that is, there is no gear-box. With velocities mostly in the order of 1–1.4 m/s and an expected best λ of 3.5, the rotational speed of the turbine and generator is expected to be mainly in the range of 10–15 rpm.

The generator design for the Söderfors station is based on the laboratory prototype described earlier [8]. To ease production, the outer diameter of the stator was reduced to 1800 mm. In order to improve efficiency at lower rotational speeds, the axial length has been reduced by designing for a lower voltage [5]. The resulting generator has 112 poles and a line to line voltage of 138 V at the design point of 1.3 m/s. See Table 1 for a summary of turbine and generator data.

FIGURE 1: Aerial view of Söderfors. The turbine and generator will be installed on the downstream side of the bridge (a), the control station will be situated about 150 m from the turbine (b). A hydropower station is situated upstream from the site (c).

TABLE 1: Design data for Söderfors turbine and generator.

Turbine	
Height	3.5 m
Radius	3.0 m
No. of blades	5
Generator	
Stator outer radius	1800 mm
Stator inner radius	1635 mm
Air gap	7 mm
No. of poles	112
Magnet thickness	10 mm
Design power	7.5 kW
Design speed	15 rpm
Design line-to-line voltage (rms)	138 V

3. Control System Design

The control system is to be able to operate the turbine and generator in different ways enabling experiments for characterizing the setup as well as identifying and evaluating different control strategies. This involves being able to keep a specified rotational speed as well as allowing for implementation of different control algorithms. At the same time, extensive measurements of at least voltages, currents, and rotational speed should be possible. Water current speed should also be measured and related to the system operation, but at the time of designing the control system it was not known if continuous real-time velocity data would be available or not. Other parameters, such as mechanical torque, forces on blades, and vibrations, were of interest but considered too much of a complication for a first unit. Conversion to grid specifications and grid connection was also set aside at this stage but can be incorporated later.

3.1. Load Control. The basic theory concerning the operation of a hydro-kinetic energy converter is very similar to that for a wind power plant, which can be studied in textbooks [9, 10]. Differences relate to the higher density of water, the presence of the water surface, turbulence, and predictability of the resouce, to name some.

The turbine captures a portion of the power in the water that passes it, P_{turbine}:

$$P_{\text{turbine}} = \frac{1}{2} C_p A_{\text{turbine}} \rho v^3 \quad [\text{W}], \tag{1}$$

$A_{turbine}$ is the cross-sectional area of the turbine, v is the water velocity, ρ is the density of water, and C_p is the power coefficient describing what portion of the available power the turbine captures.

The power capture roughly depends on the relative velocity of the blades to the water, or the tip-speed ratio, λ, defined as:

$$\lambda = \frac{\Omega r}{v}, \tag{2}$$

where Ω is the rotational speed of the turbine and r, the radius of the turbine. For the five-bladed, vertical-axis turbine designed for Söderfors, a simulated power capture curve at 1.3 m/s is presented in Figure 2. While the $C_p - \lambda$ characteristics change a little with the water velocity, the power capture for this turbine is expected to be at its highest when the tip-speed ratio is around 3.5.

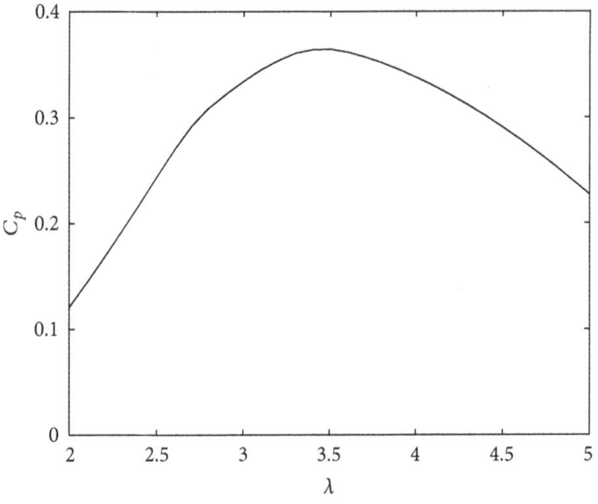

FIGURE 2: Simulated C_p-curve at 1.3 m/s for the turbine designed for Söderfors according to the streamtube model.

The generator has an efficiency of up to about 85%, according to simulations. The losses in the generator are mainly copper losses (resistive losses in the copper winding) and iron losses (mainly hysteresis and eddy current losses) in the stator steel. These losses depend on both the rotational speed of the generator and the load, or the current, withdrawn from it. For a given power, a higher rotational speed results in a higher voltage and a lower current, giving lower copper losses and higher iron losses. As seen previously [11], the iron losses tend to dominate at lower rotational speeds and the copper losses at higher rotational speeds for a similar machine, provided that the rotational speed is controlled to maintain the tip-speed-ratio where C_p is maximal.

The turbine and generator have the same shaft and thus the same rotational speed. The change of rotational speed depends on the power capture of the turbine and the power extraction from the generator to the load, P_{load}, as well as losses in bearings, seals, and the generator. With J as the moment of inertia,

$$\frac{d}{dt}\left(\frac{1}{2}J\Omega^2\right) = P_{turbine} - P_{load} - P_{losses} \quad [\mathrm{W}]. \tag{3}$$

If the power is not extracted, the rotational speed will increase, and vice versa the speed can be reduced by increasing the power delivered to the load. The change in rotational speed gives a different tip-speed-ratio λ and thus a different power capture coefficient, C_p.

Generally, the variations in water speed should be followed by the generator and turbine, giving a high energy yield. For example, if the C_p characteristics of the turbine and the water speed are known, a specific tip-speed ratio can be upheld. If the maximum C_p and corresponding tip-speed ratio, λ_0, are known, but not the water velocity, the velocity may be estimated by evaluating the absorbed power and the rotational speed, for example, from (1) we have the following

$$v = \sqrt[3]{\frac{2P_{turbine}}{C_p^{max} A\rho}} \quad [\mathrm{m/s}], \tag{4}$$

where $P_{turbine} = P_{load} + P_{losses}$ if the rotational speed is constant. Using (2), we see that the power handled by the generator should be proportional to the rotational speed cubed, Ω^3. Or, we can express the electromagnetic torque as follows

$$M_e = \frac{r^3}{2\lambda_0}\rho C_p^{max} A\Omega^2 \quad [\mathrm{Nm}]. \tag{5}$$

If the C_p is unknown, a control scheme called maximum power point tracking, MPPT, may be employed, where the load is regularly perturbated in order to find an operating point where the output power is at a maximum [10].

From a system perspective, the best way to control the turbine and generator is not easily defined. On the one hand, maximizing energy output is an attractive feature. At the same time, all equipment, including the grid connection, needs to be dimensioned for the maximum output power. If this power is seldom reached, maximizing energy output may be expensive. Instead, it may be cost-efficient to install for a smaller power and obtain a higher degree of utilization, as suggested by Clarke et al. [12].

Another aspect is the wear of the equipment. Different operating conditions result in different mechanical loads on the machine as discussed by Hu and Du [13]. Control algorithms in relation to mechanical loads can be analyzed, as in wind power [14], and allowed to influence the design of turbines and generators. Fault ride through is also an issue that has been addressed in wind power [15], and much work may be applicable also in the case of hydro-kinetic energy.

Also note that the highest power capture for the turbine may not correspond to the highest system power output due to the characteristics of the losses in the generator and other electrical components, as well as frictional losses, as they may vary with rotational speed and load. With a generator with low rotational speed, an increase in rotational speed may result in an increase in efficiency due to increased voltage, which outweighs the decrease in power capture, to some degree. This corresponds to operation at a tip-speed ratio higher than that which gives maximal C_p, which in turn means that when stopping the turbine, the power will initially increase.

Yet another issue relates to the cost of determining different parameters. Finding the C_p-characteristics for the turbine, or the equivalent for the whole system, involves initial full-scale tests over a range of velocities. Measuring water velocity and providing real-time data to a control system introduce an extra complexity of the system in the form of measurements, but also transfer of the data.

The variations in water speed at different sites have different characteristics; tides are highly predictable, local air pressure and precipitation may have a strong effect on some sites. Different control strategies may be more or less appropriate at a specific site.

Given these issues related to the control of a hydrokinetic energy converter, the control system for the Söderfors station should be able to extract power from the generator so as to maintain a specified rotational speed, thus enabling the implementation of various control strategies. The control system should also enable a controlled stopping of the turbine and generator.

3.2. Start-Up Theory. The control system is also required to be able to start the turbine. The turbine may be able to self-start under certain conditions, but it is not expected to do so generally. Instead, the turbine can be accelerated using the generator as an electric motor. At a sufficient rotational speed, the turbine will extract power from the water current, and the control system can switch from start-up to load control. A similar electric start system, though using an auxiliary winding, has been constructed and tested for use with a vertical-axis wind turbine. The energy needed to start the turbine was approximately equal to the energy produced during 3 s nominal operation [16].

During start-up, the permanent magnet machine can be used as a brushless DC motor (BLDC), for which there are established means of operating [17]. A DC-source feeds an inverter which in turn is connected to the three phases of the motor. Knowing the position of the rotor, one can determine which phases to deliver current to in order to accelerate the motor. The rotational speed will then be proportional to the DC-voltage, but by using pulse width modulation (PWM), a lower effective voltage can be obtained. Increasing the duty cycle (on time) of the PWM, the voltage, and thus the rotational speed of the rotor and turbine, can be increased gradually.

The position of the rotor can be determined using Hall effect sensors placed near the rotor. Three sensors positioned $120^\circ e$ (electrical degrees) apart give six different states that can be used to control the inverter. It is also possible to determine the rotor position without sensors mounted in the generator by analyzing the back emf from the machine. For the Söderfors setup, it was decided to install sensors in the machine, thus enabling start-up with sensors, but not excluding the possibility of designing a sensorless start.

3.3. Emergency Brake and Standstill. In the event that the turbine and generator need to be halted abruptly, an emergency brake should be present. A short circuiting of the generator could serve as an emergency brake but may harm the turbine due to the high torque involved. A dampened short circuit, essentially, providing the generator with a high load (low Ohm), should also be able to brake the turbine quickly, but somewhat more mildly.

Once at standstill, a short circuit of the generator should prevent it from self-starting.

3.4. Control System Topology. The topology for the control system, given that no power is delivered to the grid, is illustrated in Figure 3. The three phases of the generator can be connected to four different parts of the control system.

For starting the generator and turbine, a DC-source in the form of a three-phase transformer and a diode rectifier feeds an inverter. The load control comprises a rectifier, a capacitor bank, a resistive load, and an IGBT (insulated gate bi-polar transistor) to determine the amount of power spent in the load. The load IGBT is controlled with a PWM signal, and the duty cycle (on time) determines how much power is extracted for a given voltage. The emergency brake is a Y-connected load.

3.5. Simulations. Each branch (a, b, c, and d) in Figure 3 has been separately implemented in Matlab Simulink (Mathworks, http://www.mathworks.com/). Simulations were used to dimension components in the control system and to gain some knowledge of how the system would respond.

In Simulink, the existing permanent magnet synchronous machine was used to simulate the generator. No iron losses are included in this model and were not added at this point. Frictional losses in bearings and seals have also been neglected. See Table 2 for a summary of input data used. Figure 4 shows the Simulink model representing load control of the Söderfors generator and turbine. The turbine C_p is implemented as a lookup table. The water speed is specified, and the duty cycle is regulated to reach a desired rotational speed. The cable for power transmission between the generator and the control station is added as $0.1\,\Omega$ resistance per phase.

4. Realization of Control System

In general, the control system is intended to be easy to become familiar with. It is not viewed as a final product, rather a first version of itself, allowing for further development, and tests not yet concieved. At an early stage, it was decided to let the submerged turbine and generator be connected to an on-land control station via an enclosure mounted on the bridge railing about 40 m from the generator, to enable convenient deployment, as well as provide space for extra equipment that may be needed close to the water.

Most of the control system, described in more detail below, is collected in the enclosure shown in Figure 5. The power cable from the generator is connected (a). Contactors and fuses connect and disconnect the different branches of

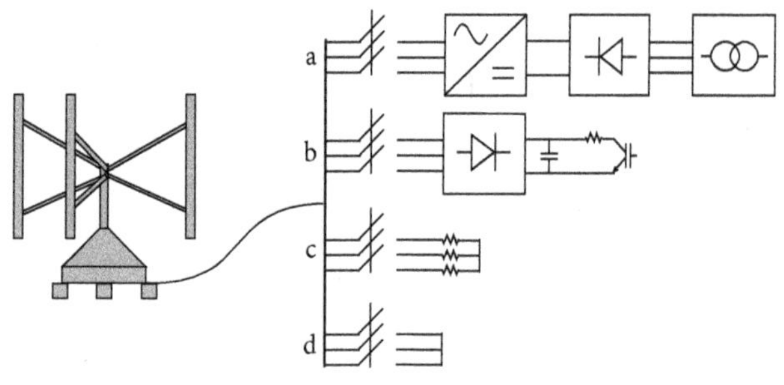

Figure 3: Schematic illustration of the control system. The turbine and generator have a common shaft. The three phases of the generator can be connected to different parts of the control system. (a) shows startup (right to left) where a DC source feeds an inverter which is controlled to start the machine, (b) is the regular controlled load, (c) is an emergency brake, and (d) is the parking brake.

Table 2: Permanent magnet synchronous machine in Simulink.

Back EMF waveform	Sinusoidal
Mechanical input	Torque
Stator phase resistance	0.335 Ω
Inductances ($L_d = L_q$)	3.5 mH
Flux linkage	1.29 Vs
Inertia	2445 kgm^2
Initial rotation	1 rad/s

the control system (b). A diode rectifier is mounted on a heat sink (c), capacitors are mounted in a separate enclosure (d), and two IGBT's (for redundancy) control the load. The inverter used for start-up is mounted on a heat sink (f). The short circuit that serves as a parking brake is manually operated (g).Voltage and current measurements are collected in three printed circuit boards mounted in zink boxes (h). The CompactRIO (i) is the interface between hardware and software.

LabView and a CompactRIO from National Instruments (http://www.ni.com/) were chosen as software and interface to hardware. The CompactRIO includes a field programmable gate array (FPGA) for small, fast data operations, a real-time processor, and a chassi that can contain four or eight modules. There is a number of different types of modules for digital and analog input and output, relays, and so forth providing the interface needed for both measurements and IGBT drivers. The alternative to this would have been to use a microprocessor and to construct an interface between hardware and software. This solution would perhaps cost less but would be more difficult to build upon and develop.

4.1. Measurement System. Measurements are necessary for the control system but are also wanted for analysis of the experimental setup. These two functions are different: for control of the system, the required sample rate may be lower, data needs to be analyzed quickly, and there is no need to save long data series.

Voltages and electrical currents from the generator are measured on-land, as well as voltage and currents in the load. From the generator line-to-line voltages of the generator, as well as line-to-neutral voltage of one phase are measured. The load (DC) voltage is measured as well as the voltage from the DC source for start-up. Currents from the generator, from the start system, to the capacitors and to the load are measured. Currents are sensed using current transducers (HAL 100 and HAL 200 from LEM (http://www.lem.com/)). Voltage division by approximately 37 is applied to all voltages. Signals from the current transducers and divided voltages are measured using C-series module NI 9205. Sampling rate can be decided by the user.

Rotational speed is measured using the signal from one of the Hall sensors used for start-up (see below).

As the only visual contact with the generator and turbine, a network camera (AXISM5014) has been installed in the generator, as well as LEDs for light. The network camera can pan, tilt, and zoom and is powered over ethernet.

Torque measurements on the shaft were considered too much of a complication for a first prototype in an aquatic setting. However, torque measurements were possible during lab tests prior to deployment. While designing the control system, it was not known if real-time water current measurements would be available. Should measurements become available, it should be possible to access them via the CompactRIO, or from a computer file.

The main control program consists of two loops: a monitoring loop for processing and storing measurements and a control loop that executes the chosen control operation. New control algorithms should be easy to implement in the control loop as new cases.

4.2. Load Control. Power from the generator is delivered to the control station in three phase cables and a common neutral. The neutral is left ungrounded and is used for measurement of phase voltage.

The three phases are fed to a rectifier and a capacitor bank to stabilize the DC level. The DC power is consumed in a resistive load via an IGBT; see Figure 3(b) in which the circuit is illustrated. The IGBT functions as a switch and is turned on and off with a pulsed signal, the duration (or duty cycle) of which can be controlled, determining the amount of power extracted from the generator and thus controlling the rotational speed of the generator rotor and turbine. See also Figure 5.

The load IGBT is chosen based on simulation results, given a hefty safety margin, and matched with the supplier's current stock, resulting in an SKM400GAL12, which includes a freewheeling diode. An extra IGBT is mounted in parallell for redundancy and can be used alternately if the primary one should falter. The IGBTs chosen are a type previously used within renewable energy projects at Uppsala University. They have performed satisfactorily, and a driver for them has been developed in house. IGBTs are controlled with digital signals from a C-series module NI 9401.

The load is a set of 1 Ω resistors (Vishay RPS 500 series) mounted on heat sinks. The nominal three-phase load for the generator is 2.5 Ω per phase, but an appropriate DC load has been estimated to be approximately 1 Ω or less (depending on

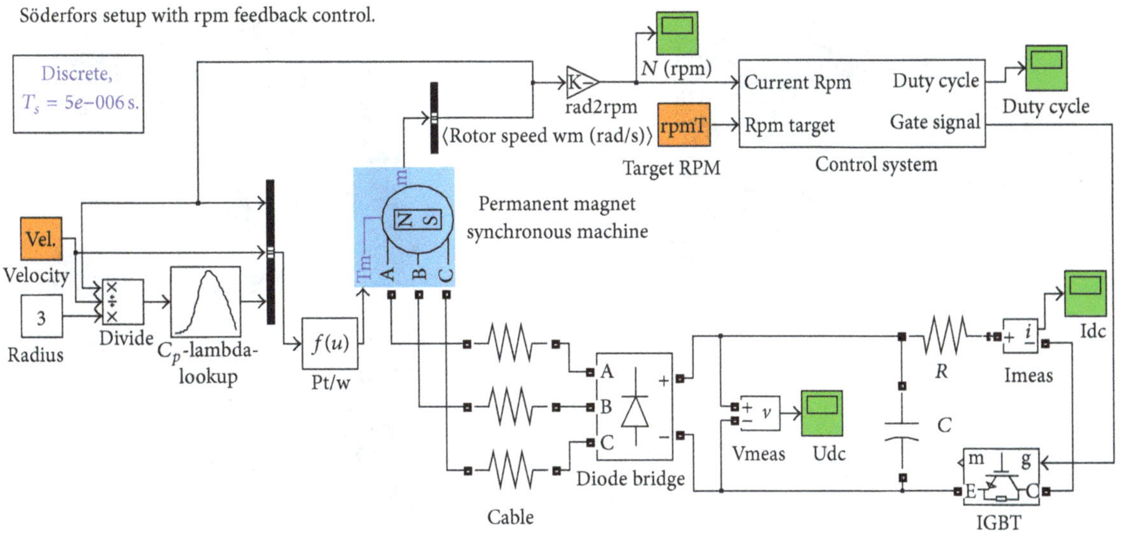

FIGURE 4: Simulink model of the Söderfors turbine and generator with control of rotational speed.

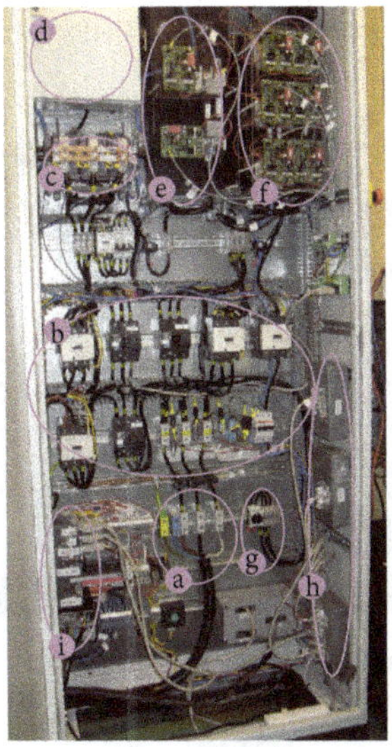

FIGURE 5: The electrical enclosure at the control station. Generator power cable (a), contactors and fuses (b), diode rectifier (c), enclosed capacitors (d), load control IGBT's (e), startup inverter (f), manually operated parking brake (g), voltage and current measurements (h), and CompactRIO (i).

the water current speed and load duty cycle used). Though variable, the default DC load is 30 resistors connected in series and parallel to 5/6 Ω. Each resistor is rated at 500 W, resulting in a load capacity of 15 kW.

The software implementation of the load control is essentially a control of the duty cycle of the IGBT. A proportional factor (which can be set by the user) is applied to the error between a desired state (e.g., rotational speed) and the actual state.

4.3. Start-Up. The start system includes three Hall latches in the generator to sense the position of the rotor: a DC source in the form of a three-phase transformer and diode bridge, a three-phase IGBT inverter, and a control program.

The three-phase transformer and diode bridge were purchased as a unit from Tramo ETV (http://www.tramoetv.se/). The transformer is connected to a standard three-phase outlet providing it with 400 V at 50 Hz. It has four taps, giving a secondary voltage of 190, 170, 150, or 130 V.

The inverter used for the start consists of three SKM300GB12 IGBTs (two in each) mounted on a heat sink, and driver circuits for each. The IGBTs are controlled by digital signals from a C-series module NI 9401 in the CompactRIO.

The Hall sensors mounted on the stator are latch type sensors, A1210, from Allegro (http://www.allegromicro.com/). Three sensors are mounted on a printed circuit board (PCB) containing supplemental electronic components according to the manufacturer's application notes. The PCB is designed to separate the sensors by 120 electrical degrees in the stator and is screwed onto a plastic mount on the stator before winding see Figure 6. The diameter of the stator is large enough that the PCB bends to fit the rounding of the stator. Four separate PCBs have been mounted in the stator for redundancy. The Hall sensor signals are read by a C-series module NI 9401.

Software implementation of the startup entails reading the signals from the Hall sensors, identifying which IGBTs to have on, and giving on signals to those IGBTs. A PWM block allows a gradual increase of rotational speed.

4.4. Emergency Brake and Standstill. An emergency brake which can be described as a damped short circuit has also been implemented. Via a mechanical emergency button or at a signal given from the computer program, a three-phase load of 0.5 Ω per phase is applied. Simulations in Simulink have shown that this load should, in an extreme case, be able to stop the turbine within 2 seconds.

At standstill, the generator can be short circuited via a switch in the main enclosure at the control station, as well

Figure 6: Hall sensors on a PCB screwed to a plastic mount attached to the stator.

as in the bridge enclosure. The turbine is expected to remain fairly still (under 1 rpm) when short circuited at the control station and even more still when shorted in the bridge enclosure.

4.5. On-Site Installation. The on-site installation comprises a submerged part, that is, the turbine and generator, and an on-land control station with the equipment described above. These two parts need to be connected by power and data cables. In order to facilitate the installation of the equipment, an electrical enclosure is mounted on the bridge railing, about 40 m from where the generator will be placed. The power and signal cables from the generator have been provided with weather proof connectors that are easily attached to the bridge enclosure. The bridge enclosure also provides a point close to the underwater site where the generator can be short circuited, signals can be amplified if necessary, power over ethernet injected, and so forth; see Figure 7. (Acoustic Doppler Current Profilers (ADCPs) are used for measuring water velocity.)

Cables from the bridge enclosure follow the bridge to land where they are buried until they reach the measurement station, slightly more than 100 m away. In the station, the power cables are screwed into terminals in the main enclosure, and signal cables are provided with standard connectors. This way, the different parts of the whole setup can be installed separately and also disconnected if necessary. Figure 8 shows the interior of the control station.

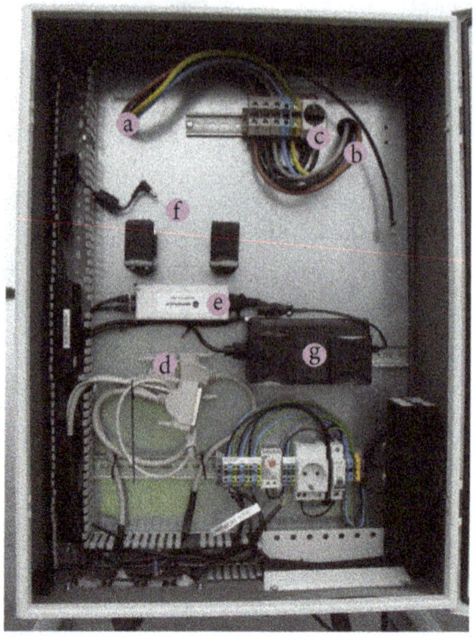

Figure 7: Interior of the bridge enclosure. Power from the generator (a) and to the control station (b) can be manually short circuited in (c). Hall sensor and other signals are relayed in (d). (e) provides power for the network camera. A computer for communication with ADCPs will be mounted (f) and the ADCPs need power (g).

Figure 8: The main enclosure (a), the regular load (b), and the emergency stop load (c) in the control room.

5. Laboratory Tests and Simulations

The generator for the Söderfors setup was assembled in the laboratory at the university. There, it was also possible to test the performance of the generator and some functions of the control system.

5.1. Laboratory Test of Load Control. The generator was connected to a motor drive that could be torque controlled. In this setup, it was possible to test if the load control system could keep a specified rotational speed and a specified load voltage. Between about 4 and 15 rpm the control system was able to hold a fairly constant rotational speed when the torque was varied. Switching frequencies for the load IGBT was tested from 10–500 Hz. In order to reduce voltage transients over the rectifier, two snubber capacitors of 100 nF each (BFC238670104 http://www.vishay.com/) were added over the IGBT.

Keeping a fixed duty cycle was also tested. Figure 9 shows the line to line voltage and current in one phase of the generator while keeping a fixed duty cycle of 10.5%. Figure 10 shows the voltage and current through the load for the same time period.

5.2. Laboratory Test of Start-Up. Start-up of the generator for Söderfors was tested using laboratory voltage supplies, since the DC source had not yet been decided on. The generator shaft was not connected to anything during this test, that is, neither to the turbine or the motor drive, and the seal on the shaft had not yet been mounted. This means that the start-up situation was simplified in that the friction was reduced, and no torque was delivered from the turbine. However, the start

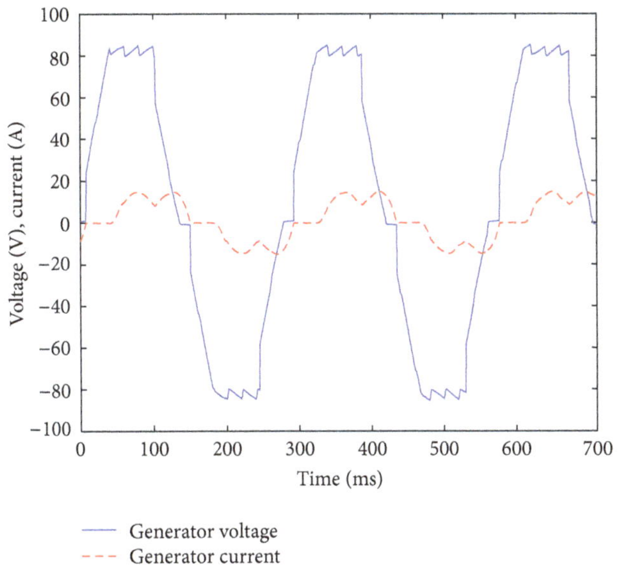

FIGURE 9: Generator voltage (line-to-line) and current while maintaining a 10.5% duty cycle.

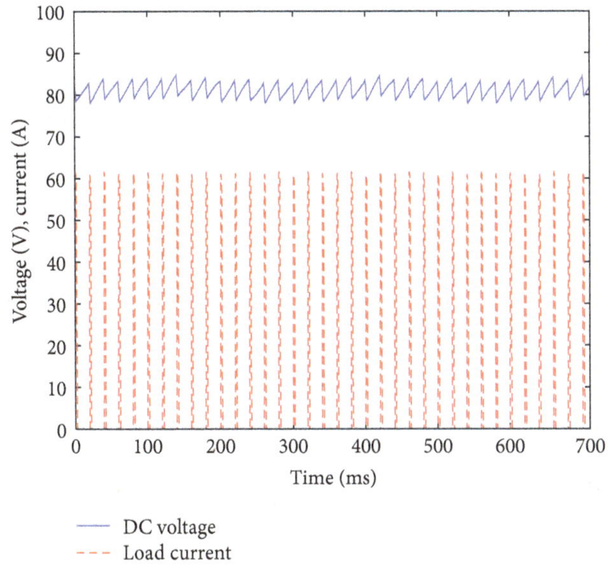

FIGURE 10: Load current and voltage while maintaining a 10.5% duty cycle.

system was able to rotate the generator smoothly up to 13 rpm (the voltage supplies could not provide higher voltage).

After the rectified transformer had been decided on and delivered, a new start-up test was performed using a similar generator [8] still in the lab. The new DC source does not have the controlled current limitation in it as the laboratory voltage supplies do. This situation gave opportunity to trim the over-current protection on the IGBT drivers and provided some practical experience. After some adjustments, the start-up system functioned well.

5.3. Simulation Results. The implementation of the system in Simulink was used for dimensioning components, as mentioned earlier. It was also used to characterize the expected performance of the system. The following figures are a result of a parameter sweep of water velocity and target rpm, stepping 1 rpm and 0.2 m/s, as well as a sweep of $\lambda = 3.5$ for the same velocities. The system is initiated with a rotational speed of 1 rad/s and simulated for 40 s. Data from the last 10 s are averaged. Where the tip-speed ratio is far from 3.5, it is difficult to reach the target rotational speed. At tip-speed-ratios above 4, the control system is not able to keep a constant rotational speed, rather the speed will oscillate around the target. Also, the control system is not dimensioned for maximum power capture at water speeds as high as 2 m/s, so as the water speed increases, the control system becomes strained.

In Figure 11 the output power is plotted against the tip-speed ratio, λ, for different water speeds. As can be seen in the figure, the maximum power output for a given water speed is reached not necessarily at $\lambda = 3.5$. Close to the design water velocity, the system power is maximized at approximately the same λ as the turbine power capture is maximized at, but for other velocities, a higher λ maximized the power output. As mentioned earlier, iron losses are not included in this simulation and may affect the result. However, we see that maximizing the system power output is a different operation than maximizing the turbine power capture. A higher rotational speed will increase the voltage output, thus decreasing the current and, consequently, the copper losses. An increase in rotational speed will increase iron losses, but as shown previously [11, 18], the copper losses dominate in this type of machine at higher water velocities.

Figure 12 shows the power capture of the turbine for $C_p = 0.36$, the maximum output power for each velocity, and the power output when operating at tip-speed ratio $\lambda_0 = 3.5$. The difference between the maximum power output and the power output at λ_0 increases with the water velocity. This is due to the increased effect of copper losses, and thus the increased benefit of reducing them.

Figure 13 shows the duty cycle (or on time) of the load versus rotational speed for different water velocities. The water velocity is expected to be mostly around 1–1.4 m/s, and the load control system appears to handle these velocities well with a maximum duty cycle of about 0.5. At 2 m/s, the full load capacity of 5/6 Ω is used and is not sufficient to control the turbine. With a smaller load resistance, it may be possible to operate at $\lambda \approx 3.5$. The current load is rated at 15 kW, which is possibly sufficient at 1.8 m/s given that there are additional losses in the system not incorporated in the simulations.

6. Conclusions

A control system for an experimental setup for hydrokinetic energy conversion has been designed and realized. It performs satisfactorily in a laboratory environment but has yet to prove itself when controlling equipment in an aquatic setting.

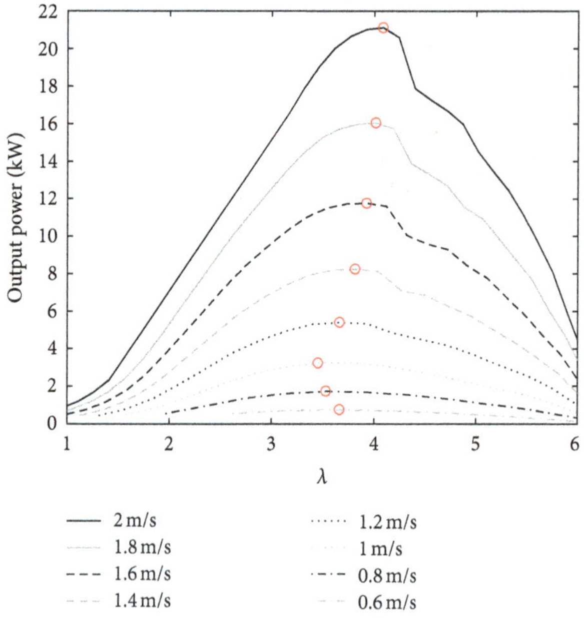

FIGURE 11: Power output of the system versus tip-speed ratio. The maximum power output for a certain water velocity, marked with a red circle on each curve, is at a higher tip-speed ratio than that which maximizes the turbine C_p.

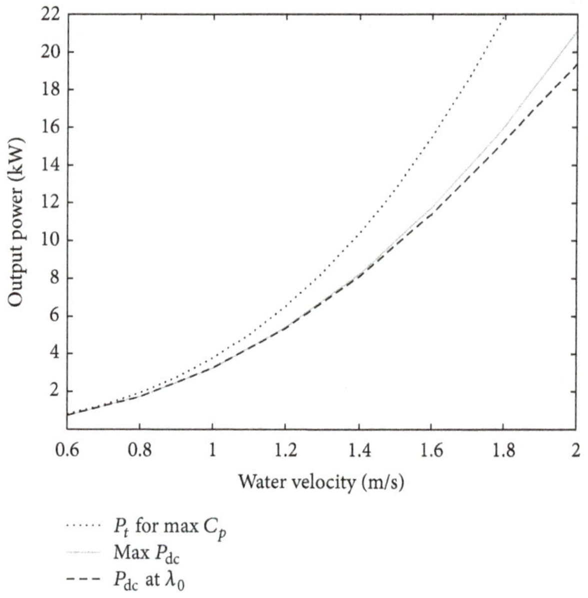

FIGURE 12: Power capture of the turbine at maximum C_p, maximum output power for each velocity, and power output for $\lambda = 3.5$.

Simulations show that maximizing the energy output of the system may coincide with a higher rotational speed than that resulting in maximum turbine power, especially at higher water velocities. The load control system for Söderfors can handle maximizing the power output within the typical velocity range of 1–1.4 m/s.

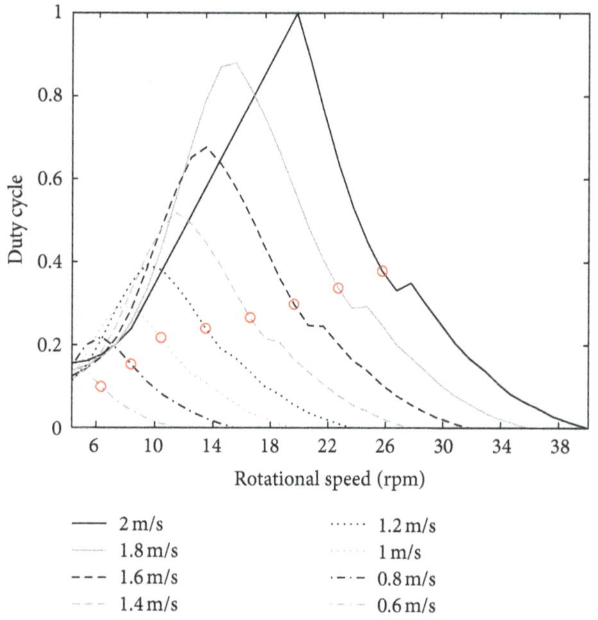

FIGURE 13: Duty cycle of the load versus rotational speed for different water velocities. The operating point with maximum power output is marked in each curve with a red circle.

Acknowledgments

This work was carried out with support from The J. Gust. Richert Foundation for Technical Scientific Work, STandUp for Energy, The Swedish Research Council (Grant no. 621-2009-4946), Vattenfall AB, and Ångpanneföreningen's Foundation for Research and Development. Thanks to the rest of the hydro-kinetics research group for your efforts and general support. Mårten Grabbe designed and built the generator. Anders Goude modeled the turbine. Emilia Lalander and Staffan Lundin characterized the Söderfors site. Thanks to Jon Kjellin, Fredrik Bülow, and Rickard Ekström for supportive discussions.

References

[1] K. Thomas, *Low speed energy conversion from marine currents [Ph.D. thesis]*, Uppsala University, 2007.

[2] K. Yuen, *System aspects of marine current energy conversion [Licentiate thesis]*, Uppsala University, 2008.

[3] M. Grabbe, *Marine current energy conversion—resource and technology [Licentiate thesis]*, Uppsala University, 2009.

[4] E. Lalander, *Modelling the hydrokinetic energy resource for in-stream energy converters [Licentiate thesis]*, Uppsala University, 2010.

[5] M. Grabbe, K. Yuen, A. Goude, E. Lalander, and M. Leijon, "Design of an experimental setup for hydro-kinetic energy conversion," *International Journal on Hydropower and Dams*, vol. 16, no. 5, pp. 112–116, 2009.

[6] K. Yuen, S. Lundin, M. Grabbe, E. Lalander, A. Goude, and M. Leijon, "The Söderfors project: construction of an experimental hydrokinetic power station," in *Proceedings of the 9th European Wave and Tidal Energy Conference (EWTEC '11)*, pp. 1–5, Southampton, UK, September 2011.

[7] E. Lalander and M. Leijon, "Numerical modeling of a river site for in-stream energy converters," in *Proceedings of the 8th European Wave and Tidal Energy Conference (EWTEC '09)*, pp. 826–832, Uppsala, Sweden, 2009.

[8] K. Thomas, M. Grabbe, K. Yuen, and M. Leijon, "A low speed generator for energy conversion from marine currents—experimental validation of simulations," *Proceedings of the IMechE A: Journal of Power and Energy*, vol. 222, no. 4, pp. 381–388, 2008.

[9] J. Manwell, J. McGowan, and A. Rogers, *Wind Energy Explained—Theory, Design and Application*, John Wiley & Sons, Chichester, UK, 2002.

[10] I. Munteanu, A. I. Bratcu, N.-A. Cutululis, and E. Ceangă, *Optimal Control of Wind Energy Systems, Towards a Global Approach*, Advances in Industrial Control, Springer, 2008.

[11] K. Yuen, K. Thomas, M. Grabbe et al., "Matching a permanent magnet synchronous generator to a fixed pitch vertical axis turbine for marine current energy conversion," *IEEE Journal of Oceanic Engineering*, vol. 34, no. 1, pp. 24–31, 2009.

[12] J. A. Clarke, G. Connor, A. D. Grant, and C. M. Johnstone, "Regulating the output characteristics of tidal current power stations to facilitate better base load matching over the lunar cycle," *Renewable Energy*, vol. 31, no. 2, pp. 173–180, 2006.

[13] Z. Hu and X. Du, "Reliability analysis for hydrokinetic turbine blades," *Renewable Energy*, vol. 48, pp. 251–262, 2012.

[14] E. A. Bossanyi, "The design of closed loop controllers for wind turbines," *Wind Energy*, vol. 3, no. 3, pp. 149–163, 2000.

[15] G. Mokryani, P. Siano, A. Piccolo, and Z. Chen, "Improving fault ride-through capability of variable speed wind turbines in distribution networks," *IEEE Systems Journal*, 2012.

[16] J. Kjellin and H. Bernhoff, "Electrical starter system for an H-rotor type VAWT with PM-generator and auxiliary winding," *Wind Engineering*, vol. 35, no. 1, pp. 85–92, 2011.

[17] T. L. Skvarenina, *The Power Electronics Handbook*, Industrial Electronics, CRC Press, 2002.

[18] S. Eriksson and H. Bernhoff, "Loss evaluation and design optimisation for direct driven permanent magnet synchronous generators for wind power," *Applied Energy*, vol. 88, no. 1, pp. 265–271, 2011.

Adaptive Impedance Control to Enhance Human Skill on a Haptic Interface System

Satoshi Suzuki and Katsuhisa Furuta

Department of Robotics and Mechatronics, School of Science and Technology for Future Life, Tokyo Denki University, 5 Asahi-Chou, Senju, Adachi-Ku, Tokyo 120-8551, Japan

Correspondence should be addressed to Satoshi Suzuki, ssuzuki@fr.dendai.ac.jp

Academic Editor: Lili Ma

Adaptive assistive control for a haptic interface system is proposed in the present paper. The assistive control system consists of three subsystems: a servo controller to match the response of the controlled machine to the virtual model, an online identifier of the operator's control characteristics, and a variable dynamics control using adaptive mechanism. The adaptive mechanism tunes an impedance of the virtual model for the haptic device according to the identified operator's characteristics so as to enhance the operator's control performance. The adaptive law is derived by utilizing a Lyapunov candidate function. Using a haptic interface device composed by a *xy*-stage, an effectiveness of the proposed control method was evaluated experimentally. As a result, it was confirmed that the operator's characteristics can be estimated sufficiently and that performance of the operation was enhanced by the variable dynamics assistive control.

1. Introduction

An impedance control is a key technology of the force/motion control for any mechanical systems such as an active vehicle suspension, a power steering system, a machining and handling by manipulators, and a tele-operation system. Since dynamics of such controlled mechanism can be adjusted by changing the virtual impedance model, this method is effective to adapt to an ever-changing environment and conditions. Also a biological system has acquired similar strategy of the variable impedance control in the course of an evolution. It is known that an impedance of a musculoskeletal system is changed dynamically during walking, running, and moving the hand [1]. Therefore, variable impedance methods have been studied for artificial legs/orthosis [2] and material-handling machines [3, 4]. Parameters of those impedance control methods are, however, often tuned empirically and intuitively; hence, the system designers have to adjust them according to individuals. Due to this issue, users sometimes have to adapt themselves to the controlled machine when the tuning condition given by the designer is not adequate for the user. To resolve this paradox, the following approach is ideal; the control characteristics of each user are identified during the operation, and then control of the machine is adjusted adaptively according to the identified user's characteristics. While several similar approaches concerning online variable impedance control are reported, troublesome processes such as a training phase [5] or an empirical tuning for different types of motion [6] are required. Since these approaches are not real adaptive control, a realization of an automatic tuning mechanism without intervention of system designers is expected. Therefore, the present paper presents a design procedure of true adaptive variable impedance control for an assisting system which is considered with the following properties based on the previous method presented in [7]:

(a) adaptivity to control characteristics of individual user,

(b) derivation of adaptive law for variable impedance tuning based on an adaptive control theory.

Main purpose of the control design proposed here is a development of an adaptive tuning law of the machine dynamics, of which parameters are fixed in an ordinary mechatronics system, in order to enhance manipulation performance of whole of a human-machine system. And, the purposes of the present study are as follows:

(c) experimental evaluation,

(d) confirmation of the benefit and issue.

Item (a) is realized with an on-line identification based on an assumed model of the human controller. Concerning item (b), an adaptive control law to adjust impedance parameters of the virtual model is derived to ensure the stability and performance of the whole human-machine system. Item (c), evaluation, was performed using a haptic interface device through a point-to-point operation task. Issues and analysis denoted at item (d) are discussed based on the results of the experiment.

This paper is organized as follows. In Section 2, a concept of the adaptive impedance control and its background are mentioned. The haptic interface system and task which were used in the experiment are explained there. Section 3 explains a procedure of the presented assistive control, and its theoretical proof is given there. Section 4 shows results of the experiment and analysis to confirm the effectiveness of the presented method. Last Section 5 presents the conclusion.

2. Human Assistive System

2.1. Concept of Adaptive Impedance Control. In order to design an assistive mechanism for user's manipulation in a human-in-the-loop system, an adequate human modeling and a feasible assistive control are required. Human modeling has been studied in the field of control engineering from its early beginnings such as a linear servo control model [8], a PID-base time-variant model [9], and an optimal control model [10]. Those models can express a human behavior well for each assumed situation; however, it is inadequate to explain the learning process of a user from the beginner phase to the expert phase. To find adequate model which can treat a human adaptability, it is adequate to refer the voluntary motion control of a musculoskeletal system. The reason is that such model is formulated to explain a process of a human development, and most popular model is a feedback-error-learning model [11, 12]. This model can be utilized to explain the learning of an external unknown dynamics since such an external system can be thought as an extension of our body. On the learning process of the external dynamics, a delay has to be considered because it concerns the stability and performance of whole human-machine system. The delay arises certainly at the visual processing and at the neural transmission between a brain and sensory receptor/muscles. Such undesirable effect given by the delay is compensated by an internal feedback compensation using an efferent neuron and by a delay compensation mechanism which is explained by the Smith predictor theory [13]. Additionally, as shown in Kleinman's research of the dynamics of a pilot [10], a human (i.e., pilot) has a high ability to compensate the delay in the response of a vehicle. Hence, if a time-delay effect inside the

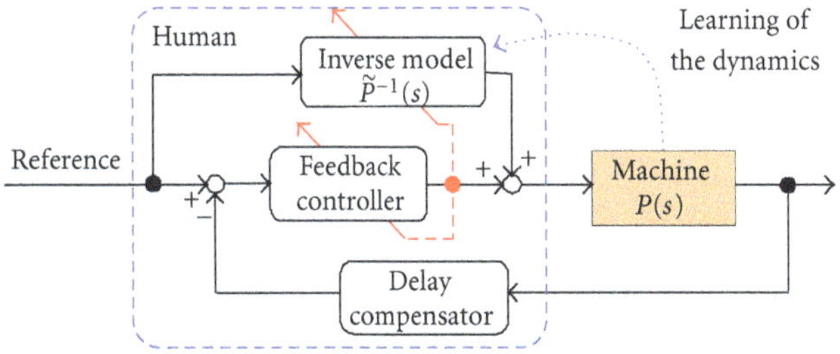

FIGURE 1: Human model inside a human-machine system.

machine side system does not change, it is expected that the delay in the human-machine system can be compensated relatively easily. Therefore, the block diagram shown in Figure 1 appears adequate for designing of a human-machine assist system. The model mentions that human learns the unknown dynamics of the machine and uses the identified model as an inverse model for the manipulation of the machine.

The concept shown in Figure 1 indicates that difference of the machine's dynamics affects indirectly to the learning of the operator. If the operated machine can change its own dynamics characteristics so as to be learned by individual operator without difficulty, performance of the operation would be enhanced. Therefore, in order to enhance the operator's performance, as shown in Figure 2, an original dynamics of the machine is replaced to a virtual dynamics model from the operator's side by making a local loop feedback with a virtual internal model control. In short, the impedance of the virtual dynamics model is modified so as to decrease an error which relates with each task performance. To summarize this discussion, the following three functions are required to realize aforementioned adaptive impedance control.

Step 1. Virtual internal model (VIM) control.

Step 2. Online identification of the operator.

Step 3. Adaptive mechanism to tune the VIM.

The VIM control for Step 1 is realized by making a local servo system that tracks the output of a virtual impedance model. The servo control input law is designed using Linear-Quadratic Regulator (LQR). Identification for Step 2 is performed by assuming a parametric model of operator's control characteristics. Concerning Step 3, the adaptive mechanism is designed by changing the impedance parameters of the VIM obtained at the Step 3 after derivation of the adaptive law of the VIM model based on a Lyapunov-like function. Details of this process are explained in Section 3.

2.2. Experimental System. A haptic interface system, which is shown in Figure 3, was used to evaluate the adaptive assistive control presented in Section 3. The haptic device consists of a two degree-of-freedom planer xy-stage, produced by NSK corporation, and a real-time CG monitor programmed by visual C++. The xy-stage is driven by two linear direct drive

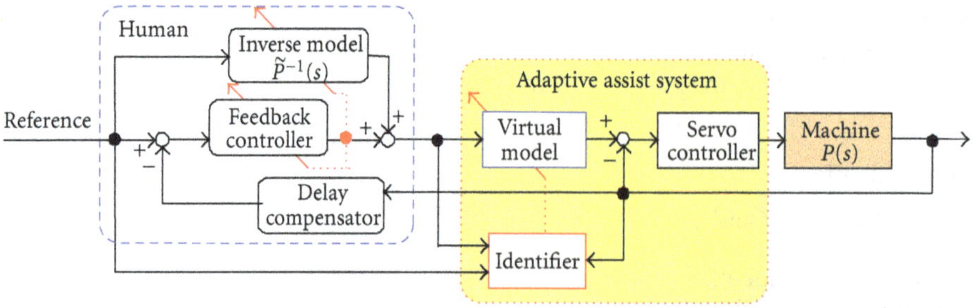

FIGURE 2: Structure of human assistive system with the adaptive impedance control.

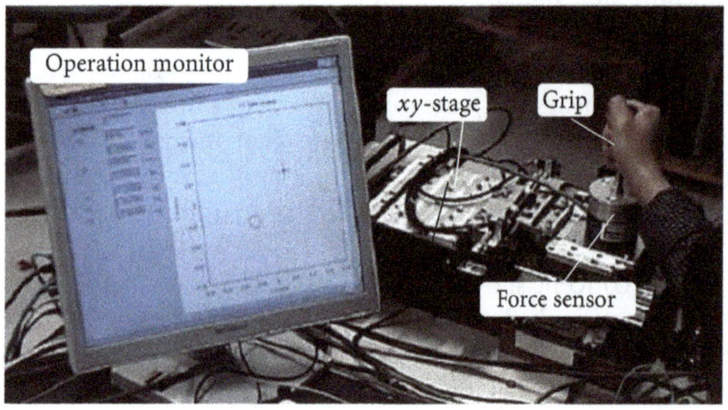

FIGURE 3: Haptic interface devices.

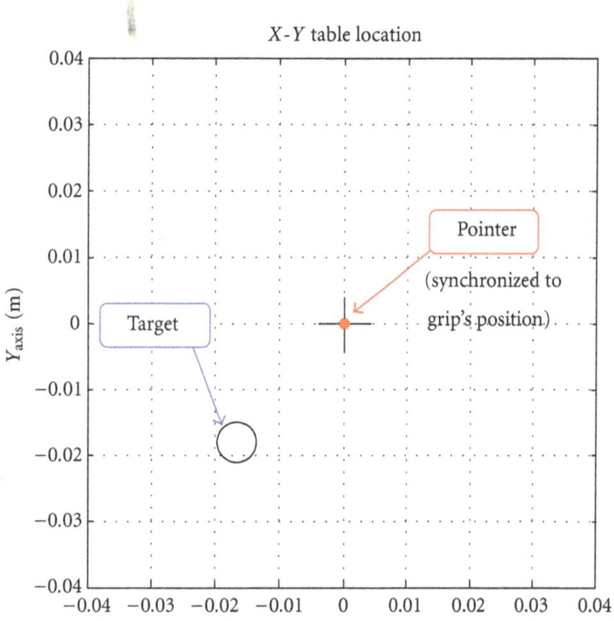

FIGURE 4: Operation monitor of the haptic interface system.

motors. The operator moves a grip attached to the xy-stage and a pointer displayed on the CG monitor (see Figure 4) is also moved according to the position of the grip. Operator's hand force is measured by a 6-axis force sensor embedded between the grip and the stage. The x- and y-axes of the stage do not effect each other because of a mechanically independent design. Computations of the control of the stage and the CG displaying are executed by a PC/AT 3 GHz computer under real-time scheduling control. The control interval is 2 ms, and the movable range is about 62 mm in both x and y directions.

2.3. Task of Manipulation. Since a point-to-point (PTP) manipulation is a popular task both in a daily life and in an industrial situation, the PTP-task was adopted for verification of the presented method. The PTP tasks were repeated by changing the target's position at random so as to keep the distance constant from each last target to the next one. As soon as the target is displayed on the monitor, the operator moves the pointer to the target by manipulating the grip of the xy-stage. When position of the pointer is kept inside the target circle for 3 seconds, one PTP motion (one trial) is finished, and then a new next target circle is displayed at random. To reduce the fatigue of the participant, ten-second rest was given to the participant after every five trials.

3. Design of Adaptive Impedance Control

3.1. Virtual Internal Model Control (Step 1). A procedure to apply the virtual internal model (VIM) control [14] to the xy-stage is mentioned in this section. Any mechanical mechanism includes nonlinearity caused by friction, variances of viscosity, and unknown dynamics; hence, it is difficult to apply a linear system control theory to actual machine without any nonlinear compensation. Since VIM is effective to suppress inherent characteristics of mechanical components such as frictions, an adaptive control for linear systems can produce an effect. Although the haptic system used in the experiment has two degree-of-freedom motions, controllers of the x- and y-axes can be designed separately thanks to the mechanical independence; hence, subscripts of x and y are omitted in later explanation. Variables and parameters of the haptic device model are shown in Figure 5 and Table 1.

The block diagram of the virtual internal model control is shown in Figure 6 and the related variables and parameters are summarized in Table 2. Dynamic equations of the stage and the virtual model are expressed as follows:

$$m_p\ddot{x}_p + d_p\dot{x}_p = f_h + f_a, \tag{1}$$

$$m_r\ddot{x}_r + d_r\dot{x}_r = f_h. \tag{2}$$

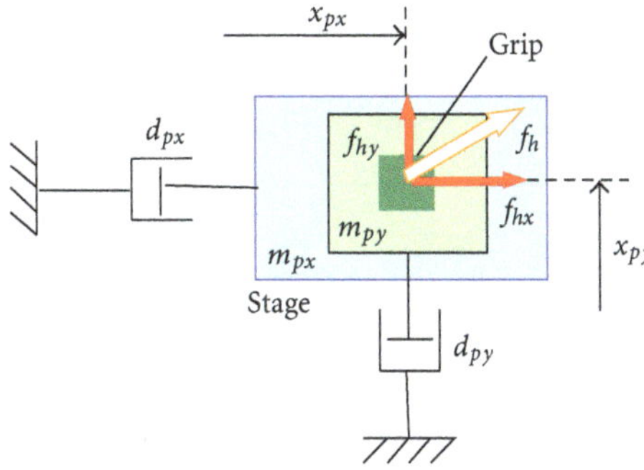

FIGURE 5: Model of the xy-stage.

TABLE 1: Parameters and variables of haptic interface device.

Variables/parameters	Unit	Meanings
x_{p*}	[m]	Position of a grip
f_{h*}	[N]	Force from an operator
f_{a*}	[N]	Force from an actuator
m_{p*}	[kg]	Mass of a stage
d_{p*}	[Ns/m]	Viscosity of a stage

$*$: x or y for x- and y-axis.

TABLE 2: Parameters and variables for virtual model control.

Variables/parameters	Unit	Meanings
x_r	[m]	Position of a virtual model
e_r	[m]	Error (=: $x_r - x_p$)
m_r	[kg]	Mass of a virtual model
d_r	[Ns/m]	Viscosity of a virtual model

Defining an error as $e_r := x_r - x_p$, (1) and (2) are transformed into

$$\frac{d}{dt}\begin{bmatrix} e_r \\ \dot{e}_r \end{bmatrix} = \begin{bmatrix} 0 & 1 \\ 0 & 0 \end{bmatrix}\begin{bmatrix} e_r \\ \dot{e}_r \end{bmatrix} + \begin{bmatrix} 0 \\ 1 \end{bmatrix} u$$
$$u := \frac{-d_r \dot{x}_r + f_h}{m_r} + \frac{d_p \dot{x}_p - f_h - f_a}{m_p}. \quad (3)$$

Minimizing the error defined by (3) makes the stage conform to a response of the virtual model described by (2). To compensate steady-state error, the integral variable $\int e_r$ is taken into consideration in the state vector as follows:

$$\frac{d}{dt}z = Az + Bu, \quad (4)$$

where

$$A := \begin{bmatrix} 0 & 1 & 0 \\ 0 & 0 & 1 \\ 0 & 0 & 0 \end{bmatrix}, \quad B := \begin{bmatrix} 0 \\ 0 \\ 1 \end{bmatrix}, \quad z := \begin{bmatrix} \int e_r \\ e_r \\ \dot{e}_r \end{bmatrix}. \quad (5)$$

The control law is calculated using an LQR method with the quadratic criterion:

$$J = \int_0^\infty \left(z^T Q z + u^T R u \right) dt, \quad (6)$$

where a positive semi-definite $Q \in R^{3\times 3}$ and a positive definite $R \in R^{1\times 1}$ are weighting matrices. The input is given as

$$u = -Fz, \quad F := R^{-1}B^T P, \quad (7)$$

where P is a symmetry positive-definite matrix of a Riccati Algebraic Equation given by

$$PA + A^T P - PBR^{-1}B^T P + Q = 0. \quad (8)$$

Since (7) is expanded as

$$-Fz = \frac{-d_r \dot{x}_r + f_h}{m_r} + \frac{d_p \dot{x}_p - f_h - f_a}{m_p}, \quad (9)$$

a final form of the control law is obtained as follows:

$$f_a = \frac{m_p}{m_r}\left(f_h - d_r \dot{x}_r\right) + d_p \dot{x}_p - f_h - m_p F z. \quad (10)$$

For the actual apparatus used in the experiment, parameters of VIM were specified as $m_{rx} = m_{ry} = 50$ [kg] and $d_{rx} = d_{ry} = 50$ [Ns/m] to intentionally obtain a slightly difficult manipulation feeling as a training test with consideration of input range of the actuators. The weighting matrices in (6) for the LQR servo design were decided as $Q = \mathrm{diag}(150, 1.12 \times 10^7, 5100)$ and $R = 1$. As a result, the feedback gain matrix was obtained as $F = [10.9, \ 3007.9, \ 100.7]$.

3.2. Online Identification of Human Control Characteristics (Step 2). Human control characteristics are complex because various kinds of compensators, such as an oculomotor control, a proprioceptive control, and a neuromuscular control, are related to each other [13]. There is, however, a fairly large body of data that can be explained by a linear model plus time delay [15] when an operation condition is limited. One of most famous models supporting such linear model assumption is a crossover model. This model insists that a frequency transfer function of a skilled operator in a man-machine system adapts to make the total system kept unchanged under a variation of the controlled system dynamics. In other words, human changes own control characteristics so that a closed-loop transfer function of whole human-machine system becomes a first-order system at a wide frequency band and the human plays a role of simple linear model. Also in previous study of the present authors, an identification analysis of the skilled operator's frequency characteristics showed an existence of the cross-over model through a juggling task using a haptic test device [16]. Hence, whole system relating a voluntary motion is simplified into the three components in this study: a linear controller inside a brain, a neuromuscular dynamics, and reaction time delay. After a learning of the machine dynamics is sufficiently finished, the human can be considered as a simple feedback controller which moves the grip to the target position by watching the monitor in case of the PTP task. Finally, a block diagram of a visual voluntary motion control is assumed as a feedback model as shown in Figure 8. In the figure, r is a reference position for a pointer, e_h is an error between the

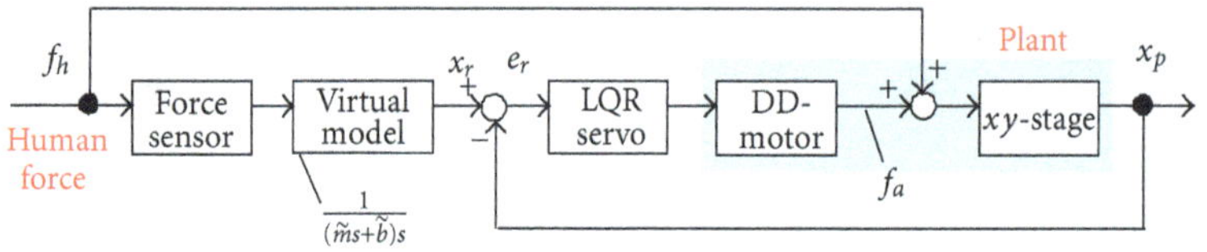

Figure 6: Block diagram of virtual internal model control.

target and the present pointer, and u_h is an input computed by a brain controller. Here, the plant block is a virtual xy-stage of which impedance property is adjusted by the VIM control. The neuromuscular dynamics can often be approximated by a first-order lag [17] and a simplest human controller is a PD controller [9]; hence, the human transfer function, $G'_h(s)$, is assumed in this study as

$$G'_h(s) = \frac{K_d s + K_p}{Ts+1} e^{-Ls}, \quad (11)$$

where K_p, K_d, T, and L are a proportional gain of the human brain controller, the differential gain of it, a time constant of the neuromuscular system, and reaction time delay, respectively. As discussed in Section 2.1, compensation of the delay factor is necessary not only for voluntary motion control in a human but also for an adequate human-machine system, and a human has an excellent ability to compensate the delay effect. And an influence of the delay to the control characteristics of the whole human-machine system depends on a response speed of the machine and the task condition. Therefore, in the present study, the response delay of participants was investigated as a preliminary experiment using the VIM control which was designed at Section 3.1. Participant aged 22 years was requested to execute the PTP manipulation hundred times. The time that the pointer begins moving just after the new target circle was displayed on the monitor was counted as the response delay. Figure 7 shows the change of the measured time delay. The dots represent measured values, and the solid lines express an approximated third-order polynomial fitting curve from the measured data. This graph shows no conclusive relationship between time delay and the number of trials, and the value is almost constant at about 0.4 second. Additional nine participants showed same tendency, and significant difference between individuals was not confirmed. For this reason, it was expected that the simple data shift would be sufficient to compensate the delay effect in the identification for the present study. Therefore, the time delay factor described in (11) was omitted for the identification by shifting the measured data for the 0.4 second as a rest time, and the following model was considered for later process:

$$G_h(s) = \frac{K_d s + K_p}{Ts+1}. \quad (12)$$

Applying a bilinear transformation

$$s \simeq \frac{2}{\Delta} \cdot \frac{1-z^{-1}}{1+z^{-1}} \quad (13)$$

to (12) yields the following discrete impulse transfer function $G_h[z]$:

$$G_h[z] = \frac{b_1 z^{-1} + b_0}{a_1 z^{-1} + 1}, \quad (14)$$

$$a_1 := \frac{-2T+\Delta}{2T+\Delta}, \quad (15)$$

$$b_0 := \frac{2K_d + K_p\Delta}{2T+\Delta}, \quad (16)$$

$$b_1 := \frac{K_p\Delta - 2K_d}{2T+\Delta}, \quad (17)$$

where Δ is a sampling interval. From (15)–(17), following equations are derived:

$$T = \frac{\Delta}{2}\frac{1-a_1}{1+a_1},$$
$$K_p = \frac{2T+\Delta}{2\Delta}(b_0+b_1), \quad (18)$$
$$K_d = \frac{2T+\Delta}{4}(b_0-b_1).$$

If a_0, b_0, and b_1 are identified from the input/output response data, characteristic parameters of the human controller can be derived using (18). These parameters are used in a design of the next variable dynamics assistive controller.

3.3. Variable Dynamics Assist Control (Step 3). An assistive control proposed in this paper changes dynamics of the internal model on-line depending on operator's characteristics. A block diagram of the assistive control is shown in Figure 9(a). In the figure, r, y, e, v, and f are a positional reference, a position of the stage, the error, an output of a brain controller, and a force generated by the hand, respectively. It is assumed that (a) K_p, K_d, and T are time-slowing changing parameters and that (b) parameters of a virtual machine $\tilde{m}$ and $\tilde{b}$ can be tuned, because the assumption (b) is realized by the VIM control, that is, $\tilde{m}$ and $\tilde{b}$ are adjustable parameters in this scheme. Figure 9(a) expresses a general human-machine system that includes a human controller $(K_p + K_d)/(Ts + 1) =: C$ and a plant $1/(\tilde{m}s + \tilde{b})s =: P$ for virtual haptic interface device. It can be considered conversely that the system consists of a plant C changing slowly the parameters (K_p, K_d, and T) and the controller P having directly variable coefficients ($\tilde{m}$ and $\tilde{b}$), as shown in Figure 9(b).

Note that the output y of new controller P cannot be changed arbitrarily and that only tuning of the controller's

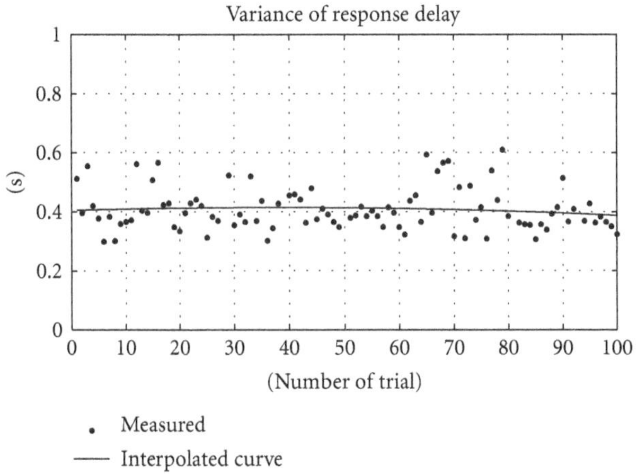

FIGURE 7: Variance of response delay.

coefficients is possible. Moreover, transformation of the block diagram shown in Figure 9(b) yields a general feedback form, as shown in Figure 9(c). In the following, in order to avoid misunderstanding owing to habits, characters for variables x and u are used instead of f and e, respectively. Then, the following equations are obtained:

$$x(s) = \frac{K_p + K_d s}{Ts+1} u(s), \tag{19}$$

$$u(s) = \frac{1}{\left(\widetilde{m}s + \widetilde{b}\right)s} e'(s), \tag{20}$$

$$e'(s) := r'(s) - x(s), \tag{21}$$

$$r'(s) := \left(\widetilde{m}s + \widetilde{b}\right)s \cdot r(s). \tag{22}$$

The purpose of the PTP task is a tracking such that $y \to r$ in the original block diagram shown in Figure 9(a). This means that $e \to 0$ (in Figure 9(b)), that is, $u \to 0$ (in Figure 9(c)); then (20) indicates that $e' \to 0$ as $t \to \infty$. First, choosing a Lyapnov candidate V as $V := (1/2)e'(t)^2$, the condition of convergence is investigated. It can be considered that a closed-loop system shown in Figure 9(c) is almost stable under the assumptions of (a) and (b); hence, it is not always necessary that $dV/dt < 0$ holds for keeping the stability. Second, an update law for $\widetilde{m}$ and $\widetilde{b}$ is derived using a Lyapunov-like analysis. If a step input is chosen for $r(t)$ for the PTP motion, the response of $r'(t)$ defined by (22) becomes almost impulse shape. The moment of $t = 0$ is, however, not important practically because the purpose of the control is an enhancement of the performance of the motion by making the tracking error small which occurs mainly by the positioning near the target position. In addition, the impulsive response converges into zero rapidly, hence, an approximation as $dr'(t)/dt \simeq 0$ holds if $t \gg 0$.

Then, the time-derivative of V can be approximated and can be transformed as follows:

$$\begin{aligned}
\frac{d}{dt}V(t) &= e'(t)\frac{d}{dt}e'(t) \simeq -e'(t)\frac{d}{dt}x(t) \quad (t>0) \\
&= -e'(t)\frac{d}{dt}\mathcal{L}^{-1}\left[\frac{K_p + K_d s}{Ts+1}\frac{1}{\left(\widetilde{m}s+\widetilde{b}\right)s}e'(s)\right] \\
&= -e'(t)\mathcal{L}^{-1}\left[\frac{K_p + K_d s}{Ts+1}\frac{1}{\left(\widetilde{m}s+\widetilde{b}\right)}e'(s)\right] \\
&= -e'(t)\mathcal{L}^{-1}\left[\frac{K_p - K_d/T}{\widetilde{b}T - \widetilde{m}} \cdot \frac{1}{s+1/T}e'(s)\right. \\
&\qquad \left. + \frac{K_p - K_d\widetilde{b}/\widetilde{m}}{\widetilde{m} - T\widetilde{b}} \cdot \frac{1}{s+\widetilde{b}/\widetilde{m}}e'(s)\right] \\
&= -e'(t)\left\{\frac{K_p - K_d/T}{\widetilde{b}T - \widetilde{m}} \cdot \phi\left(t, \frac{1}{T}\right)\right. \\
&\qquad \left. + \frac{K_p - K_d\widetilde{b}/\widetilde{m}}{\widetilde{m} - T\widetilde{b}} \cdot \phi\left(t, \frac{\widetilde{b}}{\widetilde{m}}\right)\right\},
\end{aligned} \tag{23}$$

where the function $\phi(t,\alpha)$ is defined as

$$\phi(t,\alpha) := \int_0^t e^{-\alpha(t-\tau)} \cdot e'(\tau)d\tau. \tag{24}$$

Since it is necessary for each term in (23) to be negative in order to satisfy $dV/dt < 0$ as long as possible, the following conditions are considered:

$$\frac{K_p - K_d/T}{\widetilde{b}T - \widetilde{m}} > (<)0 \quad \text{if } e'(t)\phi\left(t, \frac{1}{T}\right) > (<)0, \tag{25}$$

$$\frac{K_p - K_d\widetilde{b}/\widetilde{m}}{\widetilde{m} - T\widetilde{b}} > (<)0 \quad \text{if } e'(t)\phi\left(t, \frac{\widetilde{b}}{\widetilde{m}}\right) > (<)0. \tag{26}$$

Conversely, if parameters do not fulfill the previous inequality conditions, variable parameters $\widetilde{m}$ and $\widetilde{b}$ are tuned so as the unsatisfied condition will be recovered. Now, the following intermediate variables are introduced:

$$\begin{aligned}
\delta_1 &:= \eta_1 \cdot \left(\widetilde{b}T - \widetilde{m}\right), \\
\delta_2 &:= \eta_2 \cdot \left(\widetilde{m} - T\widetilde{b}\right), \\
\eta_1 &:= \operatorname{sgn}\left(K_p - \frac{K_d}{T}\right) \cdot \operatorname{sgn}\left\{e'(t)\phi\left(t, \frac{1}{T}\right)\right\}, \\
\eta_2 &:= \operatorname{sgn}\left(K_p - \frac{K_d\widetilde{b}}{\widetilde{m}}\right) \cdot \operatorname{sgn}\left\{e'(t)\phi\left(t, \frac{\widetilde{b}}{\widetilde{m}}\right)\right\}.
\end{aligned} \tag{27}$$

By checking signs of a numerator and a denominator of (25) and signs of $\widetilde{m}$ and $\widetilde{b}$, the following update law can be considered:

$$\begin{aligned}
\widetilde{b}[t+\Delta] &\longleftarrow \widetilde{b}[t] + k_1\sigma(\delta_1)\eta_1 \cdot |e|, \\
\widetilde{m}[t+\Delta] &\longleftarrow \widetilde{m}[t] - k_2\sigma(\delta_1)\eta_1 \cdot |e|,
\end{aligned} \tag{28}$$

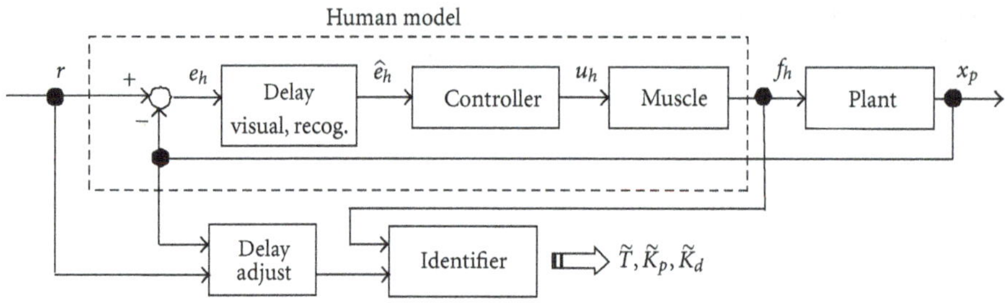

FIGURE 8: A human control model and its identification.

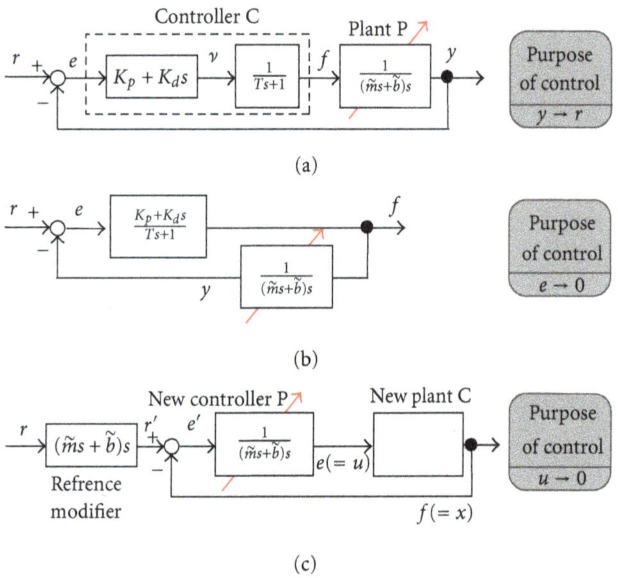

FIGURE 9: Transformation of block diagrams for variable dynamics control.

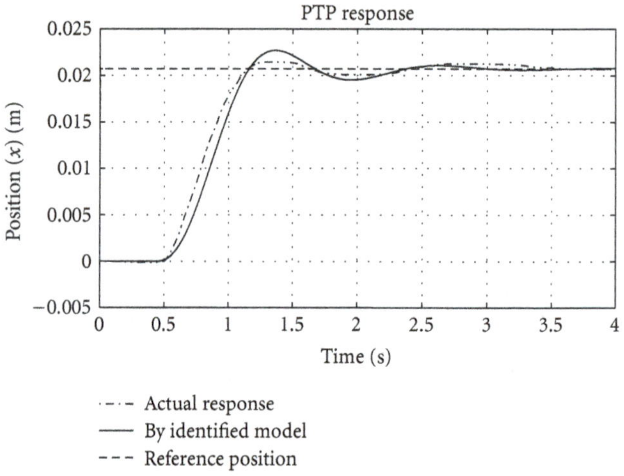

FIGURE 10: Comparison of PTP responses.

where k_1 and k_2 are positive constant parameters, Δ is a control interval, brackets in previous equations mean a discrete-time point, and a function σ is defined as

$$\sigma(\delta) = \begin{cases} 0, & \delta > 0, \\ |\delta|, & \delta < 0. \end{cases} \tag{29}$$

The other update law is derived from (26) in same manner as follows:

$$\begin{aligned} \tilde{b}[t+\Delta] &\longleftarrow \tilde{b}[t] - k_3\sigma(\delta_2)\eta_2 \cdot |e|, \\ \tilde{m}[t+\Delta] &\longleftarrow \tilde{m}[t] + k_4\sigma(\delta_2)\eta_2 \cdot |e|, \end{aligned} \tag{30}$$

where k_3 and k_4 are positive constants. Equations (28)–(30) are summarized into the following parameter update law.

$$\begin{bmatrix} \tilde{b} \\ \tilde{m} \end{bmatrix}_{[t+\Delta]} = \begin{bmatrix} \tilde{b} \\ \tilde{m} \end{bmatrix}_{[t]} + \begin{bmatrix} k_1 & -k_3 \\ -k_2 & k_4 \end{bmatrix} \begin{bmatrix} \sigma(\delta_1)\eta_1 \\ \sigma(\delta_2)\eta_2 \end{bmatrix}. \tag{31}$$

On the implementation, these parameters are updated under the following practical limit to avoid an input saturation of actual actuators:

$$\underline{b} < \tilde{b} < \overline{b}, \qquad \underline{m} < \tilde{m} < \overline{m}, \tag{32}$$

where $\underline{b}, \overline{b}, \underline{m}$, and $\overline{m}$ are constant. Here parameters k_i are chosen as they satisfy $k_1k_4 - k_2k_3 \neq 0$. Integral computation described in (24) is executed by using the following alternative online recursive computation:

$$\phi[t,\alpha] = e^{-\alpha\Delta}\phi[t-\Delta,\alpha] + e'[t]\Delta. \tag{33}$$

Since (22) cannot be computed directly, an approximation as $(\tilde{m}s+\tilde{b})s \simeq (\tilde{m}s+\tilde{b})s/(0.01s+1)^2$ is used, and the response is computed by the Eular integration with the state-space model which is derived with a controllable canonical form. K_p, K_d, and T are identified on every PTP motion and are updated according to an appropriateness of the identification result.

4. Experimental Result and Analysis

4.1. Online Identification of Human Controller. For a design of the VIM control of the xy-stage, the initial parameters were chosen as $\tilde{m}[0] = 50$ [kg], and $\tilde{b}[0] = 50$ [Ns/m]. Input information for the identification was chosen as an error

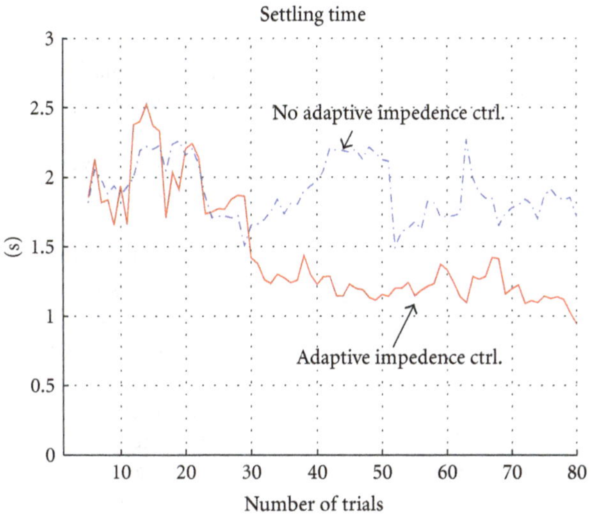

Figure 11: Evolution of settling time.

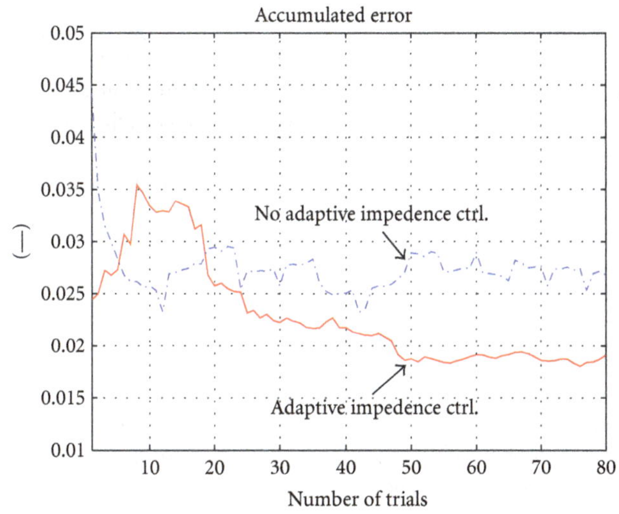

Figure 12: Evolution of accumulated error.

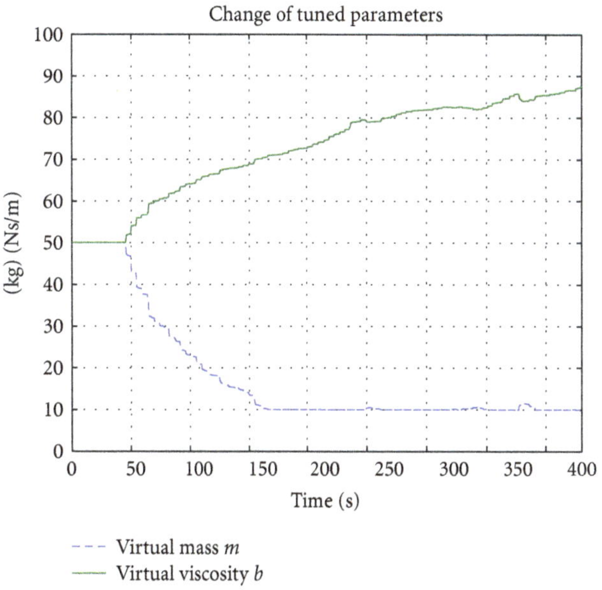

Figure 13: Evolution of tuned parameters.

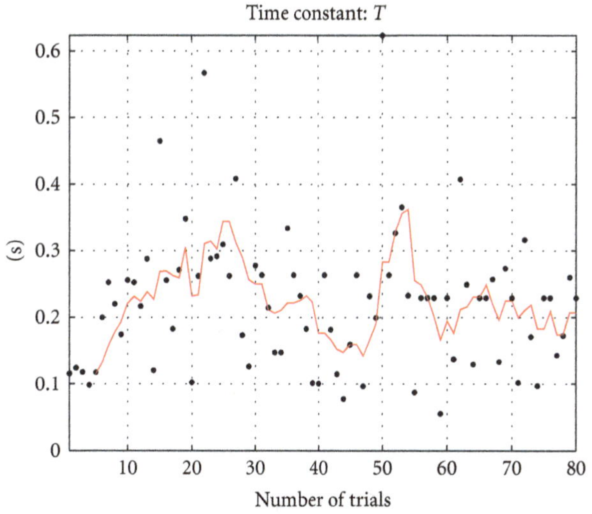

Figure 14: Change of identified T.

between the current position and the target one. Measurement value of force filtered through a 36 Hz LPF was used as output information for the identification. The time-delay effect was compensated by sifting the measured input signal at every PTP motion. Measured data was decimated by a factor of 10 for an identification; in short, an identification sampling time is 20 ms to suppress oscillation in the identified parameters. One result of the identification is shown in Figure 10. The solid curve is a simulated step response that was computed using an identified human controller model and virtual dynamics model of the stage. Those identified parameters were $\hat{K}_p = 779.0, \hat{K}_d = 288.0$, and $\hat{T} = 0.18$, and the time delay was treated as $\hat{L} = 0.406$ in the simulation. Since the response of the identified model resembles to the actual response, it can confirmed that the identification process was reasonable.

4.2. Verification of Assistive Effect. Since a key point of the proposed assistive method is to increase performance of the operator's manipulation by adjusting the machine dynamics, we investigated whether the performance of an operator who was used to the PTP operation without the adaptive control could be increased with the proposed adaptive control. From this aim, before the presented assistive control was applied to a participant, sufficient training was given to become a skilled operator using the haptic device tuned with fixed parameters which were same initial values on the assistive control. As a result of this preliminary training, it was confirmed that the performance of the participant became good and did not indicate no further improvement by checking the settling time on the PTP operation.

For the assistive control, parameters of the update law in (31) were chosen as $k_1 = k_4 = 1 \times 10^{-4}$, and $k_2 = k_3 = 2 \times 10^{-4}$. Since even the expert showed perturbation in the performance at the beginning of several trials, the normal VIM control was executed from the first trial and the adaptive

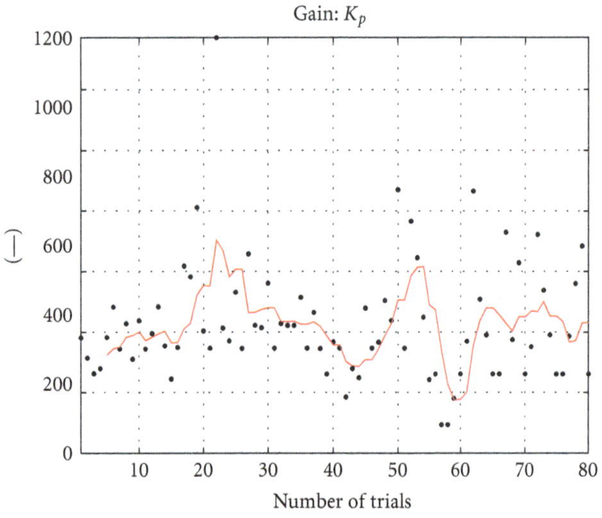

FIGURE 15: Change of identified K_p.

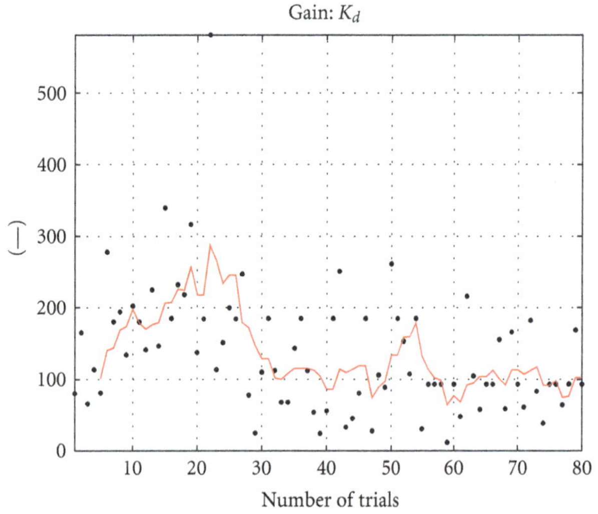

FIGURE 16: Change of identified K_d.

impedance control was activated after 50 seconds (about 10 trials). Figure 11 shows an evolution of the settling time from when the new target was displayed on the monitor till when the pointer was reached into the target circle. When three seconds passed after the pointer was kept staying inside the target circle on the monitor, it was judged that the pointer had been moved to the target by the operation. Values of the y-axis in the figure show the settling time that does not include three seconds. The solid line shows the result of the adaptive impedance control, and the dotted line shows the other result obtained by nonadaptive impedance control before the participant did not yet try the adaptive impedance control. Each line shows the evolution of trend computed by the moving average computation against five PTP tasks. While the nonadaptive impedance control case shows roughly steady state of 1.8 seconds after 50 trials, the other adaptive impedance control case shows a decrease to about 1.1 seconds. In short, speed of the PTP motion was improved by the adaptive impedance control.

Figure 12 shows an evolution of the accumulation errors $\int |e(t)|dt$ till each settling time at each trial. Similarly, the solid and dotted lines show the results of the adaptive impedance control and nonadaptive control, respectively. It can be confirmed that the accumulation error was also decreasing in case of the adaptive control. Both Figures 12 and 11 demonstrated effects of the presented method.

Figure 13 shows change of tuned parameters $\widetilde{m}$ and $\widetilde{b}$. At the beginning of trials, these values were constant because the assistive control was activated after 50 seconds. After 150 seconds, $\widetilde{m}$ was saturated at the lower limit that was specified for the safety. The reason of this nonconvergence is that the update law (31) cannot guarantee to stop the update of the parameters since the law was designed so as to make the tracking error be zero. This practical issue can be avoided by introducing a dead-zone against small error against the update law.

Finally, transitions of the identified parameters of the operator's control model, T, K_p, and K_d, are shown in Figures 14, 15, and 16, respectively. Identified parameter varies during till 30 trials. Transitions of their moving averages are comparatively flat at period of 30–80 trials except rapid change due to large outlier in 53rd trial. Although it is difficult to find tendency of change of the identified characteristics, it was confirmed that their moving averages of parameters T, K_p, and K_d are almost constant after 60 trials of when the tracking error keeps small in Figures 11 and 12. Their constant values do not differ much from their initial values which are ones before the activation of the adaptive control law. In short, it can be considered that total performance was increased by changing the machine side mainly without imposition of large change in human side. This supposition is not authentic since it is not demonstrated by statistical analysis with sufficient number of participants. These are future work.

Results of the experiment, however, showed that the proposed assistive control approach works well, and it can be said that the total performance of whole human-machine system can be enhanced by changing the dynamics of the machine itself.

5. Conclusion

For a force-feedback haptic interface system, an adaptive impedance assistive control to enhance the manipulation performance was proposed. The strategy consists of an identification of an operator's control characteristics and an adaptive online tuning of the dynamic property of the machine. The tuning is executed by changing impedance parameters of a virtual internal model for the machine. The adaptive law of the tuning was derived by utilizing a Lyapunov stability concept. Using a haptic 2-DOF interface device, it was demonstrated that proposed adaptive impedance assistive control worked effectively. The tuning law, however, cannot guarantee a convergence to a steady state without reaching to the safety limit yet since the presented algorithm was designed by only focusing on an enhancement of the performance of human manipulation without consideration of the machine limit such as an input saturation and frequency bandwidth. This practical issue will be resolved by introducing a tradeoff computation between a performance and requirement of the machine side such as an energy consumption. In the present study, however, a basic strategy for the design of human assistive system could be shown; hence, we would like to treat such practical issues in future.

Acknowledgments

This work is supported by the Grant-in-Aid for 21st Century (Center of Excellence) COE Program in Ministry of Education, Culture, Sports, Science and Technology. Preparation of the experimental system and the experiment were supported by Keiichi Kurihara and other participants who embraced the authors requests kindly. The authors are grateful to all of them for supporting this work.

References

[1] T. A. McMahon, "Mechanics of locomotion," *International Journal of Robotics Research*, vol. 3, no. 2, pp. 4–28, 1984.
[2] A. B. Joaquin and H. Herr, "Adaptive control of a variable-impedance ankle-foot orthosis to assist drop-foot gait," *IEEE Transactions on Neural Systems and Rehabilitation Engineering*, vol. 12, no. 1, pp. 24–31, 2004.
[3] Y. Yamada, H. Konosu, T. Morizono, and Y. Umetani, "Proposal of skill-assist: a system of assisting human workers by reflecting their skills in positioning tasks," in *Proceedings of the IEEE International Conference on Systems, Man, and Cybernetics*, vol. 4, pp. 11–16, October 1999.
[4] H. Konosu, I. Araki, and Y. Yamada, "Practical Development of Skill- Assist," *Journal of The Robotics Society of Japan*, vol. 22, no. 4, pp. 508–514, 2004 (Japanese).
[5] M. Uemura, K. Kanaoka, and S. Kawamura, "Power assist system for sinusoidal motion by passive element and impedance control," in *Proceedings of the IEEE International Conference on Robotics and Automation (ICRA '06)*, pp. 3935–3940, Orlando, Fla, USA, May 2006.
[6] V. Duchaine and C. M. Gosselin, "General model of human-robot cooperation using a novel velocity based variable impedance control," in *Proceedings of the 2nd Joint EuroHaptics Conference and Symposium on Haptic Interfaces for Virtual Environment and Teleoperator Systemsv (WHC '07)*, pp. 445–451, March 2007.
[7] S. Suzuki, K. Kurihara, K. Furuta, F. Harashima, and Y. Pan, "Variable dynamic assist control on haptic system for human adaptive mechatronics," in *Proceedings of the 44th IEEE Conference on Decision and Control, and the European Control Conference (CDC-ECC '05)*, pp. 4596–4601, Seville, Spain, December 2005.
[8] A. Tustin, "The Nature of the Operator's Response in Manual Control and its Implications for Controller Design," *Journal of the Institution of Electrical Engineers*, vol. 94, no. 2A, pp. 190–202, 1947.
[9] J. R. Ragazzini, "Engineering aspects of the human being as a servo-mechanism," in *Proceedings of the Meeting of the American Psychological Association*, 1948.
[10] S. Baron, D. L. Kleinman, and W. H. Levison, "An optimal control model of human response part II: prediction of human performance in a complex task," *Automatica*, vol. 6, no. 3, pp. 371–383, 1970.
[11] D. M. Wolpert and M. Kawato, "Multiple paired forward and inverse models for motor control," *Neural Networks*, vol. 11, no. 7-8, pp. 1317–1329, 1998.
[12] M. Kawato, "Internal models for motor control and trajectory planning," *Motor Systems*, vol. 9, no. 6, pp. 718–727, 1999.
[13] R. C. Miall, D. J. Weir, D. M. Wolpert, and J. F. Stein, "Is the cerebellum a smith predictor?" *Journal of Motor Behavior*, vol. 25, no. 3, pp. 203–216, 1993.
[14] K. Kosuge, K. Furuta, and T. Yokoyama, "Virtual model following control of robot arms," *IEEE Robotics and Automation*, pp. 1549–1554, 1987.
[15] D. L. Kleinman, S. Baron, and W. H. Levison, "An optimal control model of human response part I: theory and validation," *Automatica*, vol. 6, no. 3, pp. 357–369, 1970.
[16] K. Furuta, Y. Kado, S. Shiratori, and S. Suzuki, "Assisting control for pendulum-like juggling in human adaptive mechatronics," *Journal of Systems and Control Engineering IMechE*, vol. 225, no. 6, pp. 709–720, 2011.
[17] A. Phatak, H. Weinert, I. Segall, and C. N. Day, "Identification of a modified optimal control model for the human operator," *Automatica*, vol. 12, no. 1, pp. 31–41, 1976.

16

Networked Control System for the Guidance of a Four-Wheel Steering Agricultural Robotic Platform

Eduardo Paciência Godoy,[1] Giovana Tangerino Tangerino,[2] Rubens André Tabile,[2] Ricardo Yassushi Inamasu,[3] and Arthur José Vieira Porto[2]

[1] *São Paulo State University, UNESP Sorocaba, SP, Brazil*
[2] *University of São Paulo at São Carlos, São Carlos, SP, Brazil*
[3] *Brazilian Agricultural Instrumentation Research Corporation, São Carlos, SP, Brazil*

Correspondence should be addressed to Eduardo Paciência Godoy, epgodoy@sorocaba.unesp.br

Academic Editor: Yang Shi

A current trend in the agricultural area is the development of mobile robots and autonomous vehicles for precision agriculture (PA). One of the major challenges in the design of these robots is the development of the electronic architecture for the control of the devices. In a joint project among research institutions and a private company in Brazil a multifunctional robotic platform for information acquisition in PA is being designed. This platform has as main characteristics four-wheel propulsion and independent steering, adjustable width, span of 1,80 m in height, diesel engine, hydraulic system, and a CAN-based networked control system (NCS). This paper presents a NCS solution for the platform guidance by the four-wheel hydraulic steering distributed control. The control strategy, centered on the robot manipulators control theory, is based on the difference between the desired and actual position and considering the angular speed of the wheels. The results demonstrate that the NCS was simple and efficient, providing suitable steering performance for the platform guidance. Even though the simplicity of the NCS solution developed, it also overcame some verified control challenges in the robot guidance system design such as the hydraulic system delay, nonlinearities in the steering actuators, and inertia in the steering system due the friction of different terrains.

1. Introduction

Agribusiness is an activity of great importance to Brazil's economy and is responsible for more than 30% of the Brazilian gross domestic product (GDP). This economy sector has a great interest in providing solutions for sustainable agricultural development through the development and transfer of technology. The application of management techniques such as the precision agriculture (PA) aims at better utilization of the cultivated area and opens opportunities for technological development applied to the agricultural sector. References [1, 2] discussed about the new technologies and recent trends which will be required by the future in the agricultural area and the green farm concepts.

New agricultural practices related to PA have enhanced the importance in the research of embedded sensors and communication networks [3, 4] for the study of spatial variability and for the application of inputs using variable rate technology (VRT). New technologies and devices for real-time data acquisition and actuation have been released to equip agricultural machinery to support and automate these practices [5, 6]. The use of PA makes possible the culture management that seeks to maximize productivity considering spatial variability of the area, as opposed to traditional management, which applies the same amount of input for the whole area, may be resulting in better use of the chemical products, increase of productivity and reduced costs for the producer [7].

The use of robots as autonomous agricultural vehicles has an interesting potential as a valuable technological tool for PA, bringing the advantage of applying several robotic control theories for applications in various other areas [8]. This recent trend of development of mobile robots and autonomous vehicles for application in specific tasks is driven mainly by the requirement to improve the efficiency and provide better results (soil compactation reduction and machine operator absence) when compared with the use of traditional large tractors and machinery [9].

A current related trend is on the design of specific robotic platforms for autonomous vehicles and agricultural mobile robots [10, 11]. Recent applications of mobile robots have used distributed architectures based on fieldbus networks [12, 13]. Fieldbus-distributed control systems have replaced the traditional centralized control systems because of several benefits such as reduced cost and amount of wiring, increased reliability and interoperability, improved capacity for system reconfiguration, and ease of maintenance [14]. Although the fieldbus-networked control systems offer several advantages over traditional centralized control systems, the existence of communication networks makes the design and implementation of these solutions more complex. Networked control systems (NCSs) impose additional problems in control applications with fieldbus: delays, jitter, bandwidth limitations, and packet losses [15]. And the network delays may be constant, random, or time varying [16, 17]. Between the several fieldbuses, the controller area Network (CAN) protocol is the most common technology applied on embedded electronics and also in agricultural robots [18].

A common approach when designing NCS is to analyze and model the delays and missing data aiming to develop more robust control strategies [19, 20]. A control strategy is required to handle the network effects improving the performance and guaranteeing the stability of the NCS. An increasing research effort has been devoted to the PID control design for NCS. Reference [21] develops a robust $H\infty$ PID controller for NCS such that load and reference disturbances and also measurement noise can be attenuated with a prescribed level by incorporating this information in the closed loop state model. The theory of robust static output feedback (SOF) control for NCS is investigated and employed to design a remote PID controller for motor systems in [22]. In this paper the effectiveness of the proposed strategy is proved using simulations for a case study of a NCS subject to network delays and missing data.

Based on this research focus, the project "Agribot: Development of a Modular and Multifunctional Robotic Platform for Data Acquisition in Precision Agriculture" currently funded by FINEP (research and projects financing) in Brazil is developing an agricultural mobile robot called Agribot. This research and development project is a partnership among the University of São Paulo at São Carlos, the Brazilian Agricultural Instrumentation Research Corporation (Embrapa Cnpdia), and the company Máquinas Agrícolas Jacto.

In this paper a simple networked control system (NCS) solution is presented using the CAN network, for the four-wheel hydraulic steering and guidance of this agricultural robot. The correct steering control of the wheels is required for a suitable robot guidance and movement and needs to overcome some verified control challenges such as the hydraulic system delay, nonlinearities in the steering actuators, and inertia in the steering system due to the friction of different terrains. The NCS control strategy developed, centered on the robot manipulators control theory, and is based on the difference between the desired and actual position and considering the angular speed of the wheels. Experimental results with the robot demonstrated that the NCS solution for the distributed control was simple and efficient, providing suitable steering performance for the robot guidance.

This paper is organized as follows. After this introduction and literature review in Section 1, Section 2 presents a description of the agricultural robotic platform developed focusing on the mechanical structure and electronic devices. Section 3 presents the details of the wheel steering system and discusses the challenges faced to develop its control. A description of the NCS solution developed in this paper is given in Section 4, focusing on the NCS architecture, the Agribot kinematic model, and the NCS control strategy proposed. The results are resumed in Section 5, and finally some conclusions are outlined in Section 6.

2. Agribot Description

The project aims to develop a modular robotic platform able to move around in typical Brazilian agricultural environments with the main purpose of in-field data acquisition and new technologies research for sensing in agricultural area. Its main features are the robustness, mobility, high operating capacity, and autonomy consistent with agricultural needs. The robotic platform will feature a multifunctional structure to allow the coupling of different data acquisition modules to study spatial variability through embedded sensors and portable equipment. The proposed platform is composed of two main subsystems: the robotic platform subsystem and the modules subsystems. This paper focuses on the robotic platform subsystem.

2.1. Robotic Platform Description. The robotic platform subsystem, presented in Figure 1, consists of a mechanical structure and the electronic system and is able to move effectively in adverse agricultural environments. It has been manufactured by the company *Máquinas Agrícolas Jacto* and is based on the structure of a portico agricultural mobile robot developed previously [13, 23] by the research institutions.

The mechanical structure has a rectangular portico configuration with headroom of 1.8 m (Figure 1). The frame has an adjustable gauge of 2.25 m to 2.80 m, which means that it is possible to change the distance between the wheels in function of the characteristics of the field or culture. As a mobile robotic platform designed to operate in the main Brazilian agricultural environment, in almost all the growth and postharvest cycles, it requires a versatile structure. To provide this versatility, the robotic platform system was designed in separate modules, denominated as

(i) Main Frame Module, in which the main engine, fuel tank, the tank of hydraulic oil, hydraulic pumps and hydraulic cylinders are fixed;

Figure 1: Agricultural robotic platform: Agribot.

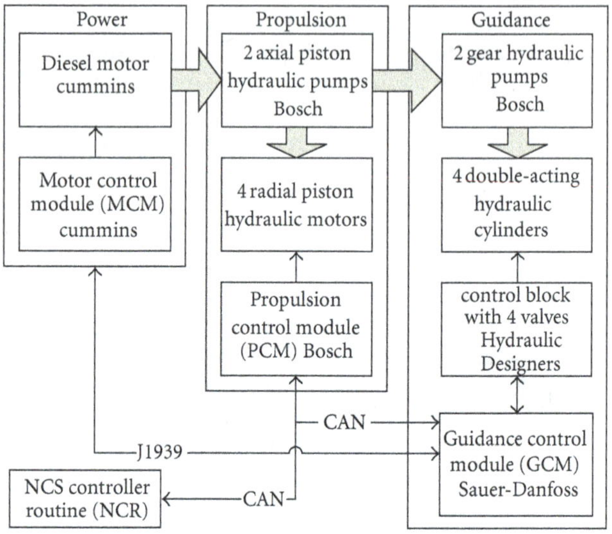

Figure 2: Electromechanical schematic view of the robotic platform.

(ii) Wheels Module, composed by a hydraulic propulsion motor, one 9.5″×24″ agricultural tire fixed directly in the hydraulic motor, steering system, pneumatic suspension system, and a telescopic rod for fixation and control the adjustment of the gauge in the main frame.

Above of the main frame a refrigerated case is located to accommodate the navigation and control systems, as well other electronic components which compose the robot. The weight of the platform is around 2800 kg.

Figure 2 presents a schematic view of the electromechanical system of the robotic platform. The main power source is a turbocharged Diesel 4-stroke engine with an electronic injection system, manufactured by *Cummins Inc.*, which provides 80 hp at 2200 RPM. The main characteristics of this kind of power source are autonomy, which in this case can reach up to 20 hours, and the ability to refuel quickly. The fuel tank has capacity of 140 liters of oil Diesel. There is also an electric power system composed by three batteries with 12 VCC and 170 Ah, connected in parallel, totaling 510 Ah, and feedbacked by one alternator fixed in the Diesel engine.

The management and the control system of the Diesel engine are owner of *Cummins Inc.* and use a SAE J1939 high layer communication protocol, based on CAN, with data transmission rate of 250 Kbit/s. The electronic control system requires the transmission of some periodic messages (by the user), otherwise the engine is turned off automatically. The input data of the system is the RPM of operation, and the output is fault alarms and some engine operating parameters.

The hydraulic propulsion system consists of two variable axial piston pumps with electronic proportional control by solenoid manufactured by *Bosch Rexroth AG*. The nominal pressure of the pump is 300 bars and maximum flow rate is 28 cm^3 per revolution, and the maximum speed at maximum flow rate is 4000 RPM. The pumps are connected in series and attached directly in the Diesel engine. The pumps feed four radial piston hydraulic motors with two speeds, also manufactured by *Bosch Rexroth AG*, and the maximum speed in 1/2 piston displacement is 465 RPM which results in a displacement speed up to 24 km/h. The nominal flow is of 470 cm^3 per revolution, and nominal and maximum torques are, respectively, 1680 Nm and 3030 Nm. The hydraulic motors are equipped with static brake with torque of the 2200 Nm and encoders for speed control.

The propulsion control system of the engines is owner of *Bosch Rexroth AG*. The electronic control system of the hydraulic system uses the CAN ISO11898 protocol with data transmission rate of 250 Kbit/s and 29-bit ID. The input data of the system are the displacement direction, speed of the motors, and static brake status, and the output data are motor speed and transmission fault alarms.

3. The Challenges of the Four-Wheel Steering Control

The use of four-wheel steering enables parallel displacement of the vehicle and facilitates maneuvering. The steering on all wheels also minimizes side slip of the wheels resulting in reduced wear on the vehicle and less damage to the culture [24]. Many works [24–26] highlight the use of robots with independent steering on all four-wheel and the development of control solutions for them. The wheel steering system of the robotic platform is described here together with the discussion of the challenges which will be faced by the NCS control development.

3.1. Details of the Wheel Steering System. The hydraulic guidance system consists of two gear pumps manufactured by *Bosch Rexroth AG*. The nominal pressure of the pump is 250 bars, maximum flow rate is 11 cm^3 per revolution, and the maximum speed is 3500 RPM. The pumps are connected in series and fixed directly after the propulsion system pumps. The pumps feed four double-acting hydraulic cylinders, which are controlled by a control block with four proportional valves with eight ways (two for each cylinder) driven by solenoid, manufactured by the *Hydraulic Designers* company. Each hydraulic cylinder is connected in a rack, which drives a pinion, converting linear motion into radial motion, allowing that the wheel turns to 133° until −133°. The position feedback of each cylinder is made by a linear potentiometer by *Gefran SpA*.

Figure 3 presents the details of a wheel module and the parts that belong to the guidance system, like cylinder, potentiometer, and motor, clarifying the operation of the robot wheel steering system.

The control system of the guidance hydraulic cylinders is done by an electronic control unit (ECU) manufactured by *Sauer-Danfoss* that was denominated as guidance control module (GCM). The GCM operates the solenoid of the *Hydraulic Designers* control block. The input data of the system are the PWM values (0–100%) that command the opening and closing of the valve of each hydraulic cylinder. The output data are the analog values read from the linear potentiometers.

The GCM is able to communicate in the CAN ISO11898 and SAE J1939 protocol, and, for this reason, this system was configured to receive and transmit the messages addressed to the Bosch (propulsion) and Cummins (engine) control modules. This fact transforms the GCM in a message center, enabling the user to exchange messages only with this module.

The schematic of the robotic platform guidance system composed by four-wheel modules is shown in Figure 4.

The GCM acts with an electrical signal on the four solenoids. The solenoids command the actuation of the four proportional valves. Each of the four valves has two ways that determine the direction of the movement of the wheel. The opening of valve releases the fluid that carries out the linear movement of hydraulic cylinders. The coupled system of racks and pinions, connected to the hydraulic cylinders, transforms the linear motion into radial motion of the wheels as explained in Figure 3.

3.2. Problems Related with the Steering System. There are some problems related to the robot wheel steering system which bring challenges to the development of the control system. Among these verified problems are the hydraulic system delay, the nonlinearities in the steering actuators, and inertia in the steering system due to the friction of different terrains. Therefore, a discussion of these problems and its influence on the control of the steering systems is presented.

The first problem is due to the hydraulic system delay. Differently of electric and pneumatic actuators that usually provide fast actuation on the controlled process, the hydraulic system used in the robot guidance has a slow response time and high inertia. These characteristics influence the system performance and consequently the controller's choice and design.

The nonlinearities in the steering actuators are another important problem. The opening action of the valves uses a PWM signal to activate the solenoids and actuate on the wheels position using the hydraulic cylinders. But there is a dead zone and a saturation limit in each cylinder, which respectively means that the cylinders do not start operating

Figure 3: Representation of a wheel module of the robot steering system.

until an approximate control signal of 30% of the PWM value, and their actions saturate at approximately 90% of the PWM value. Additionally, the change on the wheels position is not linear to the applied control signal in this range. Another issue related to the steering actuators is the difference between the areas in the two sides of the piston of the hydraulic cylinder as one side has the piston rod and the other does not. It results in a different strength for the same pressure of fluid. This fact must be considered by the control system because the same PWM control signal will provide more force in one sense of displacement than the other. This last issue also hinders the use of a single controller (usually a PID) to each wheel of the robot as the control action would be different for each side of wheel movement.

The third problem to be considered is the inertia in the steering system due to the friction between the wheel and different terrains such as earth, pasture, and asphalt. The minimum value needed for the beginning of the steering movement is not constant and depends on the amount of inertia which is being submitted to each wheel and also depends on the robot mass distribution among the wheels. Moreover, there is a difference between the inertia related to static (when the robot is stopped) and dynamic friction (when the robot is moving).

The cited problems represent challenges to the development of the steering control that must be considered and which somehow restrict the choice of the steering control strategy. On the other hand, it is desirable when designing a control system a flexible architecture and a simple algorithm which can be easily implemented and no cost computing.

4. Description of the NCS Solution

Aiming to deal with the robotic platform (Agribot) requirements and control problems, a NCS control architecture was adopted. NCS can provide efficiency, flexibility, and

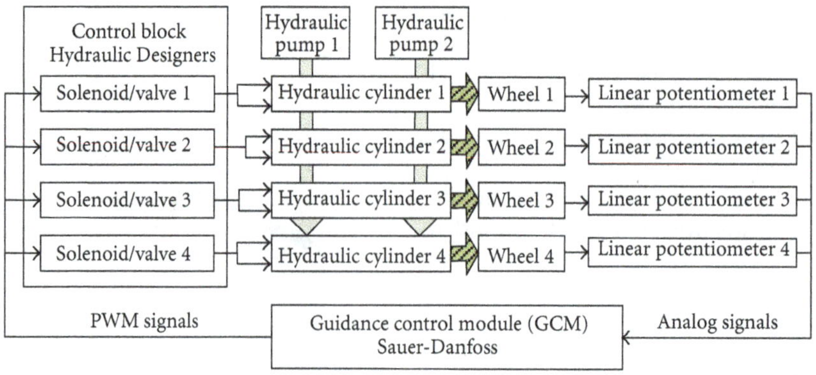

FIGURE 4: Schematic of the robotic platform guidance system.

reliability of distributed control system through distributed intelligence [15]. Furthermore, the concepts of robot manipulators control theory [27] were used as a basis for the development of the wheels steering control strategy. This chosen was driven by the fact that the robot manipulator control theory could be a simple solution to deal with the verified problems in the wheel steering control.

4.1. Architecture of the Agribot Guidance Control. In NCS, the controller, the sensor, and the plant are physically separated from each other and connected through a communication network such as CAN. The control signal is sent to the controller by a message transmitted over the network while the sensor samples the plant output and returns the information to the controller also by transmitting a message over the network. The architecture developed for the robotic platform (Agribot) follows this structure as shown in Figure 5. The NCS controller routine (NCR) was developed with *LabVIEW of National Instruments* and operates in an industrial PC that establishes communication with the guidance control module (GCM) using the CAN ISO11898 network with 250 kbit/s.

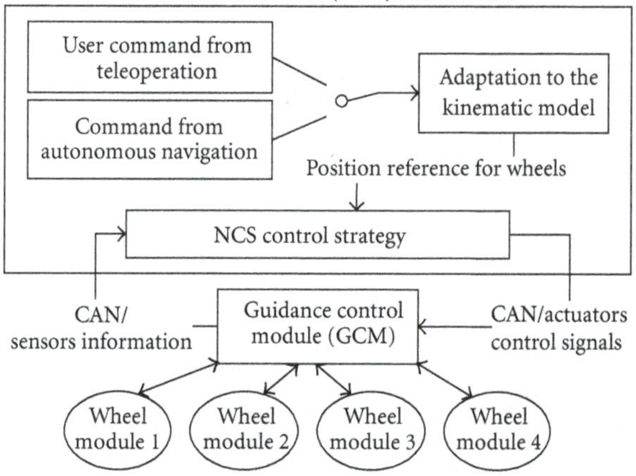

FIGURE 5: NCS architecture proposed for the Agribot Guidance Control.

In accordance with Figure 5, the guidance commands are the inputs of the Agribot kinematic model. These guidance commands can be given by the user through teleoperation or the by embedded autonomous navigation system. Some outputs of the Agribot kinematic model are the required steering position for the four-wheel of the robot. This steering position is given by the wheel angle and direction of movement. The angle of the wheel is defined in relation to the main axis from the robot, all the wheels in parallel to this axis represents the zero angle position. Steering movement from zero to left represents negative angle. Steering movement from zero to right represents positive angle. In addition, these steering positions for the wheels are used as the reference for the NCS control strategy which controls the four-wheel of the robot.

The NCR receives the CAN messages from the GCM with the sensor information (wheel angle of the potentiometer) of each wheel module, computes the algorithm of the control strategy, and sends the CAN messages to the GCM with the calculated actuators control signals (PWM values for the valves). For each of the four-wheel steering NCS that composes the robotic platform, the GCM time-driven sensors sample the wheel periodically within a sampling time of 100 ms. The threads executing in the NCR and GCM actuators are both event driven. The four closed loop NCSs in the robotic platform are sharing both limited CAN network bandwidth and NCR CPU.

4.2. Agribot Kinematic Model. The Agribot kinematic model is required for the correct guidance commands which are used as the references for the wheels steering NCS and consequently to achieve a suitable robot movement. The information is based on the dimension and the position of the wheels in relation with the center of mass (CM) of the robotic platform. To determine the CM, four scales are placed under the four-wheel with the robot. Equivalent masses are calculated for all sides, and the proportion between them gives the position of the center of mass. It is assumed for the kinematic model that the orientation of all wheels is perpendicular to the instantaneous center of rotation (ICR) and that there is no lateral sliding during the movement. The inputs of the system are the following:

(i) Turn radius (TR);

(ii) Orientation of the TR in relation to the frame (β), which assumes values to $-\pi/2$ until $\pi/2$;

(iii) Scalar velocity of the platform (V_{CM}).

The outputs are the following:

(i) Angular velocity of the platform (W_{CM});

(ii) Orientation (δ_i) and of angular velocity (rot_{W_i}) of the four-wheel.

Figure 6 presents the position of the variables of the kinematic model in relation to the robotic platform frame.

It is assumed for the kinematic model that the CM is the origin of the coordinated system. The ICR can be calculated by

$$X_{ICR} = TR \cdot \cos\left(\beta + \frac{\pi}{2}\right),$$
$$Y_{ICR} = TR \cdot \sin\left(\beta + \frac{\pi}{2}\right). \quad (1)$$

With the ICR position it is possible to determine the steering angle of the wheels. Two vectors for each wheel (Figure 6) are used for this calculation. The vector $\vec{m}$ has origin joined with the position of the wheel that desires to find the steering angle (W_i) and end in the position of the ICR. The vector $\vec{n}$ has origin in the same point of $\vec{m}$ and is oriented parallel to the frame ending in the opposite wheel (W_j). The signal of the determinant of the matrix M_i in (2), which is formed by the position of the two vectors, is used to determine the orientation of the angle. We have

$$M_i = \begin{bmatrix} (X_{ICR} - X_{W_i})(Y_{ICR} - Y_{W_i}) \\ (X_{W_j} - X_{W_i})(Y_{W_j} - Y_{W_i}) \end{bmatrix}. \quad (2)$$

The angle (γ_{W_i}) between the vectors $\vec{m}$ and $\vec{n}$ can be calculated using one of the properties of vector product as presented in

$$\gamma_{W_i} = \arccos\left(\frac{\vec{m} \cdot \vec{n}}{\|\vec{m}\| \cdot \|\vec{n}\|}\right) \cdot \frac{|\det[M_i]|}{\det[M_i]}. \quad (3)$$

Next step is converting the angle between the vectors (γ_{W_i}) to the steering angle of the wheels (δ_i). For this purpose, some logic notations are made in function of the TR and the angle γ_{W_i}, increasing or decreasing $\pi/2$ to the final result. For the case illustrated in Figure 6, the steering angle of the wheel (δ_1) is given by

$$\delta_1 = \gamma_{W_1} + \frac{\pi}{2}. \quad (4)$$

The angular velocity of the center of mass (W_{CM}) in rad/s is calculated by

$$W_{CM} = \frac{V_{CM}}{TR}. \quad (5)$$

The maximum angular velocity allowed to the platform is 0.8 rad/s. The scalar velocity of the platform will be automatically reduced for values bigger than this maximum. With the position of each wheel in relation to the CM and the position of the ICR, the radius of the patch realized for the wheel can be calculated using

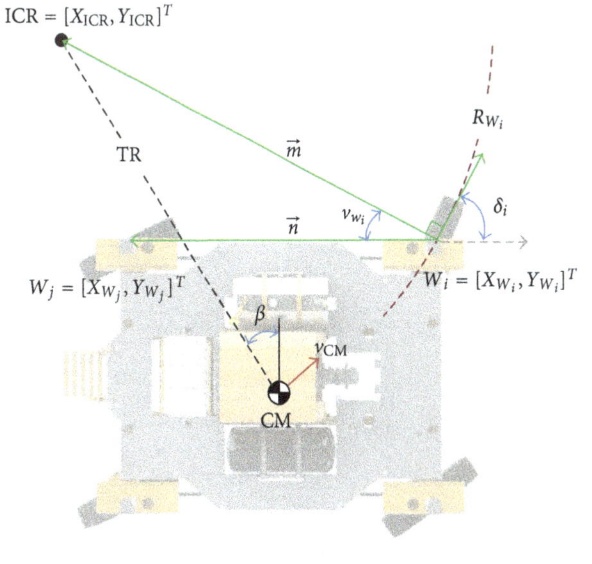

Figure 6: Variables in the Agribot kinematic model.

$$R_{W_i} = \sqrt{(Y_{ICR} - Y_{W_i})^2 + (X_{ICR} - X_{W_i})^2}. \quad (6)$$

Using the angular velocity (W_{CM}), the radius of the patch of the wheel (R_{W_i}) and the diameter of the tire (d_W), the speed of the wheel, in RPM, can be calculated by

$$\text{rot}_{W_i} = \frac{|(W_{CM} \cdot R_{W_i} \cdot V_{CM})|}{\pi \cdot d_W \cdot V_{CM}} \cdot 60. \quad (7)$$

All the calculations are made for the four-wheel. Finally, with the Agribot kinematic model, the desired angles (δ_i) for the four-wheel are used as the setpoints for the NCS steering control system.

4.3. NCS Control Strategy for Steering Control. The Agribot guidance is linked to the correct positioning control of the robot four-wheel. In the NCS architecture developed for the Agribot (Figure 5), there is one controller (NCR) responsible for the steering control of the four-wheel. Besides there is the fact that the control strategy needs to deal with some steering problems explained in Section 3.

Investigating a solution that was at the same time simple to implement and robust against the problems, in this paper a NCS control strategy centered on the robot manipulators control theory [27] was proposed. This NCS control strategy uses a position/velocity control idea [28] to calculate the control signals, based on the difference between the desired and actual position and considering the angular speed of the wheels. The fluxogram in Figure 7 details the NCS steering control algorithm developed.

The idea of the simple control strategy presented in Figure 7 is explained as follows. It is important to understand

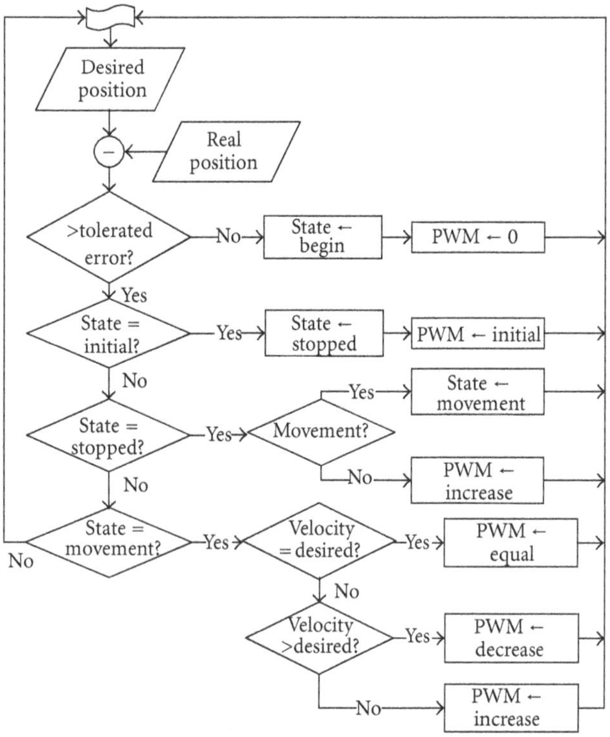

FIGURE 7: Fluxogram of the NCS control strategy developed.

that this control strategy is simultaneously applied to the steering control of the four-wheel of the platform. As a result, the actions outlined are implemented for each of the four-wheel. The steering control action begins with the comparison among the desired positions (received from the kinematic model) and the actual position of each wheel. If the difference is bigger than a maximum error tolerated (defined by the user), the control strategy will act to reduce this error. The first step is to apply an initial PWM. This value is subsequently incremented until the wheel starts its angular movement.

The wheel motion detection is done by comparing the last two readings of the wheel (angular) positioning. If the difference is bigger than the maximum tolerated, then it is considered that there is movement. After this first movement, it must be considered that the speed with which this movement is occurring. The intention is getting the slowest speed that could be constantly applied to the movement of the wheel, and keep this speed constant. To do it, a calculation is made using the previous readings of the wheel positioning, as each reading occurs every 100 ms. If the wheel speed is higher or lower than a specified value, the increment or decrement of the control signal (PWM applied to the valves) is implemented. The control algorithm runs until the error between the desired and actual position for each wheel is smaller than the error tolerated.

After some study of the robotic platform mechanical structure capabilities to deal with forces and deformations related to the requirements of its adequate guidance, the following parameters were defined for operation. A 2% of the total PWM step value to increase or decrease the control signals. An initial PWM value of 30%, in consequence to the actuators dead zone. A tolerated error is of 2 degrees for the wheel positioning and a tolerated error of 1 degree for the wheel motion detection (i.e., movement is detected if there is a difference greater than or equal to 1 degree positioning between readings).

The NCS control solution proposed in this paper for the four-wheel steering and platform guidance simplifies the control system design and implementation as only one controller can be applied to control all the wheels with a simple algorithm.

5. Results

We performed field tests using the robotic platform (Agribot) to evaluate the NCS guidance architecture and steering control strategy proposed. In the experiments, the user controls (teleoperates) the Agribot navigation and the feedback information is analyzed to check the operability and accuracy of the robot movement.

Tests were conducted on different soil types and with the robot in movement or stationary, to identify differences in the system behavior against the inertia offered by each state. The steering movement angle leaving and going from 0° to 90° represents a clockwise motion and motion from 0° to −90° a counterclockwise direction.

The defined steering control requirements for the platform guidance are slow movement speed and preferentially no overshoot on the wheel steering response, because of the great time to invert the control action on the hydraulic cylinder. Figures 8 to 11 present the accuracy of the steering wheel responses related to the setpoint required. These graphs demonstrate the steering movement of the front right wheel, front left wheel, back right wheel and back left wheel, successively.

The graphs of Figures 8 to 11 are shown separately as the setpoints, in accordance with the platform kinematic model, are different for each of the wheels. The upper graphs show the command values for the desired position and the readings values for the actual position.

At most times no steady-state error and overshoot can be verified, and an appropriate low speed was achieved for the Agribot guidance movement. These results indicate the correct design of the NCS control strategy for the wheels steering and platform guidance. The bottom graphs show the PWM value being applied by the system. These graphs of the PWM values present the dynamics of the control strategy aiming to maintain a constant speed steering (Figure 9).

Analyzing the wheel steering control graphs, it is possible to verify that the time taken to leave the actual position until the desired position changes along the platform operation and sometimes is greater in one direction of movement. This difference of values can be also seen when looking for the initial PWM values of the control signals graphs. This behavior was expected with the corrected operation of the proposed control strategy and can be explained by some facts such as the constructive characteristics of the cylinder, the different inertia related to the friction on the terrain, and the Agribot mass distribution along the wheels. These factors

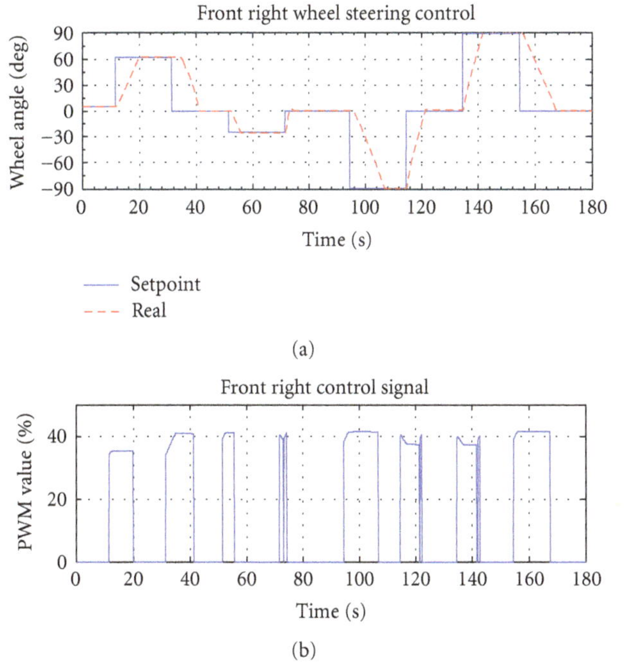

FIGURE 8: Front right wheel. (a) Wheel steering response comparing desired and real position. (b) Dynamic of PWM applied.

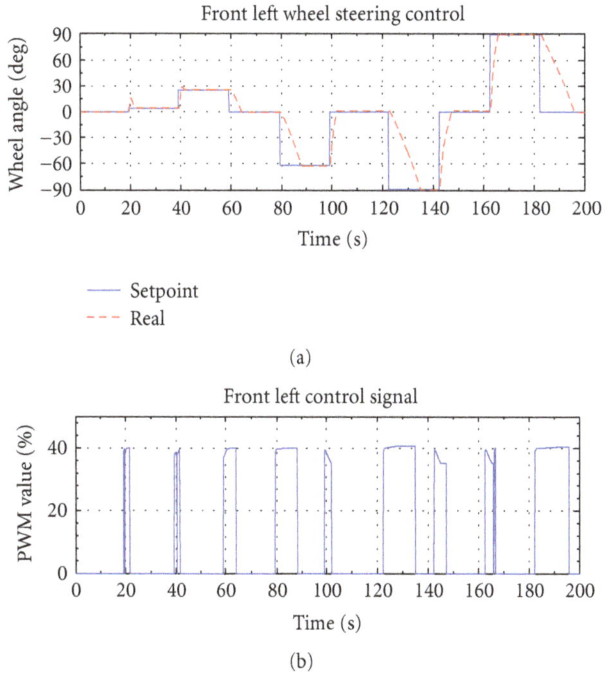

FIGURE 9: Front left wheel. (a) Wheel steering response comparing desired and real position. (b) Dynamic of PWM applied.

demand different control actions and consequently different steering outputs accordingly to the platform navigation.

Delays in the system response are, in part, attributed to the exchanging of messages through the CAN network, and in most part, attributed to the response time of the hydraulic system, which is slow and with high inertia (Figure 10).

The developed NCS control solution for the four-wheel steering and Agribot guidance simplified the control system design and implementation because only one controller was applied to control all the four-wheel. In addition, the proposed control strategy with a simple algorithm provided an ease implementing and no cost computing solution for the wheel steering control.

Despite the simplicity of the NCS control strategy, it was an effective solution to overcome the verified problems and challenges in the steering control design. However, it is important to emphasize that the control strategy could be designed for the robotic platform because of its guidance and control requirements including the slow response and movement speed.

Finally the results of the tests prove that the NCS architecture and control solution for the Agribot are reliable and satisfactory. Even though it may be desirable to improve the wheel steering control, the present level of accuracy is sufficient for the Agribot guidance, and the results indicate that the developed NCS architecture and the distributed control strategy are useful for agricultural mobile robot operations.

Future work will be done to improve the robotic platform guidance by the study of timed control solutions to synchronize the steering movements among the four-wheel. Non synchronism among the wheels was verified in the performed tests with the Agribot and is not desirable if higher navigation speed is required for the Agribot guidance.

Another interesting point would be the application of fuzzy systems [29] to deal with the Agribot steering control challenges discussed in this paper. Reference [30] attests that T-S fuzzy models can be very efficient in characterizing and dealing with nonlinearities.

6. Conclusion

In this paper a networked control system (NCS) solution was presented for the guidance of a four-wheel steering robotic platform for agricultural applications called Agribot. The solution is composed by the NCS architecture with hardware and software structure, the Agribot kinematic model used to obtain the platform guidance inputs, and the NCS control strategy, centered on the robot manipulators control theory, for the four-wheel hydraulic steering control.

The results demonstrated that the NCS solution was simple and efficient, providing suitable steering performance for the Agribot and overcoming the discussed control challenges in the robot guidance system design such as the hydraulic system delay, nonlinearities in the steering actuators and inertia in the steering system due the friction of different terrains.

The application of the NCS technology with CAN protocol provided a flexible architecture to perform the distributed control of the Agribot. The Agribot NCS architecture with distributed intelligent units (ECUs) reduced the computational load of the industrial computer and simplified the data communication among the devices of

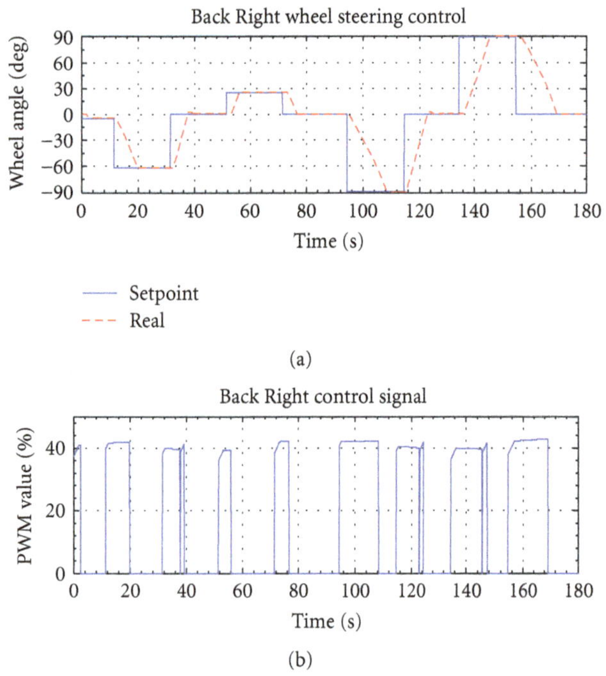

Figure 10: Back right wheel. (a) Wheel steering response comparing desired and real position. (b) Dynamic of PWM applied.

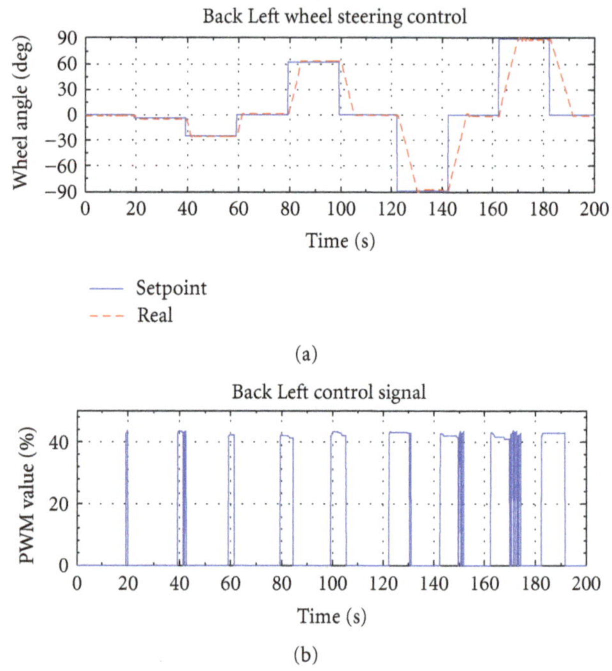

Figure 11: Back left wheel. (a) Wheel steering response comparing desired and real position. (b) Dynamic of PWM applied.

the robotic platform. Another advantage was the possibility to use only one controller to control the four-wheel of the Agribot. The NCS control strategy was developed using a simple (ease implementing and no cost computing) position/velocity control algorithm based on the difference between the desired and actual position and considering the angular speed of the wheels. This approach was proved very effective for the wheels steering control providing a proper and precise operation for the Agribot navigation.

Furthermore the NCS solution developed in this paper for the Agribot can be applied for distributed control of agricultural and other mobile robots meeting the requirements for a reliable and accurate robot movement.

Acknowledgment

This work was supported in part by FINEP (Research and Projects Financing) of Ministry of Science and Technology of Brazil, CNPq (National Council for Scientific and Technological Development of Brazil), and FAPESP (São Paulo Research Foundation). The authors also thank for Máquinas Agrícolas Jacto the collaboration and manufacturing of the mechanical structure and power system.

References

[1] T. Grift, "The first word: the farm of the future," *Transactions of the ASABE*, vol. 18, no. 1, p. 4, 2011.

[2] S. Blackmore and K. Apostolidi, "The European farm of tomorrow," *Transactions of the ASABE*, vol. 18, no. 1, p. 6, 2011.

[3] H. Auernhammer and H. Speckmann, "Dedicated communication systems and standards for agricultural applications. chapter 7, section 7.1, communication issues and Internet use," in *ASAE CIGR Handbook of Agricultural Engineering*, vol. 7, pp. 435–452, 2006.

[4] Y. Chen, X. Peng, and T. Zhang, "Application of wireless sensor networks in the field of agriculture," in *Proceedings of the ASABE Annual International Meeting*, publicationa no. 1110617, Louisville, Ky, USA, 2011.

[5] M. L. Stone, R. K. Benneweis, and J. Van Bergeijk, "Evolution of electronics for mobile agricultural equipment," *Transactions of the ASABE*, vol. 51, no. 2, pp. 385–390, 2008.

[6] J. A. Heraud and A. F. Lange, "Agricultural automatic vehicle guidance from horses to GPS: how we got here, and where we are going," in *Proceedings of the Agricultural Equipment Technology Conference*, vol. 33, ASABE publication no. 913C0109, pp. 1–67, Louisville, Ky, USA, 2009.

[7] A. K. Srivastava, C. E. Goering, R. P. Rohrbach, and D. R. Buckmaster, "Precision agriculture," in *Engineering Principles of Agricultural Machines*, P. McCann, Ed., chapter 6, pp. 123–138, ASABE, Mich, USA, 2006.

[8] T. Grift, Q. Zhang, N. Kondo, and K. C. Ting, "A review of automation and robotics for the bio industry," *Journal of Biomechatronics Engineering*, vol. 1, no. 1, pp. 37–54, 2008.

[9] S. M. Blackmore and H. W. Griepentrog, "Autonomous vehicles and robotics. chapter 7, section 7.3, mechatronics and applications," in *ASAE CIGR Handbook of Agricultural Engineering*, vol. 6, pp. 204–215, 2006.

[10] T. Bakker, K. Asselt, J. Bontsema, J. Müller, and G. Straten, "Systematic design of an autonomous platform for robotic weeding," *Journal of Terramechanics*, vol. 47, no. 2, pp. 63–73, 2010.

[11] C. Cariou, R. Lenain, B. Thuilot, and M. Berducat, "Automatic guidance of a four-wheel-steering mobile robot for accurate

field operations," *Journal of Field Robotics*, vol. 26, no. 6-7, pp. 504–518, 2009.

[12] Y. Nagasaka, H. Saito, K. Tamaki, M. Seki, K. Kobayashi, and K. Taniwaki, "An autonomous rice transplanter guided by global positioning system and inertial measurement unit," *Journal of Field Robotics*, vol. 26, no. 6-7, pp. 537–548, 2009.

[13] R. A. Tabile, E. P. Godoy, R. R. D. Pereira, G. T. Tangerino, A. J.V. Porto, and R. Y. Inamasu, "Design and development of the architecture of an agricultural mobile robot," *Engenharia Agricola*, vol. 31, no. 1, pp. 130–142, 2011.

[14] J. R. Moyne and D. M. Tilbury, "The emergence of industrial control networks for manufacturing control, diagnostics, and safety data," *Proceedings of the IEEE*, vol. 95, no. 1, pp. 29–47, 2007.

[15] R. A. Gupta and M. Y. Chow, "Networked control systemml: overview and research trends," *IEEE Transactions on Industrial Electronics*, vol. 57, no. 7, pp. 2527–2535, 2010.

[16] Y. Shi and B. Yu, "Output feedback stabilization of networked control systems with random delays modeled by Markov chains," *IEEE Transactions on Automatic Control*, vol. 54, no. 7, pp. 1668–1674, 2009.

[17] H. Zhang, Y. Shi, A. Saadat Mehr, and H. Huang, "Robust FIR equalization for time-varying communication channels with intermittent observations via an LMI approach," *Signal Processing*, vol. 91, no. 7, pp. 1651–1658, 2011.

[18] W. Baek, S. Jang, H. Song, S. Kim, B. Song, and D. Chwa, "A CAN-based distributed control system for autonomous all-terrain vehicle," in *Proceedings of the 17th IFAC World Congress*, Seoul, Republic of Korea, 2008.

[19] H. Zhang, Y. Shi, and A. Saadat Mehr, "Robust weighted H infty filtering for networked systems with intermitted measurements of multiple sensors," *International Journal of Adaptive Control and Signal Processing*, vol. 25, no. 4, pp. 313–330, 2011.

[20] H. Zhang, Y. Shi, and A. Saadat Mehr, "Robust energy-to-peak filtering for networked systems with time-varying delays and randomly missing data," *IET Control Theory and Applications*, vol. 4, no. 12, pp. 2921–2936, 2010.

[21] H. Zhang, Y. Shi, and A. S. Mehr, "Robust H_∞ PID control for multivariable networked control systems with disturbance/noise attenuation," *International Journal of Robust and Nonlinear Control*, vol. 22, no. 2, pp. 183–204, 2012.

[22] H. Zhang, Y. Shi, and A. Saadat Mehr, "Robust static output feedback control and remote PID design for networked motor systems," *IEEE Transactions on Industrial Electronics*, vol. 58, no. 12, pp. 5396–5405, 2011.

[23] E. P. Godoy, R. A. Tabile, R. R. D. Pereira, G. T. Tangerino, A. J. V. Porto, and R. Y. Inamasu, "Design and implementation of an electronic architecture for an agricultural mobile robot," *Revista Brasileira de Engenharia Agricola e Ambiental*, vol. 14, no. 11, pp. 1240–1247, 2010.

[24] T. Bak and H. Jakobsen, "Agricultural robotic platform with four wheel steering for weed detection," *Biosystems Engineering*, vol. 87, no. 2, pp. 125–136, 2004.

[25] J. Wang, Q. Wang, L. Jin, and C. Song, "Independent wheel torque control of 4WD electric vehicle for differential drive assisted steering," *Mechatronics*, vol. 21, pp. 63–76, 2011.

[26] P. F. Santana, C. Cândido, V. Santos, and J. Barata, "A motion controller for compliant four-wheel-steering robots," in *Proceedings of the IEEE International Conference on Robotics and Biomimetics*, Kunming, China, 2006.

[27] J. Roy and L. L. Whitcomb, "Adaptive force control of position/velocity controlled robots: theory and experiment," *IEEE Transactions on Robotics and Automation*, vol. 18, no. 2, pp. 121–137, 2002.

[28] I. Farkhatdinov and J. H. Ryu, "Hybrid position-position and position-speed command strategy for the bilateral teleoperation of a mobile robot," in *Proceedings of the International Conference on Control, Automation and Systems*, Seoul, Republic of Korea, 2007.

[29] T. Takagi and M. Sugeno, "Fuzzy identification of systems and its applications to modeling and control," *IEEE Transactions on Systems, Man and Cybernetics*, vol. 15, no. 1, pp. 116–132, 1985.

[30] H. Zhang, Y. Shi, and A. Saadat Mehr, "On H_∞ filtering for discrete-time Takagi-Sugeno fuzzy systems," *IEEE Transactions on Fuzzy Systems*. In press.

Modeling of Random Delays in Networked Control Systems

Yuan Ge,[1,2] Qigong Chen,[1] Ming Jiang,[1] and Yiqing Huang[1]

[1] *College of Electrical Engineering, Anhui Polytechnic University, Wuhu 241000, China*
[2] *School of Automation, Southeast University, Nanjing 210096, China*

Correspondence should be addressed to Yuan Ge; ygetoby@mail.ustc.edu.cn

Academic Editor: Mohamed Zribi

In networked control systems (NCSs), the presence of communication networks in control loops causes many imperfections such as random delays, packet losses, multipacket transmission, and packet disordering. In fact, random delays are usually the most important problems and challenges in NCSs because, to some extent, other problems are often caused by random delays. In order to compensate for random delays which may lead to performance degradation and instability of NCSs, it is necessary to establish the mathematical model of random delays before compensation. In this paper, four major delay models are surveyed including constant delay model, mutually independent stochastic delay model, Markov chain model, and hidden Markov model. In each delay model, some promising compensation methods of delays are also addressed.

1. Introduction

Communication technology and digital control have shown remarkable progress in recent years, which prompts the emergence and development of networked control systems (NCSs) [1–3]. NCSs are completely distributed feedback control systems, in which a control loop is closed via a communication network to connect sensors, controller, and actuators. Compared with traditional point-to-point control systems, NCSs reduce system wiring and cost, facilitate system diagnosis and maintenance, and increase system flexibility and reliability [4]. Thus, NCSs are becoming increasingly popular in various contexts, such as automobile, aircraft, manufacturing plant, and remote surgery [5–8].

However, the finite bandwidth and limited service in communication networks cause some new problems and challenges, such as random delays, packet losses, multipacket transmission, and packet disordering [9]. In general, packet losses mean that the delay of the packet transmission over networks is infinite. Multipacket transmission happens when the information to be transmitted is too widely scattered; that is, the delay for packing all the scattered information into one packet is too long to guarantee the packing efficiency. Packet disordering occurs when the delays of the packets transmission over networks are different, which results in that the packet sent earlier arrives at the destination node later or vice versa. Thus it can be seen that packet losses, multipacket transmission, and packet disordering are mainly caused by the existence of random delays. So, random delays are the main problem and challenge in NCSs, and the randomness of delays is generally affected by many stochastic factors (e.g., network load, nodes competition, and network congestion).

There are mainly two kinds of random delays in NCSs. One is the sensor-to-controller delay (S-C delay) in the backward network channel, and the other is the controller-to-actuator delay (C-A delay) in the forward network channel. Random delays are the major reasons for the performance deterioration and potential instability of NCSs. Conventional control theories with ideal assumptions, such as nondelayed sensing and actuation, must be reevaluated before they are applied to NCSs, which makes the analysis and design of NCSs very complex. In order to compensate for random delays in NCSs, it is necessary to establish the mathematical model of random delays before compensation. This problem has attracted strong research interests in NCSs within the control community. Generally speaking, four types of modeling methods have been proposed for the random delays in NCSs. This paper provides an overview of these methods including constant delay model in Section 2, mutually independent stochastic delay model in Section 3, Markov chain model in Section 4, and hidden Markov model in Section 5. Finally, we conclude the work in Section 6.

2. Constant Delay Model

During the early stage of NCS-related research, when the distribution characteristics of random delays are difficult to obtain, the most direct approach is to model the random delays as a constant. In this model, a receiver buffer is introduced at the controller (or actuator) node, and the buffer size is equal to the maximum S-C (or C-A) delay [10, 11]. With the constant delay model, the NCS can be treated as a deterministic system, and many deterministic control methods can be applied to the NCS. For instance, a deterministic predictor-based delay compensation methodology was proposed in [12, 13], where an observer was used to estimate the plant states and a predictor was used to compute the predictive control inputs based on past output measurements. Along this line, a first-in-first-out buffer was set for the controller to store the past output measurements, and another first-in-first-out buffer was set for the actuator to store the control inputs. All network nodes were time driven and acted synchronously. The sizes of these two buffers were, respectively, determined based on the upper bounds of S-C and C-A delays. Thus, both S-C and C-A delay were transformed to two constant values. Also, the NCS turns into a deterministic system, which is much easier to control than random delay systems. Similarly, Yu et al. [14] designed a multistep delay compensator for the NCS with dynamic noises and measurement noises.

For the NCS with constant S-C delays, Montestruque and Antsaklis [15] proposed both state-feedback and output-feedback control methodologies based on a plant model that was incorporated in the controller/actuator and was updated using the actual plant state and derived the necessary and sufficient conditions for the system stability. For the multi-input multioutput (MIMO) NCS, Lian et al. [16] established the discrete model of the NCS and designed an optimal controller, in which both S-C delay and C-A delay were constant and their sum was less than one sampling period. For the asymmetric NCS, in which sensor measurements and control inputs are transmitted by using UDP protocol and TCP protocol, respectively, Kim et al. [17] modeled such an NCS as a switched system with constant input delays and derived the sufficient conditions for the system stability by using piecewise continuous Lyapunov methods. Besides, the augmented state model methodology can also be used to analyze and design the NCS with constant delays [18, 19].

The constant delay model makes it feasible to apply conventional deterministic control methodologies to NCSs with random delays. Particularly for switched Ethernet, the random delay is relatively short and changes little, so it can be regarded as a constant delay and the deterministic control methodologies can be directly applied to the NCS closed over switched Ethernet. However, the constant delay model tends to take the upper bound of the random delay as the constant value. As a result, the random delay is artificially enlarged and then the system performance becomes degraded. Or, even worse, the system stability margin decreases so much that the system becomes unstable, which has been proved in [20, 21]. In view of this, an increasing number of researchers began to investigate stochastic control methodologies for NCSs with random delays.

3. Mutually Independent Stochastic Delay Model

Network delay tends to be stochastic since it is affected by many stochastic factors (e.g., network load, nodes competition, and network congestion), so the constant delay model and the corresponding deterministic control methodologies could hardly meet the requirement of the system performance. Thus, many people have started to study the stochastic delay model for NCSs. The stochastic delay model can be divided into two categories: the model in which delays are mutually independent and the model in which delays are probabilistically dependent. When the probabilistic dependence is unknown, the mutually independent stochastic delay model is often applied to the modeling and control of NCSs with random delays. Three main control strategies (including stochastic control, robust control, and predictive control) are surveyed in this section.

3.1. Stochastic Control. In the mutually independent stochastic delay model, each delay is treated as a mutually independent stochastic variable and its distribution can be described by a stochastic function. With this model, stochastic (optimal) control approaches are often used to investigate NCSs.

Early in the year of 1998, Nilsson et al. [22] designed an LQG-optimal stochastic controller for an NCS with mutually independent stochastic delays. In [22], the NCS was modeled as a stochastic system and the distribution of random delays was assumed to be known in advance. This assumption was relaxed by Wei et al. [23] who designed an average delay window to achieve online delay prediction for an NCS with unknown delay distribution and improved the LQG-optimal control performance of the NCS.

The delays considered in [22, 23] are all less than one sampling period. But, in real NCSs, random delays often exceed one sampling period. In view of this situation, Lincoln and Bemhardsson [24] and Hu and Zhu [25] designed the stochastic optimal controllers to guarantee the mean square exponential stability of the NCS with full (or partial) state feedback. Moreover, when the delay is infinite, Zhu et al. [26] also derived the stochastic optimal controller through solving an algebraic Riccati equation. A novel control method was presented in [27], and the maximum allowed delay bound was given based on the stochastic stability conditions.

In [28], the random delays were modeled as a linear function of the stochastic variable satisfying Bernoulli random binary distribution, and the prescribed H_∞ disturbance attenuation performance was achieved by designing an observer-based controller to guarantee the stochastically exponential stability of the closed-loop NCS. It was assumed that the construction of the observer was based on the knowledge of all the control inputs at the actuator node. The assumption was relaxed in [29] to meet the actual engineering needs. When the random delays are independent and identically distributed stochastic variables, Yu et al. [30] designed an optimal stochastic controller to guarantee the mean square exponential stability of the NCS with time-driven sensor nodes and event-driven controller/actuator nodes. For the

MIMO NCS with multiple independent stochastic delays, a delayed state-variable model was formulated and a linear quadratic regulator (LQR) optimal controller was designed to compensate for the multiple time delays in [31].

3.2. Robust Control. The stochastic delay can be transformed into the uncertainty (or disturbance) of an NCS, and then a robust controller can be designed to guarantee the robust stability and robust performance of the NCS. Different from stochastic control, robust control does not need the prior knowledge about the distribution characteristics of random delays. For instance, in [32], the random delay (S-C or C-A delay) was divided into two parts. The first part was a constant delay and the second part was an uncertain delay which was treated as the multiplicative perturbation of the NCS. And a continuous-time robust controller was designed using μ-synthesis. In [33], the NCS with asymmetric path-delays over arbitrary communication networks was investigated under the criteria of minimum H_∞ norm, and the delay-dependent switching controllers were designed via a piecewise Lyapunov function approach as well as a common Lyapunov function approach. In [34], a new discrete-time switched system model was proposed to describe the NCS with random delays, and a sufficient condition was derived for the NCS to be exponentially stable and ensure a prescribed H_∞ performance. Moreover, the obtained condition established the relations among the delay length, the delay variation frequency, and the system performances.

In [35], the H_∞ performance analysis and the H_∞ control problem of uncertain NCSs with both delays and packet losses were investigated based on a Lyapunov-Krasovskii functional method. The H_∞ performance was analyzed by introducing some slack matrix variables and employing the information of the lower bound of delays, and the H_∞ controller was designed to be of memoryless type by solving a set of linear matrix inequalities. A new Lyapunov-Krasovskii functional method, which made use of the information of both the lower and upper bounds of delays, was proposed in [36] to derive a new delay-dependent H_∞ controller that is much less conservative than the controller derived in [35] because it avoided employing model transforming and bounding technique for some cross-terms in the Lyapunov-Krasovskii function and avoided introducing slack matrix variables.

The H_∞ control problem of NCSs with both delays and packet disordering was investigated in [37], where the actuator always uses the latest arrival control law. This problem was then comprehensively investigated in [38], where the NCS was modeled as a parameter-uncertain discrete-time system with multistep delays based on matrix theory. An improved Lyapunov-Krasovskii functional method was proposed in [38] to design a less conservative H_∞ stabilization controller by solving a minimization problem based on linear matrix inequalities.

In [39], the stochastic delay was treated as a kind of uncertainty of NCSs and an H_∞ controller was designed via μ-synthesis. When the stochastic delay is unknown, bounded, and time-varying, the method to compute the maximum allowable delay bound was given in [40] and the robust stability bound was also derived there. In [41], both the deterministic information from the current timestamp and the stochastic information from the past timestamps were used to design a network-traffic-dependent H_∞ controller via introducing two algorithms: pattern generation (PG) algorithm and pattern identification (PI) algorithm. The PG algorithm classified off-line the network traffic into some separate patterns to derive a less conservative NCS, and the PI algorithm searched on-line for a pattern representing the recent network traffic to operate the designed controller. In [42], the robust control theory was combined with the network calculus theory for the delay compensation. The minimal network service curves and the maximal network traffic curves were used in the network calculus theory to estimate bounded end-to-end delays, which were translated into uncertainties in the robust controller design. For the networked cascade control system with disturbances, the criterion of its robust asymptotic stability was derived and the γ-suboptimal state feedback H_∞ controller was designed in [43].

3.3. Predictive Control. Predictive control of NCSs with random delays was investigated in [44–47]. Predictive control is suitable for complex dynamic systems with uncertain models since it is based on multistep test, rolling optimization, and feedback tuning control. In general, the setup of a networked predictive control system consists of two parts: the control prediction generator and the network delay compensator. The former generates a sequence of future control predictions to satisfy the system performance requirements using conventional control methods (e.g., PID, LQG). The latter compensates for the random network delay by choosing the latest control value from the control prediction sequences available on the plant side. In practice, predictive control has been applied into networked Hammerstein systems [48] and networked Wiener systems [49].

In [44–47], only delayed controlling data are used to derive the control predictions. However, in real systems, it is difficult to obtain the controlling data due to the existence of delays. So, an improved predictive controller, designed by using delayed sensing data, was proposed in [50]. Compared with previous results, the predictive control-based approach in [50] is easy to be implemented in practice. Besides, at any future time the predictive controller in [46] cannot know the control input of the plant in future time due to the stochasticity of network delays. So, constructing the state predictor in [46] cannot use the real but the estimative control input of the future time, which causes the problem that the error of the estimation affects the precision of the predictor and even destroys its stability. In order to decrease the conservativeness, a new design method of networked predictive control was proposed in [51] by packing the current predictive control input with history predictive inputs. By this means the future plant input is known, and the state predictor can be designed. Thus its performance and stability are not affected by the future input of the plant.

The networked predictive control approach transforms the closed-loop system to a switched system, and the stability of the switched system is guaranteed based on the condition that all subsystems of the switched system have to possess a common Lyapunov function under arbitrary switching. However, this condition is too strict in practice because some delay values may be not admissible considering the system stability. In order to overcome this limitation, an improved networked predictive control strategy was proposed in [52] by appropriately assigning the subsystem or designing the switching signal. Average dwell time method was given as an effective tool for finding such switching signal that a common Lyapunov function was not necessary for the overall system still to be stable even though there existed unstable subsystems. Furthermore, an improved predictive control design strategy, combined with the switched Lyapunov function technique, was proposed in [53], where the controller gain varied with the random delay to make the corresponding closed-loop system asymptotically stable with an H_∞-norm bound. Besides, a new predictive control scheme based on multirate Kalman Filtering was presented in [54] to compensate for the random delays and packet losses in the feedback channel of NCSs.

4. Markov Chain Model

Stochastic delays are not always mutually independent. Sometimes, there are some probabilistic dependency relationships among the delays, such as Bernoulli distribution, Markov chain. For the past few years, a lot of effort has been put on the research about Markov-based modeling and compensating methods for stochastic delays in NCSs. Since there are mainly two kinds of random delays: S-C delays and C-A delays, most existing methods can be divided into two categories: one considering S-C delays (or C-A delays) and the other one considering both S-C delays and C-A delays. The second category can be further divided into two subcategories: the method using one Markov chain to model the sum of S-C delay and C-A delay and the method using two Markov chains to model S-C delay and C-A delay, respectively.

4.1. Considering S-C Delays or C-A Delays. In [55], network only existed between the sensor and the controller, and the stochastic S-C delay was modeled as a finite-state Markov chain. Based on the Markov delay model, the NCS was modeled as a discrete-time jump linear system and a V-K iteration algorithm was proposed to design switching and nonswitching output feedback stabilizing controllers. When the S-C delay was longer than one sampling period [56], it was divided into two parts: the first part was integer times of the sampling period and the second part was less than one sampling period. The second part was modeled as a finite-state Markov chain. Thus the NCS was modeled as a discrete-time Markovian jump linear system (MJLS) and an H_∞ optimal controller was designed.

In the aforementioned two papers, the Markov states were assumed to be completely observed. However, the access to the Markov states may not be possible in some circumstances, which makes is quite possible that the associated Markov states are inaccessible to controller. For this case, to design mode-independent controllers, a mode-dependent Lyapunov function was used to derive output-feedback H_2/H_∞ controller for the NCS subject to random failures and stochastic Markovian S-C delays in [57]. In practical systems, it is difficult to obtain the exact transition probability matrix of the Markov chain of the S-C delays. To solve this problem, a polytopic-type uncertainty in the transition probability matrix of the Markov chain was taken into account in [58], where both S-C delays and packet losses were simultaneously considered. The robust fault detection problem addressed there was converted into an auxiliary robust filtering problem, and a sufficient condition for the existence of the desired robust fault detection filter was established in terms of linear matrix inequalities.

A more general NCS framework was considered in [59], where Markovian C-A delays, packet losses, and uncertainty of NCSs were simultaneously investigated. Packet losses were compensated and a sufficient condition on the existence of robust H_∞ disturbance attenuation level was given. In [60], the C-A delays were also modeled as a Markov chain. Thus, the NCS was modeled as a discrete-time MJLS and a stabilizing controller was designed by using the delay-quantization and augmented controller vector methods.

4.2. Considering Both S-C Delays and C-A Delays. Generally, S-C delays and C-A delays coexist in NCSs. In [61], both S-C delay and C-A delay were lumped together as a single delay governed by a Markov chain, and an optimal controller was designed to satisfy the prespecified performance. Besides, a method was proposed to calculate the state transition matrix of the Markov chain. This calculating method was reused in [62]. The delay from the sensor to the actuator, which is equal to the sum of S-C delay and C-A delay because the calculating time at controller is usually very short, was also modeled as a Markov chain in [63]. And a T-S model was employed to represent an NCS with all possible delays considered there. As an application of the T-S based modeling method, a parity-equation approach and a fuzzy-observer based approach were developed for the fault detection of the NCS. Under some geometric conditions, the T-S model was transformed to an output-feedback form and then an observer-based fault detection method was presented for a class of nonlinear NCSs with Markovian delays from sensor to actuator [64]. Based on the same delay model, an uncertain NCS with stochastic but bounded delays was modeled as a discrete-time jump linear system governed by a finite-state Markov chain in [65]. In order to make the system asymptotically mean square stable under a serials of stochastic disturbance such as random delays and actuator failures, an improved V-K iteration algorithm was used to successfully designed a class of reliable controllers. For the NCS with multiple faults and Markovian delays from senor to actuator, a robust fault isolation filter was developed with its parameters satisfying H_∞ disturbance attenuation and poles assignment constraints in the frame of MJLS [66]. When the delay is modeled as a Markov chain, a fixed state transition matrix is usually assigned. However, the controller designed based on the fix transition matrix may not stabilize NCSs when delays change along with network load. In order to solve the problem, an unknown and time-varying Markov chain with multiple possible transition matrices was used to model the time-varying part of delays, and then a stochastic switching controller was constituted in [67]. The switching rules were found through an intelligent algorithm based on the greedy algorithm.

In aforementioned papers (i.e., [61–67]), S-C delay and C-A delay are lumped together as a single delay and then modeled as a single Markov chain. But it is not always feasible for the two delays to be lumped together. So it is quite general to model S-C delay and C-A delay as two Markov chains. In [68], the two delays were modeled as two different Markov chains, and the resulting closed-loop system was jump linear system with two modes characterized by two Markov chains. A mode-dependent state-feedback controller was designed with its gain depending on both the current S-C delay and the previous C-A delay. However, practically the previous C-A delay is not always available because the transmission of the C-A delay value will also suffer from the S-C delay when it is transmitted through the S-C network before reaching the controller. In order to overcome the problem, the most recent available C-A delay and the current S-C delay were used to derive a two-mode-dependent controller in [69]. It is worthwhile noting that the most recent available C-A delay may not be the previous C-A delay. Therefore, the resulting closed-loop system is not a standard MJLS. Actually, it is a special MJLS with multistep delay mode jump as shown in [69]. Through calculating the transition probability matrix for the multistep delay mode jump, an output feedback controller was derived and the sufficient and necessary conditions for the stochastic stability were established. Furthermore, the mixed H_2/H_∞ control problem of NCSs with S-C delays and C-A delays governed by two Markov chains was investigated in [70], where a two-mode-dependent output feedback controller was designed based on the available S-C and C-A delay values. In [71], the S-C and C-A delays were also modeled as two Markov chains and a state-feedback controller based on both the plant mode and the delay mode was designed to guarantee the stochastic stability of NCSs.

Networked predictive control strategy has been discussed in the previous section, where the random delays are mutually independent. But, when the random delays have Markovian properties, how to achieve the predictive control of such NCSs? This problem was investigated in [72, 73], where the resulting closed-loop system eventually turned into a jump linear system with two modes, and a state-feedback model predictive controller was designed to guarantee the stochastic stability of the closed-loop system.

5. Hidden Markov Model

As mentioned in Section 1, the network-induced delay is random because it is affected by many stochastic factors (e.g., network load, nodes competition, and network congestion). All of these factors can be combined into an abstract and hidden variable which is defined as the network state. So, the distribution of delays is governed by the network state. With every packet transmission over network, the network state will jump from one mode to another within a finite state space. In general, the jump of modes satisfies Markovian property, which makes it possible to model the network state as a Markov chain. Moreover, it should be a hidden Markov chain because the network state cannot be observed directly but can be estimated through observing network delays. Therefore, the relation between the network state and the network delay is modeled as a hidden Markov model (HMM). In HMM, the network state (i.e., the state of the hidden Markov chain) determines the stochastic distribution of delays, while delays can be used as a kind of probes to infer the HMM and estimate the hidden network states [74]. Different from the Markov chain model in which the current delay is governed by the previous delay, the current delay is governed by the current network state in the HMM. From this point of view, the HMM reveals the essential generation mechanism of random delays.

To the best of our knowledge, the HMM was first introduced to NCSs by Nilsson [75]. In [75], the network state was simplified as the network load with three states ("L" for low network load, "M" for medium network load, and "H" for high network load). The transitions between different states were modeled as a Markov chain. With every state in the Markov chain, there was a corresponding delay distribution modeling the delay for the state, which was referred to as an HMM. Based on the HMM, the NCS was treated as a jump linear system and an LQG optimal controller was designed to guarantee the system stability. Following the same line as in [75], the sensor-to-actuator delays were considered as interval variables governed by a Markov chain in [76]. Based on the intervals of different mode, the corresponding plant was converted to a discrete-time Markov jump system with norm-bounded uncertainties, and an H_∞ state feedback controller was constructed via a set of linear matrix inequalities.

When S-C delays and C-A delays coexisted in an NCS, two hidden Markov chains were used in [77] to model the sensor-to-controller network state and the controller-to-actuator network state, respectively. So, there are two HMMs: one is for the relation between the S-C delay and the sensor-to-controller network state, and the other is for the relation between the C-A delay and the controller-to-actuator network state. Based on the Lyapunov-Razumikhin method, a mode-dependent state feedback controller was designed to stabilize this class of systems by solving bilinear matrix inequalities. Based on the same delay modeling method as that in [77], a dynamic output feedback controller was designed to achieve both robust stability and prescribed disturbance attenuation performance for a class of uncertain NCSs with random delays in [78], and a new stochastic Lyapunov-Krasovskii function was proposed to develop a delay-dependent criterion for determining a mode-dependent state-feedback H_∞ controller for a class of continuous-time NCSs with random delays in [79]. In order to obtain less conservative results than those in [77], free weighting matrices were introduced in [79] by using the Newton-Leibniz formula to avoid estimating some cross-terms in the Lyapunov-Krasovskii function.

Generally, the Baum-Welch algorithm (i.e., expectation maximization algorithm) can be used to estimate the parameters (e.g., initial network state distribution, network state transition probability matrix, and delay observation matrix) of the HMM of delays in NCSs (see [80, 81]). Based on these parameters, the current C-A delay can be predicted before designing the current controller [82], and the prediction can be used to design a state-feedback controller or an optimal controller to compensate for the effect of the current C-A delay on NCSs [83, 84]. In order to estimate the parameters of the HMM, the historical C-A delays were uniformly quantized to obtain the discrete observations which were used as the inputs of the estimating process [80–84]. However, in practical systems, delays seldom uniformly distribute over a delay interval. So, when quantizing the delays, some relatively centralized delays should be quantized as the same observation value, while those relatively decentralized delays should be quantized into different observation values. Based on this rule, K-means clustering was proposed in [85] to quantize the past C-A delays, which leads to more precise prediction and better compensation of the current C-A delay. A composite hidden Markov model was presented in [86] to model both S-C delays and C-A delays and the model illustrated the inner relationship between these two kinds of delays. Based on the composite HMM, the C-A delay was estimated by using S-C delays and a single mode-dependent state feedback controller was designed via solving linear/bilinear matrix inequalities.

6. Conclusions

NCS is a very important research area with wide applications. Random delay is one of the fundamental problems in NCSs. In this paper, various approaches for the modeling of random delays in NCSs are surveyed. The earliest and most simple approach is to model the random delay as a constant, and the corresponding delay compensation is usually based on deterministic control methods. Since the real network-induced delay is often timevarying or even random, an increasing number of literatures tend to adopt the stochastic delay model without or with probabilistic dependency relationships among delays. The one without probabilistic dependency mainly deals with mutually independent stochastic delays and utilizes stochastic control, robust control, predictive control, or other intelligent control strategies to compensate for the random delays. The one with probabilistic dependency mainly uses one or two Markov chains-based modeling and compensation methods for random delays. When the stochastic characteristics of delays are caused by many stochastic factors in networks (such as network load, nodes competition, and network congestion), all these factors are integrally defined as the network state, and the HMM is introduced to model the random delays. Even though the HMM is, to some extent, superior to the Markov chain model, it is imprudent to consider the HMM as the best and ultimate model for the random delays in NCSs because of, for example, the poor real-time performance of the HMM. There is still much work to do to sufficiently and completely describe the stochastic distribution of random delays in NCSs. In summary, the modeling of random delays in NCSs is still in their infancy and should deserve a lot of attention in the years to come.

Acknowledgments

This work was supported partially by the National Natural Science Foundation of China (61203034, 61172131, and 61271377), the Natural Science Foundation of Anhui Province (1308085QF120), and the Outstanding Youth Foundation of Anhui Provincial Colleges and Universities (2012SQRL086ZD).

References

[1] W. Zhang, M. S. Branicky, and S. M. Phillips, "Stability of networked control systems," *IEEE Control Systems Magazine*, vol. 21, no. 1, pp. 84–97, 2001.

[2] J. Baillieul and P. J. Antsaklis, "Control and communication challenges in networked real-time systems," *Proceedings of the IEEE*, vol. 95, no. 1, pp. 9–28, 2007.

[3] R. A. Gupta and M.-Y. Chow, "Networked control system: overview and research trends," *IEEE Transactions on Industrial Electronics*, vol. 57, no. 7, pp. 2527–2535, 2010.

[4] T. C. Yang, "Networked control system: a brief survey," *IEE Proceedings: Control Theory and Applications*, vol. 153, no. 4, pp. 403–412, 2006.

[5] M. E. M. Ben Gaid, A. Çela, and Y. Hamam, "Optimal integrated control and scheduling of networked control systems with communication constraints: application to a car suspension system," *IEEE Transactions on Control Systems Technology*, vol. 14, no. 4, pp. 776–787, 2006.

[6] Z. Ma, D. Cui, and P. Cheng, "Dynamic network flow model for short-term air traffic flow management," *IEEE Transactions on Systems, Man, and Cybernetics Part A*, vol. 34, no. 3, pp. 351–358, 2004.

[7] A. G. Martin and R. E. H. Guerra, "Internal model control based on a neurofuzzy system for network applications. a case study on the high-performance drilling process," *IEEE Transactions on Automation Science and Engineering*, vol. 6, no. 2, pp. 367–372, 2009.

[8] Y. H. Kim, L. D. Phong, W. M. Park, K. Kim, and K. H. Rha, "Laboratory-level telesurgery with industrial robots and haptic devices communicating via the internet," *International Journal of Precision Engineering and Manufacturing*, vol. 10, no. 2, pp. 25–29, 2009.

[9] L. X. Zhang, H. J. Gao, and O. Kaynak, "Network-induced constraints in networked control systems—a survey," *IEEE Transactions on Industrial Informatics*, vol. 9, no. 1, pp. 403–416, 2013.

[10] R. Luck and A. Ray, "Delay compensation in integrated communication and control systems—I: conceptual development and analysis," in *Proceedings of American Control Conference (ACC '90)*, pp. 2045–2050, San Diego, Calif, USA, May 1990.

[11] R. Luck and A. Ray, "Delay compensation in integrated communication and control systems—II: implementation and verification," in *Proceedings of American Control Conference (ACC '90)*, pp. 2051–2055, San Diego, Calif, USA, May 1990.

[12] R. Luck and A. Ray, "An observer-based compensator for distributed delays," *Automatica*, vol. 26, no. 5, pp. 903–908, 1990.

[13] R. Luck and A. Ray, "Experimental verification of a delay compensation algorithm for integrated communication and control systems," *International Journal of Control*, vol. 59, no. 6, pp. 1357–1372, 1994.

[14] Z. Yu, H. Chen, and Y. Wang, "Control of network system with random communication delay and noise disturbance," *Control and Decision*, vol. 15, no. 5, pp. 518–526, 2000 (Chinese).

[15] L. A. Montestruque and P. J. Antsaklis, "On the model-based control of networked systems," *Automatica*, vol. 39, no. 10, pp. 1837–1843, 2003.

[16] F.-L. Lian, J. Moyne, and D. Tilbury, "Optimal controller design and evaluation for a class of networked control systems with distributed constant delays," in *Proceedings of American Control Conference (ACC '02)*, pp. 3009–3014, Anchorage, Alaska, USA, May 2002.

[17] D. K. Kim, J. W. Ko, and P. Park, "Stabilization of the asymmetric network control system using a deterministic switching system approach," in *Proceedings of the 41st IEEE Conference on Decision and Control (CDC '02)*, pp. 1638–1642, Las Vegas, Nev, USA, December 2002.

[18] F.-L. Lian, J. Moyne, and D. Tilbury, "Network design consideration for distributed control systems," *IEEE Transactions on Control Systems Technology*, vol. 10, no. 2, p. 297, 2002.

[19] M. S. Branicky, S. M. Phillips, and W. Zhang, "Stability of networked control systems: explicit analysis of delay," in *Proceedings of the American Control Conference (ACC '00)*, pp. 2352–2357, Chicago, Ill, USA, June 2000.

[20] J. A. Yorke, "Asymptotic stability for one dimensional differential-delay equations," *Journal of Differential Equations*, vol. 7, no. 1, pp. 189–202, 1970.

[21] K. Hirai and Y. Satoh, "Stability of a system with variable time delay," *IEEE Transactions on Automatic Control*, vol. AC-25, no. 3, pp. 552–554, 1980.

[22] J. Nilsson, B. Bernhardsson, and B. Wittenmark, "Stochastic analysis and control of real-time systems with random time delays," *Automatica*, vol. 34, no. 1, pp. 57–64, 1998.

[23] Z. Wei, C.-H. Li, and J.-Y. Xie, "Improved control scheme with online delay evalution for networked control systems," in *Proceedings of the 4th World Congress on Intelligent Control and Automation (WCICA '02)*, pp. 1319–1323, Shanghai, China, June 2002.

[24] B. Lincoln and B. Bemhardsson, "Optimal control over networks with long random delays," in *Proceedings of the 14th International Symposium on Mathematical Theory of Networks and Systems (MTNS '00)*, pp. 84–90, Perpignan, France, June 2000.

[25] S. S. Hu and Q. X. Zhu, "Stochastic optimal control and analysis of stability of networked control systems with long delay," *Automatica*, vol. 39, no. 11, pp. 1877–1884, 2003.

[26] Q.-X. Zhu, S.-S. Hu, and Y. Liu, "Infinite time stochastic optimal control of networked control systems with long delay," *Control Theory and Applications*, vol. 21, no. 3, pp. 321–326, 2004 (Chinese).

[27] C. Ma and H. Fang, "Stochastic stabilization analysis of networked control systems," *Journal of Systems Engineering and Electronics*, vol. 18, no. 1, pp. 137–141, 2007.

[28] F. Yang, Z. Wang, Y. S. Hung, and M. Gani, "H_∞ control for networked systems with random communication delays," *IEEE Transactions on Automatic Control*, vol. 51, no. 3, pp. 511–518, 2006.

[29] W. Wang, Q.-B. Lin, F.-H. Cai, and F.-W. Yang, "Design of H-infinity output feedback controller for networked control system with random delays," *Control Theory and Applications*, vol. 25, no. 5, pp. 920–924, 2008 (Chinese).

[30] Z. Yu, H. Chen, and Y. Wang, "Research on mean square exponential stability of time-delayed network control system," *Control and Decision*, vol. 15, no. 3, pp. 228–289, 2000 (Chinese).

[31] F.-L. Lian, J. Moyne, and D. Tilbury, "Modelling and optimal controller design of networked control systems with multiple delays," *International Journal of Control*, vol. 76, no. 6, pp. 591–606, 2003.

[32] F. Göktas, *Distributed control of systems over communication networks [Ph.D. thesis]*, University of Pennsylvania, Philadelphia, Pa, USA, 2000.

[33] D. K. Kim, P. Park, and J. W. Ko, "Output-feedback $\mathscr{H}_\infty$ control of systems over communication networks using a deterministic switching system approach," *Automatica*, vol. 40, no. 7, pp. 1205–1212, 2004.

[34] W.-A. Zhang, L. Yu, and S. Yin, "A switched system approach to H_∞ control of networked control systems with time-varying delays," *Journal of the Franklin Institute*, vol. 348, no. 2, pp. 165–178, 2011.

[35] D. Yue, Q.-L. Han, and J. Lam, "Network-based robust H_∞ control of systems with uncertainty," *Automatica*, vol. 41, no. 6, pp. 999–1007, 2005.

[36] X. Jiang, Q.-L. Han, S. Liu, and A. Xue, "A new H_∞ stabilization criterion for networked control systems," *IEEE Transactions on Automatic Control*, vol. 53, no. 4, pp. 1025–1032, 2008.

[37] Y.-L. Wang and G.-H. Yang, "H_∞ control of networked control systems with time delay and packet disordering," *IET Control Theory and Applications*, vol. 1, no. 5, pp. 1344–1354, 2007.

[38] J.-N. Li, Q.-L. Zhang, Y.-L. Wang, and M. Cai, "H_∞ control of networked control systems with packet disordering," *IET Control Theory and Applications*, vol. 3, no. 11, pp. 1463–1475, 2009.

[39] Z. X. Yu, H. T. Chen, and Y. J. Wang, "Design of closed loop network control system based on H_∞ and μ-synthesis," *Journal of Tongji University*, vol. 29, no. 3, pp. 307–311, 2001 (Chinese).

[40] L. Dritsas and A. Tzes, "Robust stability bounds for networked controlled systems with unknown, bounded and varying delays," *IET Control Theory and Applications*, vol. 3, no. 3, pp. 270–280, 2009.

[41] S. H. Kim and P. Park, "Networked-based robust H_∞ control design using multiple levels of network traffic," *Automatica*, vol. 45, no. 3, pp. 764–770, 2009.

[42] N. Vatanski, J.-P. Georges, C. Aubrun, E. Rondeau, and S.-L. Jämsä-Jounela, "Networked control with delay measurement and estimation," *Control Engineering Practice*, vol. 17, no. 2, pp. 231–244, 2009.

[43] C. Huang, Y. Bai, and X. Liu, "H-infinity state feedback control for a class of networked cascade control systems with uncertain delay," *IEEE Transactions on Industrial Informatics*, vol. 6, no. 1, pp. 62–72, 2010.

[44] G. P. Liu, J. X. Mu, D. Rees, and S. C. Chai, "Design and stability analysis of networked control systems with random communication time delay using the modified MPC," *International Journal of Control*, vol. 79, no. 4, pp. 288–297, 2006.

[45] G.-P. Liu, Y. Xia, D. Rees, and W. Hu, "Design and stability criteria of networked predictive control systems with random network delay in the feedback channel," *IEEE Transactions on Systems, Man and Cybernetics Part C*, vol. 37, no. 2, pp. 173–184, 2007.

[46] G.-P. Liu, Y. Xia, J. Chen, D. Rees, and W. Hu, "Networked predictive control of systems with random network delays in both forward and feedback channels," *IEEE Transactions on Industrial Electronics*, vol. 54, no. 3, pp. 1282–1297, 2007.

[47] G. P. Liu, S. C. Chai, J. X. Mu, and D. Rees, "Networked predictive control of systems with random delay in signal transmission channels," *International Journal of Systems Science*, vol. 39, no. 11, pp. 1055–1064, 2008.

[48] Y.-B. Zhao, G.-P. Liu, and D. Rees, "A predictive control-based approach to networked hammerstein systems: design and stability analysis," *IEEE Transactions on Systems, Man, and Cybernetics, Part B*, vol. 38, no. 3, pp. 700–708, 2008.

[49] Y.-B. Zhao, G.-P. Liu, and D. Rees, "A predictive control based approach to networked wiener systems," *International Journal of Innovative Computing, Information and Control*, vol. 4, no. 11, pp. 2793–2802, 2008.

[50] Y.-B. Zhao, G. P. Liu, and D. Rees, "Improved predictive control approach to networked control systems," *IET Control Theory and Applications*, vol. 2, no. 8, pp. 675–681, 2008.

[51] Y. Guo and S. Li, "A new networked predictive control approach for systems with random network delay in the forward channel," *International Journal of Systems Science*, vol. 41, no. 5, pp. 511–520, 2010.

[52] R. Wang, G.-P. Liu, B. Wang, W. Wang, and D. Rees, "L_2-gain analysis for networked predictive control systems based on switching method," *International Journal of Control*, vol. 82, no. 6, pp. 1148–1156, 2009.

[53] R. Wang, G.-P. Liu, W. Wang, D. Rees, and Y.-B. Zhao, "H_∞ control for networked predictive control systems based on the switched Lyapunov function method," *IEEE Transactions on Industrial Electronics*, vol. 57, no. 10, pp. 3565–3571, 2010.

[54] B. Liu, Y. Xia, M. S. Mahmoud, H. Wu, and S. Cui, "New predictive control scheme for networked control systems," *Circuits, Systems, and Signal Processing*, vol. 31, no. 3, pp. 945–960, 2012.

[55] L. Xiao, A. Hassibi, and J. P. How, "Control with random communication delays via a discrete-time jump system approach," in *Proceedings of American Control Conference (ACC '00)*, pp. 2199–2204, Chicago, Ill, USA, June 2000.

[56] Q. F. Wang and H. Chen, "H_∞ control of networked control system with long time delay," in *Proceedings of the 7th World Congress on Intelligent Control and Automation (WCICA '08)*, pp. 5457–5462, Chongqing, China, June 2008.

[57] S. Aberkane, J. C. Ponsart, and D. Sauter, "Output-feedback $\mathscr{H}_2/\mathscr{H}_\infty$ control of a class of networked fault tolerant control systems," *Asian Journal of Control*, vol. 10, no. 1, pp. 34–44, 2008.

[58] X. He, Z. Wang, and D. H. Zhou, "Robust fault detection for networked systems with communication delay and data missing," *Automatica*, vol. 45, no. 11, pp. 2634–2639, 2009.

[59] J. Li, Q. Zhang, and Y. Xie, "Robust H_∞ control of uncertain networked control systems with dropout compensation and Markov jumping parameters," in *Proceedings of the 7th World Congress on Intelligent Control and Automation (WCICA '08)*, pp. 7965–7969, Chongqing, China, June 2008.

[60] L. Liu, L. Shan, and C. Tong, "Delay-quantization and augmented controller vector method of networked control systems modeling," in *Proceedings of the 1st International Workshop on Education Technology and Computer Science (ETCS '09)*, pp. 278–282, Wuhan, China, March 2009.

[61] Z. Yu, H. Chen, and Y. Wang, "Research on Markov delay characteristic-based closed loop network control system," *Control Theory and Applications*, vol. 19, no. 2, pp. 263–267, 2002 (Chinese).

[62] C. Ma and H. Fang, "Research on stochastic control of networked control systems," *Communications in Nonlinear Science and Numerical Simulation*, vol. 14, no. 2, pp. 500–507, 2009.

[63] Y. Zheng, H. Fang, and H. O. Wang, "Takagi-Sugeno fuzzy-model-based fault detection for networked control systems with Markov delays," *IEEE Transactions on Systems, Man, and Cybernetics, Part B*, vol. 36, no. 4, pp. 924–929, 2006.

[64] Z. Mao, B. Jiang, and P. Shi, "Fault detection for a class of nonlinear networked control systems," *International Journal of Adaptive Control and Signal Processing*, vol. 24, no. 7, pp. 610–622, 2010.

[65] C. Yang, Z.-H. Guan, J. Huang, H. O. Wang, and K. Tanaka, "Stochastic controlling tolerable fault of network control systems," in *Proceedings of American Control Conference (ACC '08)*, pp. 1979–1984, Seattle, Wash, USA, June 2008.

[66] D. Sauter, S. Li, and C. Aubrun, "Robust fault diagnosis of networked control systems," *International Journal of Adaptive Control and Signal Processing*, vol. 23, no. 8, pp. 722–736, 2009.

[67] C. Yang, Z.-H. Guan, J. Huang, and T. Qian, "Design of stochastic switching controller of networked control systems based on greedy algorithm," *IET Control Theory and Applications*, vol. 4, no. 1, pp. 164–172, 2010.

[68] L. Zhang, Y. Shi, T. Chen, and B. Huang, "A new method for stabilization of networked control systems with random delays," *IEEE Transactions on Automatic Control*, vol. 50, no. 8, pp. 1177–1181, 2005.

[69] Y. Shi and B. Yu, "Output feedback stabilization of networked control systems with random delays modeled by Markov chains," *IEEE Transactions on Automatic Control*, vol. 54, no. 7, pp. 1668–1674, 2009.

[70] Y. Shi and B. Yu, "Robust mixed $\mathscr{H}_2/\mathscr{H}_\infty$ control of networked control systems with random time delays in both forward and backward communication links," *Automatica*, vol. 47, no. 4, pp. 754–760, 2011.

[71] J. Wang, C. Liu, and H. Yang, "Stability of a class of networked control systems with Markovian characterization," *Applied Mathematical Modelling*, vol. 36, no. 7, pp. 3168–3175, 2012.

[72] J. Wu, L. Zhang, and T. Chen, "Model predictive control for networked control systems," *International Journal of Robust and Nonlinear Control*, vol. 19, no. 9, pp. 1016–1035, 2009.

[73] Y. Xia, G.-P. Liu, M. Fu, and D. Rees, "Predictive control of networked systems with random delay and data dropout," *IET Control Theory and Applications*, vol. 3, no. 11, pp. 1476–1486, 2009.

[74] W. Wei, B. Wang, and D. Towsley, "Continuous-time hidden Markov models for network performance evaluation," *Performance Evaluation*, vol. 49, no. 1-4, pp. 129–146, 2002.

[75] J. Nilsson, *Real-time control systems with delays [Ph.D. thesis]*, Lund Institute of Technology, Lund, Sweden, 1998.

[76] Y. Wang and Z. Sun, "H_∞ control of networked control systems via lmi approach," *International Journal of Innovative Computing, Information and Control*, vol. 3, no. 2, pp. 343–352, 2007.

[77] D. Huang and S. K. Nguang, "State feedback control of uncertain networked control systems with random time delays," *IEEE Transactions on Automatic Control*, vol. 53, no. 3, pp. 829–834, 2008.

[78] D. Huang and S. K. Nguang, "Robust disturbance attenuation for uncertain networked control systems with random time delays," *IET Control Theory and Applications*, vol. 2, no. 11, pp. 1008–1023, 2008.

[79] Y. Liu and D. Sun, "Delay-dependent H_∞ stabilisation criterion for continuous-time networked control systems with random delays," *International Journal of Systems Science*, vol. 41, no. 11, pp. 1399–1410, 2010.

[80] F.-C. Liu and Y. Yao, "Modeling and analysis of networked control systems using hidden markov models," in *Proceedings of the 4th International Conference on Machine Learning and Cybernetics (ICMLC '05)*, pp. 928–931, Guangzhou, China, August 2005.

[81] Y. Zhang and S. Wang, "Delay-loss estimation and control for networked control systems based on hidden Markov models," in *Proceedings of the 6th World Congress on Intelligent Control and Automation (WCICA '06)*, pp. 4415–4419, Dalian, China, June 2006.

[82] S. Cong, Y. Ge, Q. Chen, M. Jiang, and W. Shang, "DTHMM based delay modeling and prediction for networked control systems," *Journal of Systems Engineering and Electronics*, vol. 21, no. 6, pp. 1014–1024, 2010.

[83] Y. Ge, S. Cong, and W. W. Shang, "Compensation for the stochastic delays in networked control systems," *Journal of Graduate University of Chinese Academy of Sciences*, vol. 30, no. 2, pp. 179–186, 2013.

[84] Y. Ge, S. Cong, and W. W. Shang, "DTHMM based optimal controller design for networked control systems," *Journal of University of Science and Technology of China*, vol. 42, no. 2, pp. 161–169, 2012.

[85] Y. Ge, S. Cong, and W. W. Shang, "K-means based delay quantization and prediction in networked control systems," in *Proceedings of the 31st Chinese Control Conference (CCC '12)*, pp. 5921–5926, Hefei, China, July 2012.

[86] A. Ghanaim and G. Frey, "Modeling and control of closed-loop networked PLC-systems," in *Proceedings of the American Control Conference (ACC '11)*, pp. 502–508, San Francisco, Calif, USA, July 2011.

18

Application of Neuro-Wavelet Algorithm in Ultrasonic-Phased Array Nondestructive Testing of Polyethylene Pipelines

Reza Bohlouli, Babak Rostami, and Jafar Keighobadi

Faculty of Mechanical Engineering, University of Tabriz, 29 Bahman, Tabriz, 5166614766, Iran

Correspondence should be addressed to Babak Rostami, rostami_babak@yahoo.com

Academic Editor: Ricardo Dunia

Polyethylene (PE) pipelines with electrofusion (EF) joining is an essential method of transportation of gas energy. EF joints are weak points for leakage and therefore, Nondestructive testing (NDT) methods including ultrasonic array technology are necessary. This paper presents a practical NDT method of fusion joints of polyethylene piping using intelligent ultrasonic image processing techniques. In the proposed method, to detect the defects of electrofusion joints, the NDT is applied based on an ANN-Wavelet method as a digital image processing technique. The proposed approach includes four steps. First an ultrasonic-phased array technique is used to provide real time images of high resolution. In the second step, the images are preprocessed by digital image processing techniques for noise reduction and detection of ROI (Region of Interest). Furthermore, to make more improvement on the images, mathematical morphology techniques such as dilation and erosion are applied. In the 3rd step, a wavelet transform is used to develop a feature vector containing 3-dimensional information on various types of defects. In the final step, all the feature vectors are classified through a backpropagation-based ANN algorithm. The obtained results show that the proposed algorithms are highly reliable and also precise for NDT monitoring.

1. Introduction

The ultrasonic technique as a nondestructive testing (NDT) method has been widely used over decades to evaluate the quality of materials and equipments without causing damage in a large range of industries. In the evaluation of pressure vessels and piping, not only is UT utilized in manufacturing quality controlling, but also has been used in service monitoring and residual life prediction, such as the inspection of welded joints, monitoring of crack propagation, and evaluation of materials property deterioration. In the specific case of welded materials, the research for the development of an acceptable system for analyzing the extracted images from the welded joints has grown considerably in the last years [1–4].

One of its applications is in the gas pipelines where the usage of natural gas in residential, commercial, and industrial facilities is increased day by day. In this way, Polyethylene pipes rapidly substituted the metal pipes, because the polyethylene pipes have a high-corrosion resistance, easy to form, lighter, and cheaper than metal ones. In fact, the main reason for using PE pipes in gas distribution is that its material has a high-chemical resistance against corrosive materials in transported gas. In addition, PE pipes are easy to carry, lie down, and make connections. Because of these benefits, gas distribution companies and water and sewage organizations would change their existing systems and use the PE pipes. The demand of polyethylene (PE) pipeline is increased for gas energy transportation and electrofusion (EF) joining is an essential method to build PE pipeline [5].

It is important to note that, usually the EF joints are considered as weak points for leakage, and kind of nondestructive test is necessary. One of the main factors disturbing the reliability and accuracy of the test is the encountered noise during inspection. The most commonly used ultrasonic detector is the A-scan detector. This kind of traditional UT has several disadvantages such as the need for a skilled and experienced technician to judge the defect, also the lack of permanent record, which is extremely important in the condition monitoring and in-service inspection [5].

These problems may be easily solved by the introduction of a digital ultrasonic system, which combines the computer

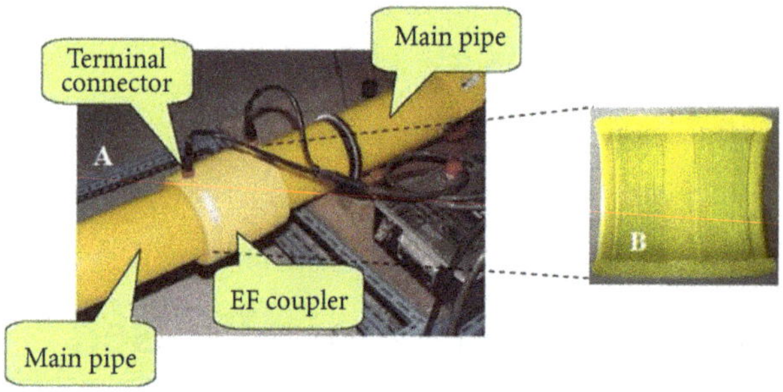

FIGURE 1: (A) Electrofusion (EF) joining of two polyethylene (PE) pipes using EF coupler. (B) Cross-section of EF coupler including heating wires.

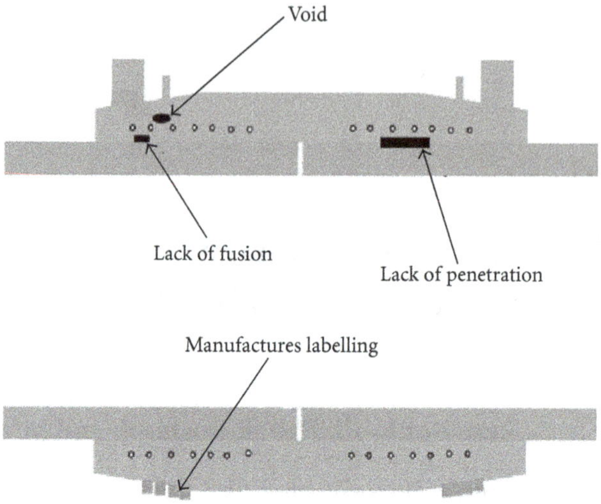

FIGURE 2: General structure and location of joint flaws.

and digital signal processing (DSP) technology into UT instrument [6–9]. Electrical, pulse, ringing, and structure noises are the most commonly encountered noises, which reduce the quality of extracted images for NDT evaluations. But in order to improve the process of image processing, wavelet transforms are used significantly [10–12]. The wavelet transform, multiresolution analysis, and other space frequency or space scale approaches are now considered as standard tools by the researchers in signal processing, and many applications have been proposed. The theory of wavelet analysis has been well studied in [13], where images are represented by wavelet reconstructing toolbox.

Bhuiyan et al. have successfully used wavelet transforms to significantly improve identification and classification rates on fingerprints. They showed that wavelet contains features that are more pronounced for higher accuracy in recognizing fingerprints [10].

Also in [11, 12] the output of wavelet transformation to obtain some features is used as an input to the artificial neural network (ANN) classifier for pattern recognition.

The aim of this paper is to present a new combined intelligent algorithm considering digital image processing, neural networks, and mathematical morphology techniques for improving the quality of extracted images from NDT evaluations. Features are extracted from Haar wavelet decomposition of the JPG images. The simulations are done using MATLAB platform. Obtained results show that the proposed algorithms to ultrasonic signals are highly reliable and precise for quality of NDT testing and monitoring.

2. Electrofusion Joining of PE Pipe

EF joining is one of the widespread PE pipe weld methods. An estimated annual use of EF joining was over 15 million in 1993 [1]. This technique makes possible joining of preassembled pipes to be carried out with minimum equipment. Electrofusion method is a system that welds pipes together through fittings whose internal surfaces are covered with special resistance wires (as shown in Figure 1). Welding is performed through melting plastic material with heating coils that reach a high temperature as a result of the stress applied to the sockets on fittings by an electrofusion machine. The electrofusion welding process can be described in three stages: initial heating and fitting expansion, heat soaking to create the joint, and finally joint cooling. The application of the electrofusion process is shown in Figure 1.

Usually, wound-heating wires exist between the coupler and main pipes. The distance between adjacent wires is usually very close, for example, 1$\sim$3 mm. The fundamental ideas of joining process are to heat the wires and to melt the inside and outside surface of coupler and main pipes respectively, and then to consolidate fusion area. Welding faults (bad preparation of the pipe, poor cleaning, pipe surface being not scraped, pipe and fitting being badly clamped, or not respecting of fusion time) can generate defects. Figure 2 shows a cross-section view of the coupling. The position of the heating wires and possible flaw locations are shown.

3. Proposed Algorithm

In order to improve the raw UT images for observation and accurate analysis, a combined algorithm is proposed considering various image preprocessing, mathematical morphology (such as dilation and erosion operations), wavelet feature extractor, and artificial neural network. Flowchart of proposed algorithm is shown in Figure 3.

The starting point of this algorithm is collecting and inserting the raw UT images. In this way, all electrofusion (EF) joints of polyethylene pipes must be tested by ultrasonic method. Then extracted images will be prepared and saved as JPEG pictures for computer evaluations.

Once the template and test images are resized, threshold values for the grayscale images are determined to convert the images into binary ones. Grayscale images with levels between 0 and 255 are converted into binary images. It is assumed that the initial threshold is equal to 0.55. This value will be updated during this algorithm if it cannot find an appropriate result.

Then binary images will be fed into the image processing step including preprocessing sections such as noise reduction and segmentation based on the image processing toolbox

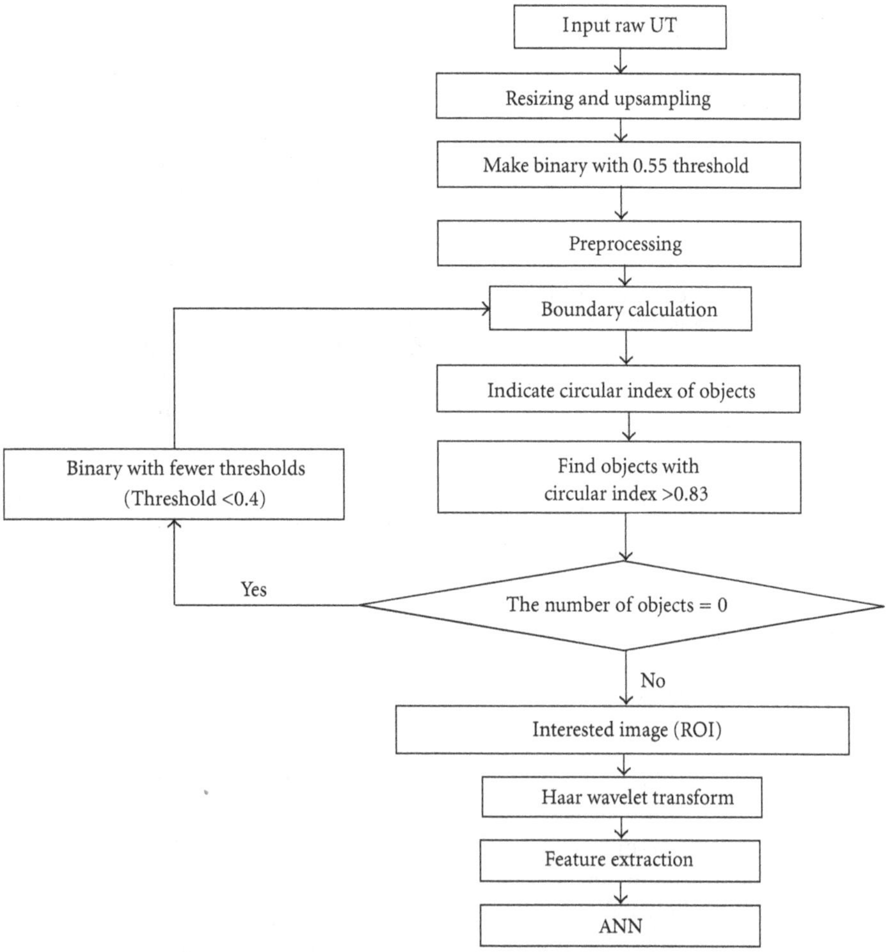

FIGURE 3: Schematic diagram of proposed algorithm.

of MATLAB software [14]. Preprocessing involves applying radiometric and geometric corrections to a raw image data. Several levels of corrections have been defined. For the preprocessing step, all noises and unwanted objects will be removed from the document's image. This leads to an easier and more effective process. Afterward, more improvement will be made bymorphological operations.

Mathematical morphology [15] is a mathematical tool for analyzing image on morphological basis. Its basic idea is to use structuring elements of certain morphology to measure and extract corresponding morphology in the image for making analysis and recognition. The application of mathematical morphology helps to simplify image data, so the basic morphologies of image can be maintained and unrelated structure can be removed. Mathematical morphology consists of four basic steps including morphology corrosion, morphology expansion, opening operation, and closing operation. In this paper, we focus on the dilation and erosion operations of the mathematical morphology. Given $f(x, y)$ is a grey-scale image on domain Z. $g(x, y)$ is a structuring element on domain Z hence, the formula for grey-scale expansion operation of image $f(x, y)$ based on $g(x, y)$ is as follows:

$$(f \Theta g)(x, y) = \min(f(x+i, y+i) - g(i, j)), \quad (i, j) \in D_g, \ (x, y) \in D_f. \tag{1}$$

Accordingly, the erosion of $f(x, y)$ by $g(x, y)$, denoted by $f \Theta g$, is defined as

$$(f \oplus g)(x, y) = \max(f(x+i, y+i) + g(i, j)), \quad (i, j) \in D_g, \ (x, y) \in D_f. \tag{2}$$

Many other morphological operations are based on these two basic operations [15].

The operations of erosion, dilation, opening, closing, and others can extract many types of information about a binary image. Morphological reconstruction can be applied for restoring the lost image information and also for segmenting object image as shown simulations [16]. Then the prepared ROI section from the images will be used by wavelet transform. Finally, all features like mean and standard deviation of data will be trained by a multilayer neural network. Details regarding this algorithm and its simulation are presented in the next section.

4. Simulation Results

The experimental results and performance evaluation of the proposed method are described in this section. Real time ultrasonic array technique was applied to obtain the ultrasonic images of the cross-section of electrofusion joints.

Ultrasonic array transducers in this simulation have 96 array elements and the center frequencies of the transducers were 7.5 MHz for the high-resolution application. To monitor the real time image, a PC monitor is connected to the system. Also 3D-FFT (Haar wavelet) was used in feature extraction processes of ROI images and simulations were performed on 2 GHz PC by using MATLAB.

4.1. Image Resizing. Tests showed that the best performance will be achieved by increasing the dimension of input pictures to twice their initial size. Because the raw image are small and by increasing the dimension more details such as pixels and the edges of regions will appear significantly. The results of this part are presented in Figure 4.

4.2. Noise Reduction and Morphological Application. Noise reduction is the process of removing noise, holes, and unwanted details from images. Noises in binary images mainly consist of isolated pixels of opposite value in image objects or background (also called salt-and-pepper noise); small holes on objects and small spots on the background; line merging and splitting. In the simulated case of this paper, a combined binary filter is used for erosion or dilation of objects, removal of noise in an image, detect edges, and smoothing the image.

This filter consists of two steps. In first step, a separate program is applied to indicate and remove all holes with areas less than 30 (This threshold is determined by text). Then, in the second step, morphological method is applied for more improvement. The result of noise reduction based on first step of binary filter is presented in Figure 5. It is clear that unwanted areas are removed from the images.

But in case of morphological application, designing a structural element is necessary. Mathematical morphology regards an image as a set and uses another smaller set which is called as structuring element to probe the image.

This apparent geometric description of set theory makes mathematical morphology more suitable for visual information processing. Mathematical morphology is originally proposed for binary images, and its basic theory is developed in this application. In this case, a 5×5 structuring element (Figure 6) is designed for morphological application. So in some cases, the results of this application are shown in Figure 7.

Based on the proposed algorithm in Figure 3, the important step of this algorithm is to extract a high-quality ROI for further examination by wavelet transform. So it is necessary to trace and indicate the exterior boundaries of objects, as well as boundaries of holes inside the objects, in the binary image. So nonzero pixels will belong to an object and 0 pixels constitute the background. These boundaries for each region in binary images are specified using MATLAB functions. In the next step, circular index of each obtained object from previous step should be indicated. Circular index is defined by

$$\text{Circular Index} = \frac{4 \times \pi \times S}{P^2}, \tag{3}$$

where S and P are the area and perimeters for each object, respectively.

To imply this matter, an acceptable value for circular index (circular threshold) should be found for a sufficient comparison with current situation of each object.

Tests and simulations on database showed that the best value for this circular threshold is around 0.83, so acceptable circular objects will be selected and simultaneously the performance of this algorithm will be sufficient enough. In this way, all regions with a circular index higher than 0.83 will be shown as heating wires as a very important factor for final evaluation. The simulation result of this step, which is about boundaries and circular object indication, are shown in Figure 8. As shown in this figure, circular index for all objects is determined.

Based on presented information in Figure 8, there are various objects with circular index higher than 0.83 and ROI will be selected around these objects.

Extracted ROI is shown in Figure 9. This region will be included by the exact places of the EF joints in polyethylene pipelines.

4.3. Data Extraction by Wavelets Transform. Wavelet transform exploits both the spatial and frequency correlation of data by dilations (or contractions) and translations of mother wavelet on the input data. It supports the multi resolution analysis of data that can be applied to different scales according to the details required, which allows progressive transmission and zooming of the image without the need of extra storage [17, 18].

The implementation of wavelet compression scheme is very similar to the subband coding scheme: the signal is decomposed using filter banks (Figure 10). The output of the filter banks is downsampled, quantized, and encoded. The decoder decodes the coded representation, up-samples and recomposes the signal [17].

Wavelet transform divides the information of an image into approximation and detail subsignals. The approximation of sub-signal shows the general trend of pixel values, and other three detail sub-signals show the vertical, horizontal and diagonal details or changes in the images.

In this paper 3D wavelet is used, and by utilizing the ROI (from previous section), all information about images will be extracted by wavelet transform. The results of using wavelet are shown in Table 1, where Fi is a matrix for saving the extracted data and all data will indicate the entire extracted features from the images. In fact, in this simulation 5-level HAAR wavelet is used and it is clear that Fi matrix includes from 5 rows and each row will be about each level extraction for wavelet. Also in this matrix, it is clear that we have 6 columns and each column will include features such as: the mean and standard deviation of data for horizontal, vertical, and diagonal details, respectively.

4.4. Final Evaluation by ANN. Using the extracted feature vector representations from previous section, neural classifier is trained and tested to recognize and classify the scenes.

Neural networks are based on models of biological neurons and form a parallel information processing array

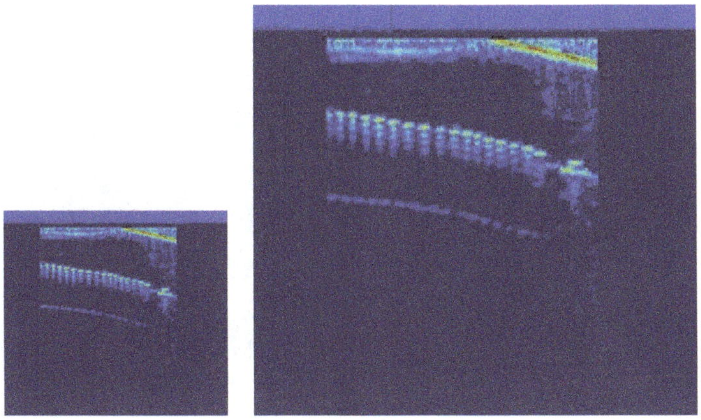

(a) (b)

FIGURE 4: Image resizing.

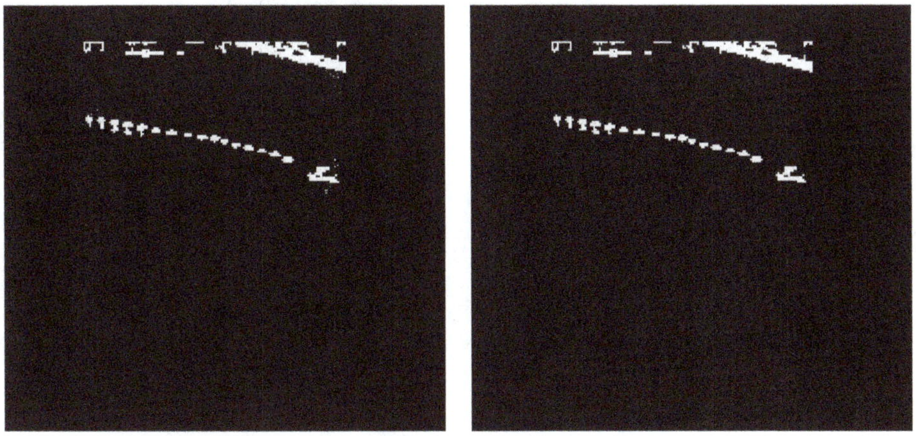

(a) (b)

FIGURE 5: Noise reduction for kind of sample image. (a) before (b) after.

0	0	1	0	0
0	1	1	1	0
1	1	1	1	1
0	1	1	1	0
0	0	1	0	0

FIGURE 6: Structuring element in morphological application.

based on a network of interconnected elements [19, 20]. A Multilayer perceptron (MLP) networks is used in this paper. This type of network is trained using a process of supervised learning in which the network is presented with a series of matched input and output patterns and the connection strengths or weights of the connections are automatically adjusted to decrease the difference between the actual and desired outputs [19]. The structure of this kind of network is shown is Figure 11.

The introduced ANN was trained by the main features of images for different joints. So part of the data will be used for training phase, and in the next step trained network will be tested. Figure 12 depicts the converging training graph of neural classifier for Haar wavelet features, respectively. This network is trained after 4000 epoch and the error is acceptable. After that, a testing set by other images will be used to check the classification performance and its accuracy.

In fact, the trained network is used for testing, and in this way 10 practical images are applied for this evaluation. Based on the simulation results, 9 images are recognized correctly. It is assumed that the image with correct joint is 1 and each image with incorrect joint is -1.

Based on used testing images we have 7 images with correct joint and 3 incorrect joint.

y_net1 = sim (net1, test)

Columns 1 through 7

-0.8467 -0.9962 -0.9756 0.9336 0.8449 0.9450 0.5027

Columns 8 through 10

-0.2186 0.0566 0.9777

In this section, based on the previous information, the network is trained. Next, the results of testing network are presented in Figure 12 and Table 2. Note that the desired

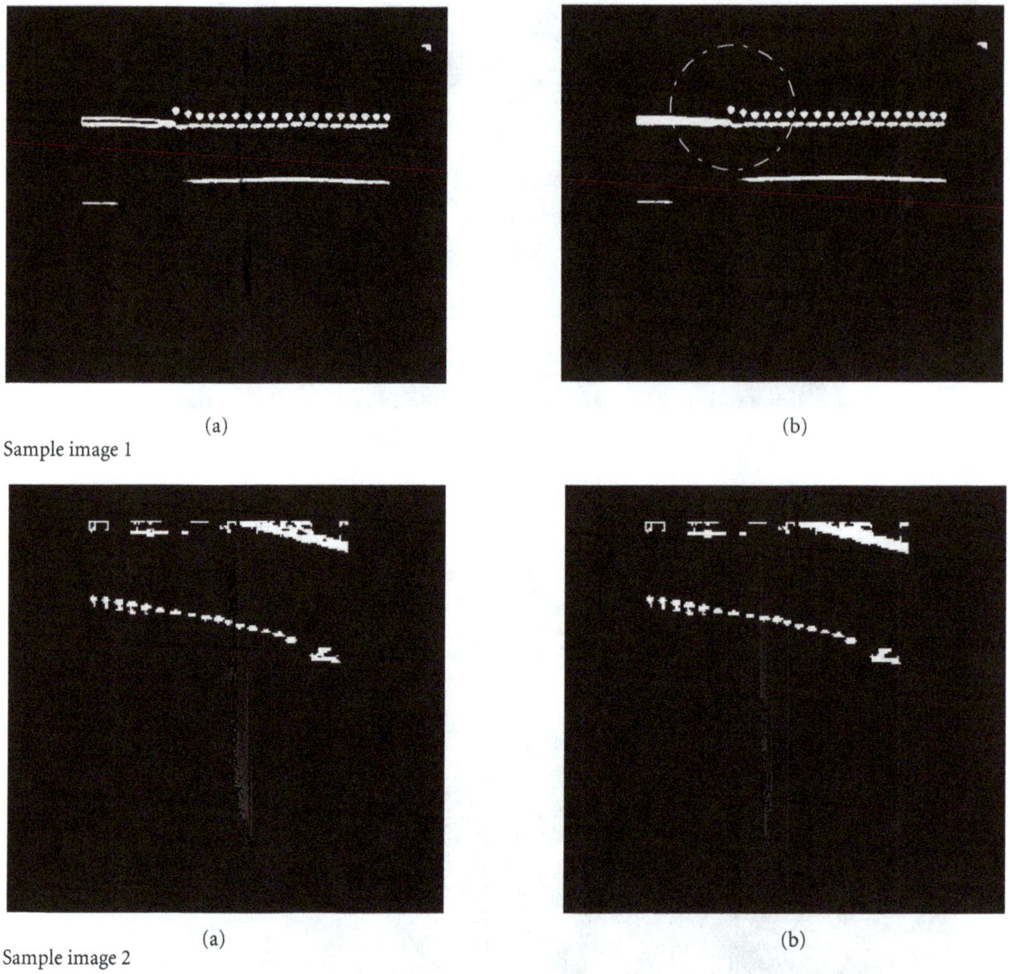

Figure 7: The effect of morphology on two sample images. (a) Before morphology application, (b) after morphology.

Metrics closer to 1 indicate that the object is approximately round

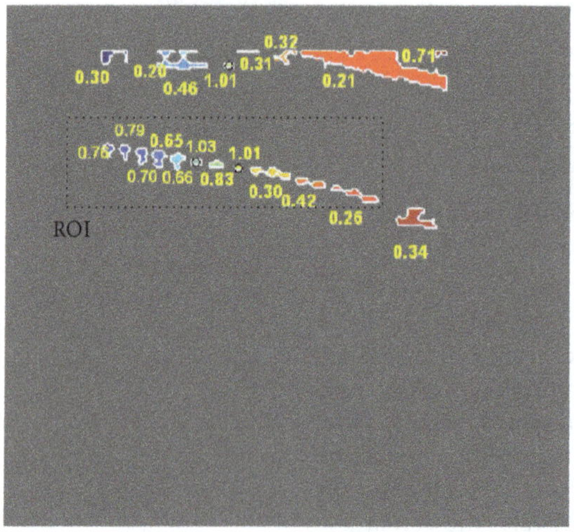

Figure 8: Circularity of objects.

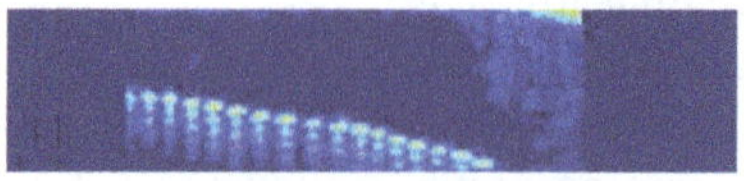

Figure 9: Extracted ROI with improved quality.

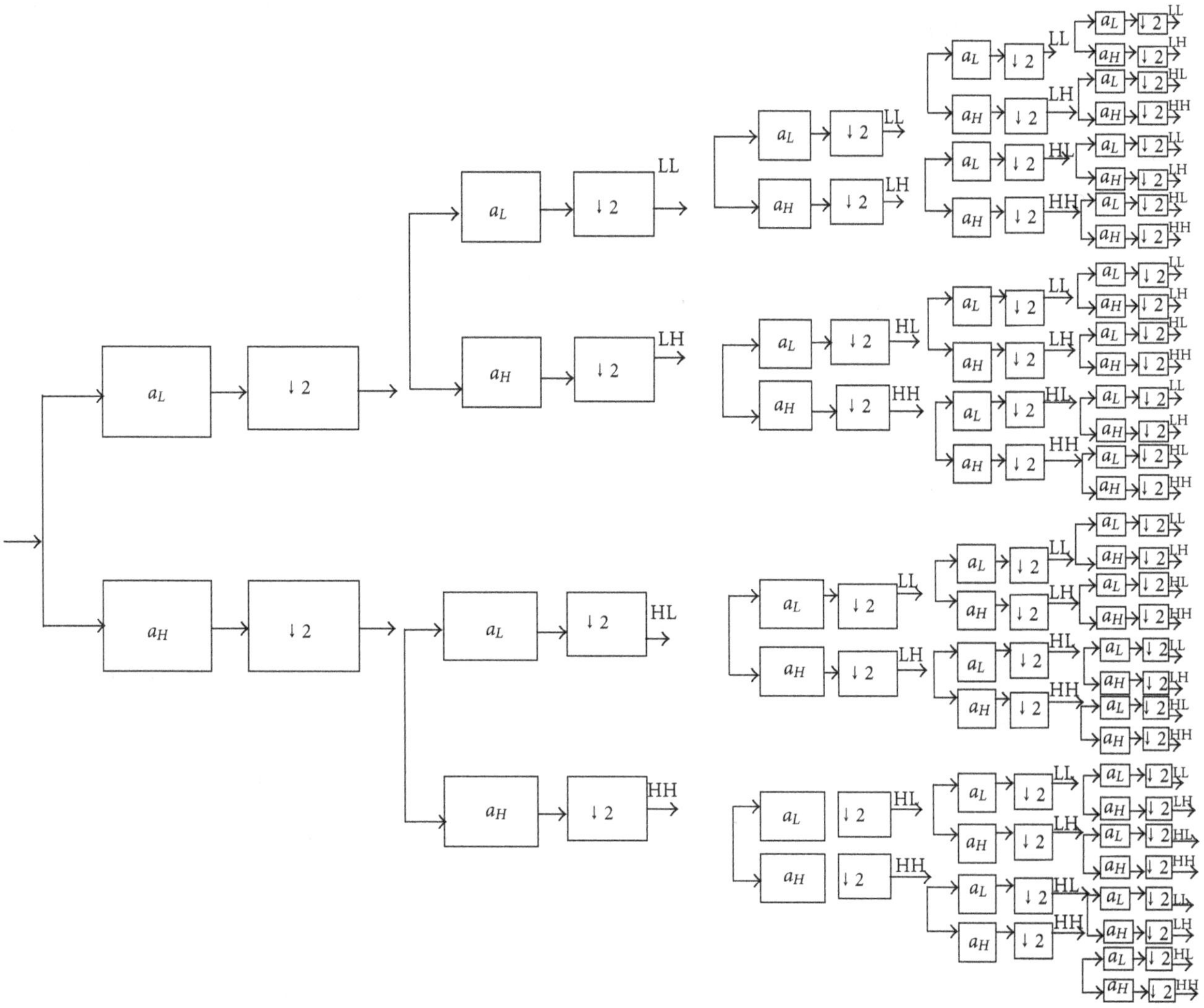

Figure 10: Example of five-level in wavelet transform.

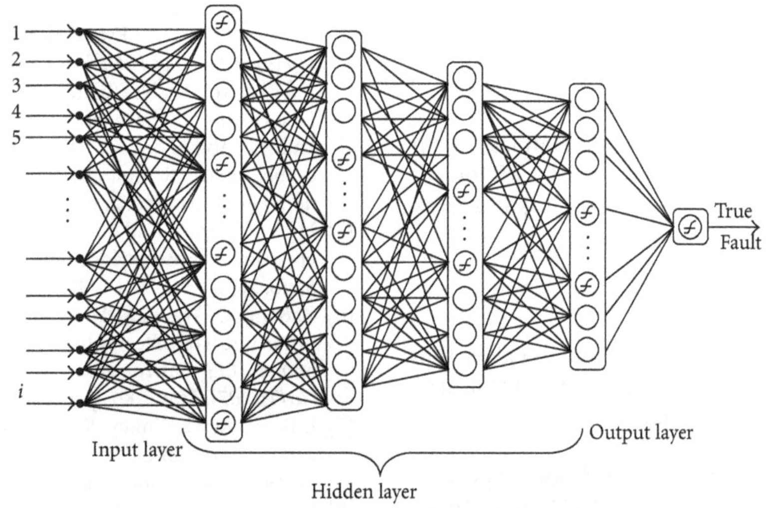

Figure 11: A typical multilayer perceptron network.

Table 1: Fi matrix.

14.569306864	7.0399609375	14.752532451	5.9073046875	20.791458954	8.6166406250
22.728265859	10.038593750	21.102602900	8.1773437500	24.808366126	10.330781250
18.682656381	8.1012500000	23.328353275	8.6918750000	25.175306545	11.230625000
34.302752032	15.230000000	26.909669279	12.005000000	42.179155356	20.607500000
30.981330318	15.340000000	35.931600284	17.530000000	54.869266111	34.900000000

Table 2: Simulation results in case study.

Q	Classification of correct and incorrect joint				
	Actual		ANN		
	Value	Type	Value	Type	
1	−1	Disjoint	−0.8467	Disjoint	OK
2	−1	Disjoint	−0.9962	Disjoint	OK
3	−1	Disjoint	−0.9756	Disjoint	OK
4	1	Joint	0.9336	Joint	OK
5	1	Joint	0.8449	Joint	OK
6	1	Joint	0.9450	Joint	OK
7	1	Joint	0.5027	Joint	OK
8	1	Joint	−0.2186	Disjoint	—
9	1	Joint	0.1566	Joint	OK
10	1	Joint	0.9777	Joint	OK

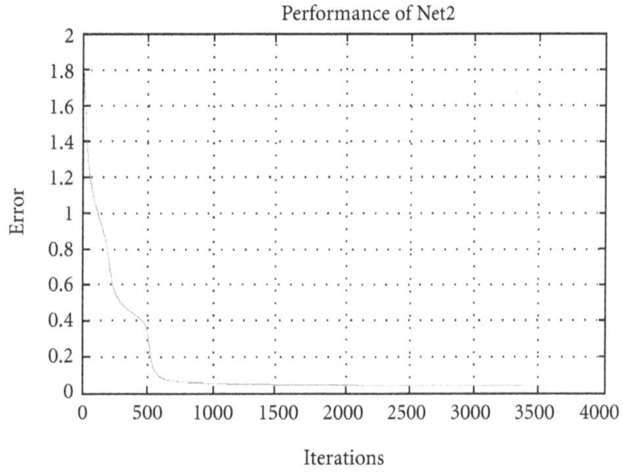

Figure 12: Training of network.

outputs for training of networks are −1 and 1. In fact, when the estimated outputs of ANN are positive, the PE joint is correct but for the negative values the PE joint is in a weak condition.

5. Conclusion

Based on practical experiments, a new combined algorithm based on a digital image processing, wavelet transform, artificial neural networks, and mathematical morphology techniques is presented.

This algorithm is applied and tested for improving the quality of the raw UT images from NDT evaluations. Obtained results show that the proposed algorithms to ultrasonic signals are highly reliable and also precise for the NDT testing and monitoring. It should be noted that extracted ROI by this method is applied for final evaluation by ANN and wavelet transform.

References

[1] J. Shi, J. Zheng, and W. Guo, "Formation mechanism and micro-structure of the Eigen-line in electro-fusion joints of polyethylene pipes," in *Proceedings of the American Society of Mechanical Engineers, Pressure Vessels and Piping Division*, pp. 303–313, July 2008.

[2] D. J. Hughes, E. L. Heeley, and C. Curfs, "A non-destructive method for the measurement of residual strains in semi-crystalline polymer components," *Materials Letters*, vol. 65, no. 3, pp. 530–533, 2011.

[3] H. Shin, Y. Jang, J. Kwan, and E. Lee, *Nondestructive Testing of Fusion Joints of Polyethylene Piping By Real Time Ultrasonic Imaging*, vol. 10, NDT.net, 2005.

[4] H. Kasban, O. Zahran, H. Arafa, M. El-Kordy, S. M. S. Elaraby, and F. E. Abd El-Samie, "Welding defect detection from radiography images with a cepstral approach," *NDT and E International*, vol. 44, no. 2, pp. 226–231, 2011.

[5] H. A. Mehrabi and J. Bowman, "Electrofusion welding of cross-linked polyethylene pipes," *Iranian Polymer Journal*, vol. 6, no. 3, pp. 195–203, 1997.

[6] Z. Song, Q. Wang, X. Du, and Y. Wang, "A high speed digital ultrasonic flaw detector based on PC and USB," in *Proceedings of the IEEE Instrumentation and Measurement Technology (IMTC '07)*, May 2007, paper no. 4258079.

[7] J. Brase, R. McKinney, K. Blaedel, J. Oppenheimer, S. Wong, and J. Simmons, "An automated ultrasonic testing bed," *Materials Evaluation*, vol. 42, no. 12, pp. 1619–1625, 1984.

[8] I. L. Iznita, S. A. Vijanth, P. A. Venkatachalam, and S. N. Lee, "Computerized segmentation of sinus images," in

Proceedings of the Innovative Technologies in Intelligent Systems and Industrial Applications (CITISIA '09), pp. 125–128, July 2009.

[9] Z. Li, X. Xu, X. Zhu, and C. Sui, "Computer system of ultrasonic signal sampling and analyzing," *Nondestructive Testing*, vol. 17, no. 6, pp. 154–156, 1996.

[10] M. I. H. Bhuiyan, M. O. Ahmad, and M. N. S. Swamy, "Spatially adaptive thresholding in wavelet domain for despeckling of ultrasound images," *IET Image Processing*, vol. 3, no. 3, pp. 147–162, 2009.

[11] A. M. Martínez, "Recognizing imprecisely localized, partially occluded, and expression variant faces from a single sample per class," *IEEE Transactions on Pattern Analysis and Machine Intelligence*, vol. 24, no. 6, pp. 748–763, 2002.

[12] S. Mitra and Y. Liu, "Local facial asymmetry for expression classification," in *Proceedings of the IEEE Computer Society Conference on Computer Vision and Pattern Recognition (CVPR '04)*, pp. II889–II894, June 2004.

[13] M. Vetterli and J. K. Evic, *Wavelets and Subband Coding*, Prentice-Hall, Englewood Cliffs, NJ, USA, 1995.

[14] MathWoks Incorporation, Matlab. Neural Networks Toolbox, V. 7. 2.

[15] Y. Q. Fu and Y. S. Wang, "Algorithm for edge detection of gray-scale image based on mathematical morphology," *Journal of Harbin Engineering University*, vol. 10, no. 5, pp. 685–687, 2005.

[16] J. Xu, "Morphological representation of 2-D binary shapes using rectangular components," *Pattern Recognition*, vol. 34, no. 2, pp. 277–286, 2001.

[17] J. S. Walker, *A Primer on Wavelets and Their Scientific Applications Edition*, Chapman & Hall/CRC, New York, NY, USA, 2nd edition, 2008.

[18] F. Bettayeb, T. Rachedi, and H. Benbartaoui, "An improved automated ultrasonic NDE system by wavelet and neuron networks," *Ultrasonics*, vol. 42, no. 1–9, pp. 853–858, 2004.

[19] S. Haykin, *Neural Networks and Learning Machines*, Pearson Education, Upper Saddle River, NJ, USA, 3rd edition, 2009.

[20] T. H. Martin, B. D. Howard, and H. B. Mark, *Neural Network Design*, PWS, Boston, Mass, USA, 1995.

Pilot-Induced Oscillation Suppression by Using L_1 Adaptive Control

Chuan Wang, Michael Santone, and Chengyu Cao

Department of Mechanical Engineering, University of Connecticut, Storrs, CT 06269-3139, USA

Correspondence should be addressed to Chuan Wang, chuan.wang@uconn.edu

Academic Editor: Lili Ma

Despite significant technical advances, pilot-induced oscillation (PIO) continues to occur in both flight tests and operational aircrafts. Such a phenomenon has led to significant research activities that aim to alleviate this problem. In this paper, the L_1 adaptive controller has been introduced to suppress the PIO, which is caused by rate limiting and pure time delay. Due to its architecture, the L_1 adaptive controller will achieve a desired response with fast adaptation. The analysis of PIO and its suppression by L_1 adaptive controller are presented in detail in the paper. The simulation results indicate that the L_1 adaptive control is efficient in solving this kind of problem.

1. Introduction

The high performance demands of modern aircrafts, especially highly maneuverable military jets, require the implementation of advanced control systems. The use of a modern electronic flight control system could provide great potential for improvement in the aircraft's performance. In spite of all the improvements, a significant handling quality problem arose with the introduction of electronic flight control system: pilot-induced oscillation, also known as pilot in-the-loop oscillation.

Pilot-induced oscillations are described as pilot-aircraft dynamic couplings, which could lead to instability in the systems [1]. Both previous and current research has attempted to explain, predict, and avoid these oscillations. Almost every modern aircraft has experienced PIO, which is well known by the public for the catastrophic event it caused, such as the YF-22 [2] and Olympic Airways Falcon 900. The occurrence of such events has led to significant research activities that are intended to alleviate the negative effects due to PIO. Despite the research efforts made, PIOs continue to occur, and reports of PIOs on operational aircrafts are increasing.

The focus of this paper is to suppress the pilot-induced oscillation caused by both rate limiting and pure time delay by using the L_1 adaptive controller. The effects of rate limiting and other system nonlinearities are considered to be the main factors that result in the occurrence of PIOs [3–5]. Some of the existing methods, to some extent, could handle the PIO caused by rate limiting, but not necessarily work with the presence of pure time delay in the actuator model. The L_1 adaptive control is known for its fast adaptation and smooth control implementation due to its powerful control architecture [6, 7]. With proper design, the L_1 adaptive controller will make the system respond in the desired manner.

The objective of this paper is to design the L_1 adaptive controller to make the inner loop of the system respond according to a given first-order system, while suppressing the PIO phenomenon caused by both rate limiting and pure time delay in the pilot dynamic model. Section 2 gives a brief introduction of pilot-induced oscillation and pilot model. In Section 3, the design of an inner loop adaptive controller to suppress the PIO and to track desired response is given. The inner loop adaptive controller cannot handle the adversely high pilot command inputs because they are the commands to the inner closed loop system. The inner loop controlled by L_1 adaptive controller solely tracks the command inputs from the pilots.

2. Pilot-Induced Oscillation

Pilot-induced oscillations can be regarded as a closed-loop instability of the pilot-aircraft loop, and it often happens when the pilot proves to be unable to adapt himself to a sudden change of the vehicle dynamics during a high demanding flight task. PIOs are complex interactions between the pilot and the aircraft dynamics.

2.1. Types of PIOs. According to the report by National Research Council (NRC) Committee in 1997 [8], PIOs can be separated into three categories.

(i) Category I PIO: Linear pilot-vehicle system oscillation. These PIOs result from linear phenomena such as excessive time delay and excessive phase loss between the pilot's control input and the aircraft response. They are the simplest to model, understand, and prevent. Before the introduction of fly-by-wire technology, almost all PIOs were this type, and a large amount of research has been conducted on them. But they are currently the least common in operational flight. The main causes of Category I PIO are excessive lags due to time delays and various digital filter dynamics in the flight control system. These effects lead to a high frequency phase roll-off in the frequency response. There are roughly two groups of design criteria for preventing Category I PIO. One is according to the flight control stability, and the other is the handling quality requirements. In the first group of design criteria, the open loop aircraft frequency response needs to be checked as well as the phase and magnitude margin. If the phase rate of the system is too high, a small increase in frequency will result in a strong additional phase delay. This is usually related with time delay in the whole system.

(ii) Category II PIO: Quasi-linear events with some nonlinear contribution, such as rate or position limiting. For the most part, these PIOs can be modeled as linear events, with an identifiable nonlinear contribution that may be treated separately. Figure 1 shows two typical positions where rate-limiting blocks are installed in the flight control system. The block after the pilot model is installed to prevent the system from receiving a high input rate by the pilot. The other block after the controller is used to protect the actuator against overload. In the case of rate limiting, nonlinear system theory should be utilized for the analysis of the flight system, such as the Lyapunov theory and the phase plane method.

The majority of aircraft crashes due to PIOs are caused by Category II PIO through activating actuator rate limits. When the rate limiter is saturated, phase lag occurs and the aircraft dynamics change suddenly, which cause PIO and threaten the flight safety. Rate limiting adds additional phase lag, increasing the delay between the pilot input and the aircraft response. This tends to make the pilot compensate with faster responses, often worsening the situation. Rate limiting also reduces the gain, which the pilot interprets as a lack of control response and therefore makes larger command inputs, again making the situation worse.

(iii) Category III PIO: Nonlinear PIOs with transients. Such events are difficult to recognize and rarely occur but are always severe. Mode switching or rapid changes in effective vehicle characteristics could be reasons for this type of PIO. Fortunately, these events that result in nonoscillatory divergence and loss of control of the aircraft are rare.

2.2. Rate Limiting. Rate limiting of the actuator is one of main factors that lead to pilot-induced oscillation. Actuator rate limiting occurs when pilot input command error requires a higher rate than the actuator can actually provide. A simplified model of a rate-limited actuator is shown in Figure 2.

Rate limiting adds additional phase lag, increasing the delay between the pilot input and aircraft response. This tends to make the pilot compensate with faster responses, which will often worsen the situation.

2.3. Pilot Modeling. With rate limiting considered, another important factor that causes the PIO phenomenon is the pilot model, which is also regarded as the weakest point in the analysis due to its high nonlinearity and complexity. However, there are several specific pilot behavioral patterns [9] that could be used to analyze the cause of these PIOs. The pilot model is the source factor that distinguishes severe PIO problems from most aircraft feedback control design problems. The difference resides in unique human properties related to the adaptive characteristics of the human pilot.

Pilots exhibit peculiar transitions in the organizational structure of the pilot vehicle system. These transitions can involve both the pilot's compensation and effective architecture of the pilot's control strategy. The human pilot dynamics can be roughly separated into the following types.

(i) Compensatory behavior: essentially, the pilot has generated a lead to cancel out the lag in the aircraft model. However, the higher frequency lags of the pilot model can be approximated at the lower frequencies by the pure time delay.

(ii) Pursuit behavior: the introduction of the pursuit behavior permits an open-loop control in conjunction with the compensatory closed-loop error correcting action. The pilot model of both compensatory behavior and pursuit behavior will be superior to that where only compensatory operations are possible.

(iii) Precognitive behavior: this kind of pilot model gives a higher level of control performance. Based on the knowledge of the system dynamics, the pilot model will generate proper control signals at the right time so as to result in machine outputs that are almost as desired. This operation also appears in company with compensatory behavior as well as pursuit behavior. Most highly skilled movements will automatically fall into this category.

3. PIO Suppression by L_1 Adaptive Control

3.1. System Modeling. In most PIO cases that have happened in recent years, rate- and position-limiting phenomena appear. Hence, numerous research works have been conducted to analyze rate and position limiting [10–12]. In this situation, PIOs can be explained by limit cycles occurring in a nonlinear system where the nonlinearities cause a sustained,

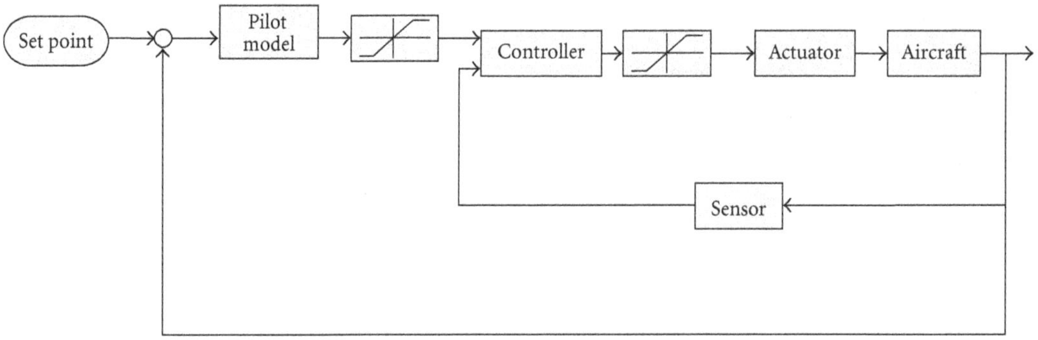

FIGURE 1: Rate limiting in flight control systems.

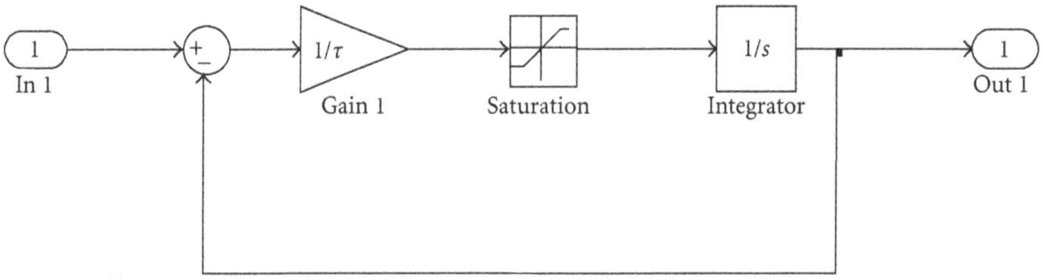

FIGURE 2: Rate limit actuator dynamic model.

constant amplitude oscillation. In those studies, rate limiting of the actuator is considered as one possible cause for PIOs. There are some other factors, such as relatively high pilot command gain and lag in the pilot model, contributing to the onset of PIOs. Rate limiting is commonly met in aviation practice. Hence, if an adaptive inner loop can compensate for the nonlinearities caused by rate limiting and eliminate the limit-cycle oscillations, it can be considered that the PIO is suppressed for such commonly seen situations. We apply the L_1 adaptive controller on the F-14 model taken from [10] in order to test this idea. A model of the Grumman Aircraft Company (GAC) F-14 was developed for this investigation. The approximated model in [10] is compared with data of a real F-14 that has severe PIOs during flight tests. The rate-limiting phenomenon we considered is due to sudden off-nominal conditions such as hydraulic pressure failure as in the F-14 example. Therefore, it is hard to estimate the nonlinearity caused by rate limiting in advance, like most other research work has done. Adaptive control is needed for such situations.

The control structure is shown in Figure 3. It is a longitudinal dynamic model. The system states are $[\alpha\, \theta\, q]$, and control input is the elevator. The L_1 adaptive inner loop controller is a pitch-rate augmentation system, which provides the desired dynamics for the $[\alpha\, q]$ subsystem. The outer loop takes the θ feedback signal and injects it into the pilot model. The pilot model outputs a pitch rate command for the inner loop to achieve the pitch attitude control (the θ angle). The pilot model is given by

$$p(s) = K(s+\beta)e^{-\tau s}, \tag{1}$$

which describes the compensatory behavior of the pilot.

3.2. Delay Margin Detection. Consider the Pilot model described by (1); we assume the desired response of the inner loop is given by

$$g_c(s) = \frac{1}{T_p s + 1}. \tag{2}$$

With the Pilot model, we could derive the transfer function of the closed-loop as

$$g(s) = \frac{p(s)g_c(s)}{p(s)g_c(s)+1} = \frac{K(s+\beta)e^{-\tau s}}{T_p s + 1 + K(s+\beta)e^{-\tau s}}. \tag{3}$$

In the PIO problem concerned, the pure time delay deviates between 0.2 and 0.3 second. The time delay term can be approximated by $e^{-\tau s} = (1-(\tau s/2))/(1+(\tau s/2))$ according to Pade approximation. Equation (3) can be further written as

$$\begin{aligned} g(s) &= \frac{K(s+\beta)1 + \tau s/2}{\left(T_p s+1\right)(1+\tau s/2) + K(s+\beta)(1-\tau s/2)} e^{-\tau s} \\ &= \frac{(K\tau/2)s^2 + (1+\tau\beta K/2)s + \beta K}{\left(\left(T_p - K\right)\tau/2\right)s^2 + \left(T_p + K - \tau/2 - \tau\beta K/2\right)s + 1 + K\beta} e^{-\tau s}, \end{aligned} \tag{4}$$

according to Routh stability criterion, we can get the following inequalities:

$$\begin{gathered} T_p > K, \\ 1 + K\beta > 0, \\ T_p > K\tau\beta/2 - \tau/2 - K. \end{gathered} \tag{5}$$

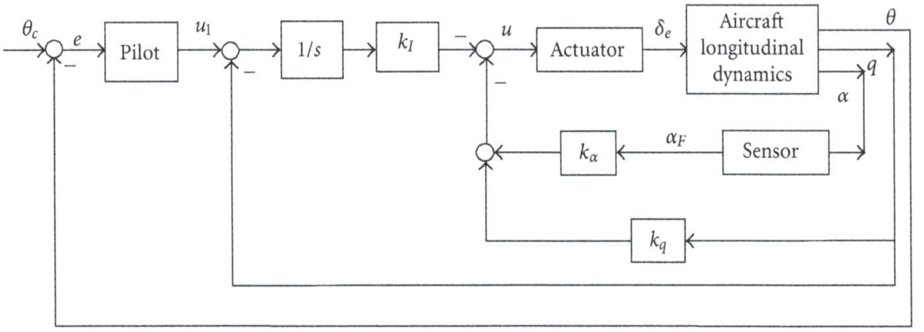

FIGURE 3: Control structure.

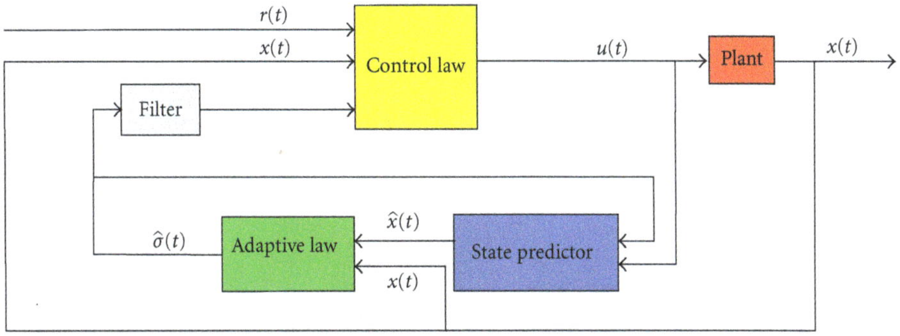

FIGURE 4: L_1 adaptive controller architecture.

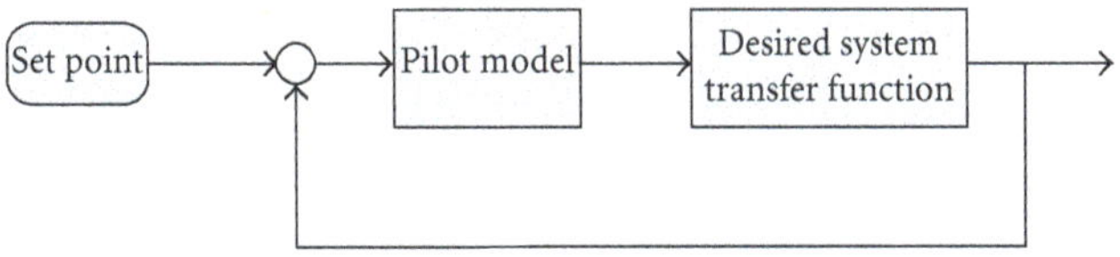

FIGURE 5: PIO delay margin detection architecture.

If the parameter T_p chosen is satisfied with unequal conditions in (5), the system will achieve the desired response with stability condition. From many PIO studies, the onset of PIO often was accompanied by high pilot command inputs. If some off-nominal conditions happen, such as hydraulic failure of the actuators in F-14 PIO accidents, the rate limiting comes into play. When the L_1 inner loop adaptive controller is applied to this model, it can compensate for the unknown effects caused by rate limiting and time delay, no matter whether they are linear or nonlinear. The design of the L_1 adaptive controller does not need a priori information for the rate limits of the actuator.

3.3. L_1 Adaptive Control Design to Achieve Desired Response. Consider the L_1 architecture given in Figure 4, the L_1 adaptive control [13] could be divided into three parts: the adaptive law, the state predictor, and the control law.

State Predictor Design. The state predictor of the aircraft longitudinal dynamic system in Figure 3 is designed as follows:

$$\dot{\hat{x}}(t) = A_m\hat{x}(t) + Bu(t) + \hat{\sigma}_x(t), \qquad \hat{x}(0) = x_0, \tag{6}$$

where $A_m \in R^{3\times3}$, $B \in R^{3\times1}$,

$$\hat{x}(t) = \begin{bmatrix} \hat{\theta}(t) \\ \hat{q}(t) \\ \hat{\alpha}(t) \end{bmatrix}, \qquad \hat{\sigma}_x(t) = \begin{bmatrix} \hat{\sigma}_1(t) \\ \hat{\sigma}_2(t) \\ \hat{\sigma}_3(t) \end{bmatrix}. \tag{7}$$

Matrix A should be Hurwitz to make sure of the stability of the model. $\hat{\sigma}$ can be divided into two parts, the model matched part and the model unmatched part

$$\hat{\sigma} = B\hat{\sigma}_m + \overline{B}\hat{\sigma}_{um}, \tag{8}$$

where $\overline{B}$ is the null space of B. $\hat{\sigma}_m$ is the matched part, and $\hat{\sigma}_{um}$ indicates the unmatched part. $\hat{\sigma}_m$ and $\hat{\sigma}_{um}$ can be calculated as follows:

$$\begin{bmatrix} \hat{\sigma}_m \\ \hat{\sigma}_{um} \end{bmatrix} = \begin{bmatrix} B & \overline{B} \end{bmatrix}^{-1} \hat{\sigma}. \tag{9}$$

Adaptive Law Design. Assume the system dynamics is as follows:

$$\dot{x}(t) = Ax(t) + Bu + \sigma(t). \tag{10}$$

Given any $T > 0$, we have

$$\Phi_x(T) = \int_0^T e^{A_m(T-\tau)} d\tau. \tag{11}$$

Letting $\tilde{x}(t) = \hat{x}(t) - x(t)$, the adaptive law for $\hat{\sigma}_x(t)$ is given by

$$\begin{aligned} \hat{\sigma}_x(t) &= \hat{\sigma}_x(iT), \quad t \in [iT, (i+1)T], \\ \hat{\sigma}_x(iT) &= -\Phi_x^{-1}(T)\mu_x(iT), \quad i = 0, 1, 2, \ldots, \end{aligned} \tag{12}$$

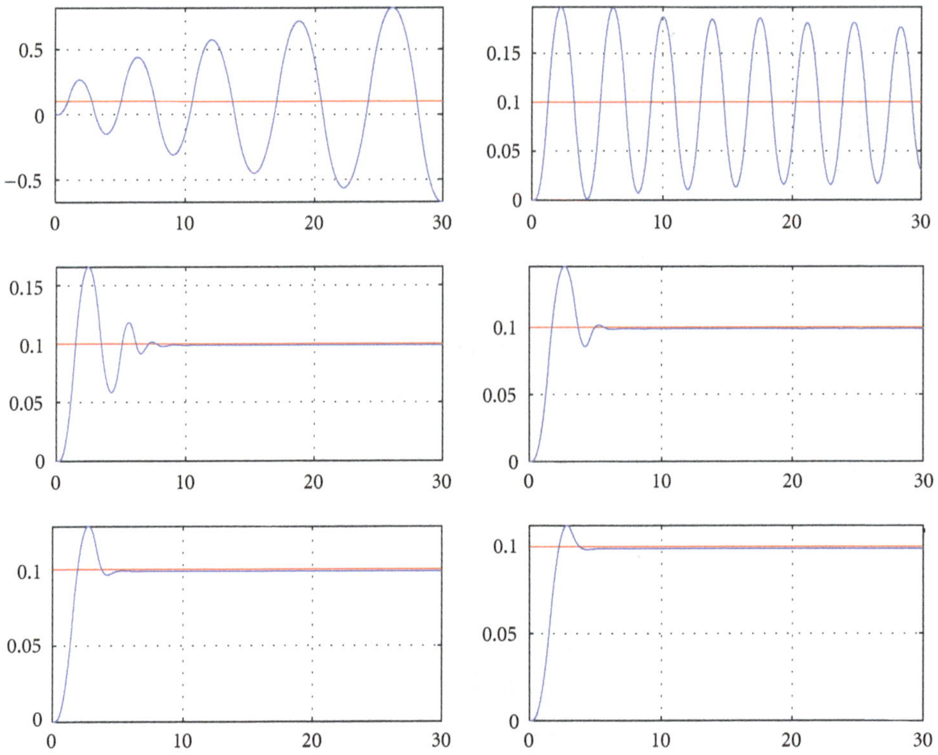

FIGURE 6: system responses with increasing T_p.

where $\Phi(T)$ is defined above, and

$$\begin{aligned} \mu_x(iT) &= e^{A_m T}\tilde{x}(iT), \quad i = 1, 2, \ldots \\ \hat{\sigma}_x(t) &= -\Phi_x^{-1}(T)e^{A_m T}\tilde{x}(t). \end{aligned} \tag{13}$$

Control Law Design. The output estimation of the system is given by

$$\hat{y} = C\hat{x}. \tag{14}$$

The control signal can be calculated by

$$\begin{aligned} u &= u_r + u_m + u_{um}, \\ u_r &= -\frac{r}{CA_m^{-1}B}, \\ u_m &= -\hat{\sigma}_m, \\ u_{um} &= \left[C(SI - A_m)^{-1}B\right]^{-1} C(SI - A_m)^{-1}\overline{B}\hat{\sigma}_{um}. \end{aligned} \tag{15}$$

u_r is the control signal used to track the reference signal with desired response designed aforementioned. u_m and u_{um} are control signals used to cancel the uncertainty part of the system, where u_m is the matched control signal and u_{um} is the unmatched one.

4. Simulation Study

In general, rate limiting results in an amplitude reduction and a significant added phase lag. With the introduction of pure time delay into the pilot model, the closed-loop system could even become unstable.

Based on the delay margin detection architecture shown in Figure 5, the time variable in the desired system response model can be obtained from the result derived in the previous section. There is another alternative method that could be used for the detection of the delay margin by searching for the optimal value of T_p in (2). With an increase in T_p, the system response will change gradually in accordance. At some critical point, the oscillation will disappear and the system response will converge to the set-point value instead. If we further increase the time variable, the damping ratio of the closed-loop system will increase as well, which will affect the system's transient response including overshoot, rise time and so forth (Figure 6).

In Figure 7, the F-14 system responses of different time constants T_p are given. If the time variable T_p is small as shown in Figure 7(a), the system responds rapidly, but the overshot is also very large. With increasing T_p, the overshoot decreases, but transient performance reduces as well. Thus, we need to build a balance between the overshoot and the rise time, which can be simply implemented by introducing an objective function as

$$\min J = \omega_1 \delta + \omega_2 T_r, \tag{16}$$

where ω_1, ω_2 are weighting coefficients and δ, T_r represent the overshoot and rise time, respectively.

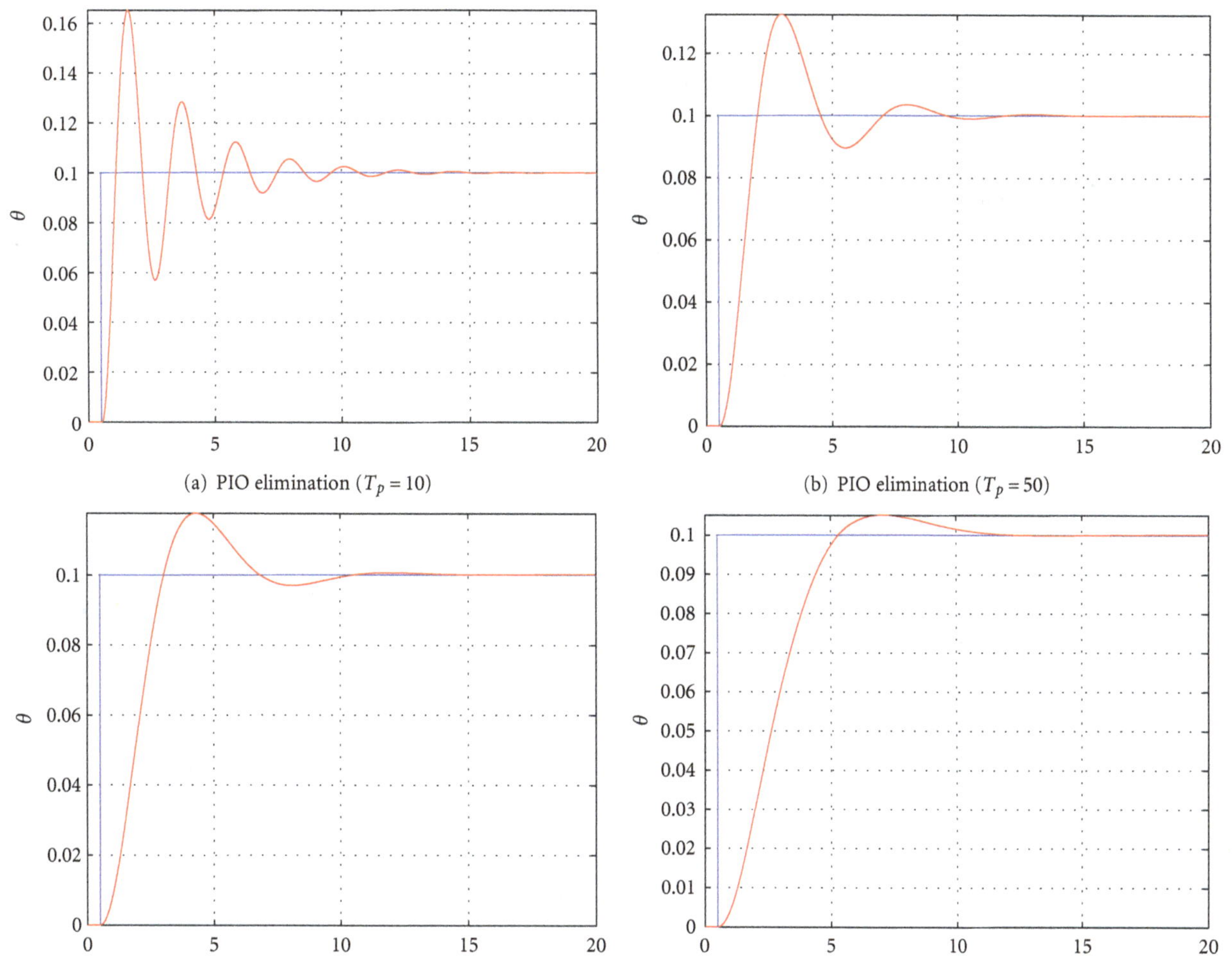

(a) PIO elimination ($T_p = 10$)

(b) PIO elimination ($T_p = 50$)

(c) PIO elimination ($T_p = 100$)

(d) PIO elimination ($T_p = 200$)

FIGURE 7: PIO elimination with different desired responses.

5. Conclusion and Future Research

This paper proposes an alternative method for suppressing the pilot-induced oscillation phenomenon caused by rate limiting and time delay in the pilot dynamic model. The delay margin is detected at the beginning, and the critical part of the method is design the L_1 adaptive controller to achieve the desired response of the system.

To further improve the performance of this method, we need to consider the following issues. First, certain criterion need to be built to determine the choice of T_p. The system response can be evaluated by stability, overshoot, and rise time, and so forth. For the robustness of this method, a time-varying pilot model should be considered.

Acknowledgment

The authors would like to thank Dr. Jiang Wang from Zona Inc. for his useful suggestions and help.

References

[1] I. Ashkenas, "Pilot-induced oscillations: their cause and analysis," Tech. Rep., 1964, DTIC Document.

[2] M. Dornheim, "Report pinpoints factors leading to yf-22 crash," *Aviation Week and Space Technology*, vol. 137, no. 19, pp. 53–54, 1992.

[3] D. G. Mitchell, B. A. Kish, and J. S. Seo, "A flight investigation of pilot-induced oscillation due to rate limiting," in *Proceedings of the IEEE Aerospace Conference*, vol. 3, pp. 59–74, March 1998.

[4] S. L. Gatley, M. C. Turner, I. Postlethwaite, and A. Kumar, "A comparison of rate-limit compensation schemes for pilot-induced- oscillation avoidance," *Aerospace Science and Technology*, vol. 10, no. 1, pp. 37–47, 2006.

[5] Y. Yildiz and I. V. Kolmanovsky, "A control allocation technique to recover from pilot-induced oscillations (capio) due to actuator rate limiting," in *Proceedinjgs of the American Control Conference (ACC '10)*, pp. 516–523, July 2010.

[6] C. Cao and N. Hovakimyan, "L_1 adaptive controller for a class of systems with unknown nonlinearities: part I," in *Proceedings*

of the American Control Conference (ACC '08), pp. 4093–4098, June 2008.

[7] C. Cao and N. Hovakimyan, "L_1 adaptive controller for nonlinear systems in the presence of unmodelled dynamics: part II," in *Proceedings of the American Control Conference (ACC '08)*, pp. 4099–4104, USA, June 2008.

[8] National Research Council (U.S.). Committee on the Effects of Aircraft-Pilot Coupling on Flight Safety, *Aviation Safety and Pilot Control: Understanding and Preventing Unfavorable Pilot-Vehicle Interactions*, National Academies Press, Washigton, DC, USA, 1997.

[9] D. McRuer, U. S. N. Aeronautics, S. A. Scientific, and T. I. Program, *Pilot-Induced Oscillations and Human Dynamic Behavior*, vol. 4683, NASA, 1995.

[10] A. Robbins, *Pilot variability during pilot-induced oscillation*, Ph.D. thesis, Virginia Polytechnic Institute and State University, 1999.

[11] M. R. Anderson, "Pilot-induced oscillations involving multiple nonlinearities," *Journal of Guidance, Control, and Dynamics*, vol. 21, no. 5, pp. 786–791, 1998.

[12] D. H. Klyde and D. G. Mitchell, "A pio case study—lessons learned through analysis," in *Proceedings of the AIAA Atmospheric Flight Mechanics Conference and Exhibit*, pp. 1–17, 2005.

[13] N. Hovakimyan and C. Cao, *L_1 Adaptive Control Theory: Guaranteed Robustness with Fast Adaptation*, SIAM, Philadelphia, Pa, USA, 2010.

20

Model Predictive Control of Uncertain Constrained Linear System Based on Mixed $\mathcal{H}_2/\mathcal{H}_\infty$ Control Approach

Patience E. Orukpe

Department of Electrical and Electronic Engineering, University of Benin, P.M.B 1154, Benin City, Edo State, Nigeria

Correspondence should be addressed to Patience E. Orukpe, patience.orukpe01@imperial.ac.uk

Academic Editor: Marcin T. Cychowski

Uncertain constrained discrete-time linear system is addressed using linear matrix inequality based optimization techniques. The constraints on the inputs and states are specified as quadratic constraints but are formulated to capture hyperplane constraints as well. The control action is of state feedback and satisfies the constraints. Uncertainty in the system is represented by unknown bounded disturbances and system perturbations in a linear fractional transform (LFT) representation. Mixed $\mathcal{H}_2/\mathcal{H}_\infty$ method is applied in a model predictive control strategy. The control law takes account of disturbances and uncertainty naturally. The validity of this approach is illustrated with two examples.

1. Introduction

Model predictive control (MPC) is a class of model-based control theories that use linear or nonlinear process models to forecast system behaviour. MPC is one of the control techniques that is able to cope with model uncertainties in an explicit way [1]. MPC has been used widely in practical applications to industrial process systems [2] and active vibration control of railway vehicles [3]. One of the methods used in MPC when uncertainties are present is to minimise the objective function for the worst possible case. This strategy is known as minimax and was originally proposed [4] in the context of robust receding control, [5] in the context of feedback and feedforward control and [6] in the context of $\mathcal{H}_\infty$ MPC. MPC has been applied to $\mathcal{H}_\infty$ problems in order to combine the practical advantage of MPC with the robustness of the $\mathcal{H}_\infty$ control, since robustness of MPC is still being investigated for it to be applied practically.

This work is motivated by the work in [7, 8] where uncertainty in the system was modeled by perturbations in a linear fractional representation. In [9], model predictive control based on a mixed $\mathcal{H}_2/\mathcal{H}_\infty$ control approach was considered. The designed controller has the form of state feedback and was constructed from the solution of a set of feasibility linear matrix inequalities. However, the issue of handling both uncertainty and disturbances simultaneously was not considered. In this paper, we extend the result of [9] to constrained uncertain linear discrete-time invariant systems using a mixed $\mathcal{H}_2/\mathcal{H}_\infty$ design approach and the uncertainty considered is norm-bounded additive. This is more suitable as both performance and robustness issues are handled within a unified framework.

The method presented in this paper develops an LMI design procedure for the state feedback gain matrix F, allowing input and state constraints to be included in a less conservative manner. A main contribution is the accomplishment of a prescribed disturbance attenuation in a systematic way by incorporating the well-known robustness guarantees through $\mathcal{H}_\infty$ constraints into the MPC scheme. In addition, norm-bounded additive uncertainty is also incorporated. A preliminary version of some of the work presented in this paper was presented in [10].

The structure of the work is as follows. After defining the notation, we describe the system and give a statement of the mixed $\mathcal{H}_2/\mathcal{H}_\infty$ problem in Section 2. In Section 3, we derive sufficient conditions, in the form of LMIs, for the existence of a state feedback control law that achieves the design specifications. In Section 4, we consider two examples that illustrate our algorithm. Finally, we conclude in Section 5.

The notation we use is fairly standard. $\mathcal{R}$ denotes the set of real numbers, $\mathcal{R}^n$ denotes the space of n-dimensional (column) vectors whose entries are in $\mathcal{R}$ and $\mathcal{R}^{n\times m}$ denoting the space of all $n \times m$ matrices whose entries are in $\mathcal{R}$. For $A \in \mathcal{R}^{n\times m}$, we use the notation A^T to denote transpose. For $x, y \in \mathcal{R}^n$, $x < y$ (and similarly $\leq$, $>$, and $\geq$) is interpreted element wise. The identity matrix is denoted as I and the null matrix by 0 with the dimension inferred from the context.

2. Problem Formulation

We consider the following discrete-time linear time invariant system:

$$\begin{aligned} x_{k+1} &= Ax_k + B_w w_k + B_u u_k + B_p p_k, \\ q_k &= C_q x_k + D_{qu} u_k + D_{qw} w_k, \\ p_k &= \Delta_k q_k, \\ z_k &= \begin{bmatrix} C_z x_k \\ D_{zu} u_k \end{bmatrix}, \\ & x_0 \text{ given}, \end{aligned} \tag{1}$$

where x_0 is the initial state, $x_k \in \mathcal{R}^n$ is the state, $w_k \in \mathcal{R}^{n_w}$ is the disturbance, $u_k \in \mathcal{R}^{n_u}$ is the control, $z_k \in \mathcal{R}^{n_z}$ is the controlled output, $A \in \mathcal{R}^{n\times n}$, $B_w \in \mathcal{R}^{n\times n_w}$, $B_u \in \mathcal{R}^{n\times n_u}$, $C_z \in \mathcal{R}^{n_{z_1}\times n}$, and $D_{zu} \in \mathcal{R}^{n_{z_2}\times n_u}$, and where $n_z = n_{z_1} + n_{z_2}$. The signals q_k and p_k model uncertainties or perturbations appearing in the feedback loop.

The operator, Δ_k, is block diagonal:

$$\Delta_k \in \mathbf{\Delta}_k = \left\{ \Delta_k = \begin{bmatrix} \Delta_{1k} & & 0 \\ & \ddots & \\ 0 & & \Delta_{tk} \end{bmatrix} : \|\Delta_{ik}\| \leq 1 \ \forall i \right\}, \tag{2}$$

and is norm bounded by one. Scalings can be included in C_q and B_p, thus generalizing the bound. Δ_k can represent either a memoryless time-varying matrix with $\overline{\sigma}(\Delta_{ik}) \leq 1$, for $i = 1, \ldots, t$, $k \geq 0$, or the constraints:

$$p_{ik}^T p_{ik} \leq q_{ik}^T q_{ik}, \quad i = 1, \ldots, t, \tag{3}$$

where $p_k = [p_{1k}, \ldots, p_{tk}]^T$, $q_k = [q_{1k}, \ldots, q_{tk}]^T$, and the partitioning is induced by Δ_k. Each Δ_k is assumed to be either a full block or a repeated scalar block, and models a number of factors, such as dynamics or parameters, nonlinearities, that are unknown, unmodeled or neglected. In this work, we only consider full blocks for simplicity.

In terms of the state space matrices, this formulation can be viewed as replacing a fixed (A, B_u, B_w) by $(A, B_u, B_w) \in (\mathcal{A}, \mathcal{B}_u, \mathcal{B}_w)$, where

$$(\mathcal{A}, \mathcal{B}_u, \mathcal{B}_w) = \left\{ \left[A + B_p\Delta_k C_q, B_u + B_p\Delta_k D_{qu}, B_w + B_p\Delta_k D_{qw} \right] \mid \Delta_k \in \mathbf{\Delta}_k \right\}. \tag{4}$$

In robust model predictive control, we consider norm-bounded uncertainty and define stability in terms of quadratic stability [11] which requires the existence of a fixed quadratic Lyapunov function ($V(\zeta) = \zeta^T P \zeta$, $P > 0$) for all possible choices of the uncertainty parameters.

In the case of norm-bounded uncertainties:

$$\begin{bmatrix} A & B_w & B_u \end{bmatrix} \in \left\{ \begin{bmatrix} A^o & B_w^o & B_u^o \end{bmatrix} + F_A \Delta_H \begin{bmatrix} E_A & E_w & E_u \end{bmatrix} : \Delta \in \mathbf{\Delta} \right\}, \tag{5}$$

where $[A^o \ B_w^o \ B_u^o]$ represents the nominal model, $\Delta_H = \Delta(I - H\Delta)^{-1}$, with

$$\Delta \in \mathbf{\Delta} := \left\{ \Delta = \operatorname{diag}\left(\delta_1 I_{q_1}, \ldots, \delta_l I_{q_l}, \Delta_{l+1}, \ldots, \Delta_{l+f}\right) : \|\Delta\| \leq 1, \delta_i \in \mathcal{R}, \Delta_i \in \mathcal{R}^{q_i \times q_i} \right\} \tag{6}$$

and where F_A, E_A, E_w, E_u, and H are known and constant matrices with appropriate dimensions. This linear fractional representation of uncertainty, which is assumed to be well posed over $\mathbf{\Delta}$ (i.e., $\det(I - H\Delta) \neq 0$ for all $\Delta \in \mathbf{\Delta}$), has great generality and is used widely in robust control theory [12].

We use the following lemma, which is a slight modification of a result in [13] and which uses the fact that $\Delta \in \mathbf{\Delta}$ to remove explicit dependence on Δ for the solution with norm bounded uncertainties.

Lemma 1. *Let $\mathbf{\Delta}$ be as described in (6) and define the subspaces*

$$\begin{aligned} \mathbf{\Sigma} &= \left\{ \operatorname{diag}\left(S_1, \ldots, S_l, \lambda_1 I_{ql+1}, \ldots, \lambda_s I_{ql+f}\right) : S_i = S_i^T \in \mathcal{R}^{q_i \times q_i}, \quad \lambda_j \in \mathcal{R} \right\}, \\ \mathbf{\Gamma} &= \left\{ \operatorname{diag}\left(G_1, \ldots, G_l, 0_{ql+1}, \ldots, 0_{ql+f}\right) : G_i = -G_i^T \in \mathcal{R}^{q_i \times q_i} \right\}. \end{aligned} \tag{7}$$

Let $T_1 = T_1^T, T_2, T_3, T_4$ be matrices with appropriate dimensions. We have $\det(I - T_4\Delta) \neq 0$ and $T_1 + T_2\Delta(I - T_4\Delta)^{-1}T_3 + T_3^T(I - \Delta^T T_4^T)^{-1}\Delta^T T_2^T < 0$ for every $\Delta \in \mathbf{\Delta}$ if there exist $S \in \mathbf{\Sigma}$ and $G \in \mathbf{\Gamma}$ such that $S > 0$ and

$$\begin{bmatrix} T_1 + T_2 S T_2^T & T_3^T + T_2 S T_4^T + T_2 G \\ T_3 + T_4 S T_2^T + G^T T_2^T & T_4 S T_4^T + T_4 G + G^T T_4^T - S \end{bmatrix} < 0. \tag{8}$$

If $\mathbf{\Delta}$ is unstructured, then (8) becomes

$$\begin{bmatrix} T_1 + \lambda T_2 T_2^T & T_3^T + \lambda T_2 T_4^T \\ T_3 + \lambda T_4 T_2^T & \lambda\left(T_4 T_4^T - I\right) \end{bmatrix} < 0, \tag{9}$$

for some scalar $\lambda > 0$. In this case, condition (9) is both necessary and sufficient.

We also use the following Schur complement result [14].

Lemma 2. *Let $X_{11} = X_{11}^T$ and $X_{22} = X_{22}^T$. Then*

$$\begin{bmatrix} X_{11} & X_{12} \\ X_{12}^T & X_{22} \end{bmatrix} \geq 0 \Longleftrightarrow X_{22} \geq 0, \quad X_{22} - X_{12}^T X_{11}^{+} X_{12} \geq 0, \quad X_{12}(I - X_{11}X_{11}^{+}) = 0, \tag{10}$$

where X_{11}^{+} denotes the Moore-Penrose pseudo-inverse of X_{11}.

We assume that the pair (A, B_u) is stabilizable and that the disturbance is bounded as

$$\|w\|_2 := \sqrt{\sum_{k=0}^{\infty} w_k^T w_k} \le \overline{w}, \tag{11}$$

where $\overline{w} \ge 0$ is know.

The aim is to find a state feedback control law $\{u_k = Fx_k\}$ in $\mathcal{L}_2$, where $F \in \mathcal{R}^{n_u \times n}$, such that the following constraints are satisfied for all $\Delta_k \in \mathbf{\Delta}_k$.

(1) *Closed-loop stability*: the matrix $A + B_u F$ is stable.

(2) *Disturbance rejection*: for given $\gamma > 0$, the transfer matrix from w to z, denoted as T_{zw}, is quadratically stable and satisfies the $\mathcal{H}_\infty$ constraint

$$\|z\|_2 < \gamma \|w\|_2, \tag{12}$$

for $x_0 = 0$.

(3) *Regulation*: for given $\alpha > 0$, the controlled output satisfies the $\mathcal{H}_2$ constraint:

$$\|z\|_2 := \sqrt{\sum_{k=0}^{\infty} z_k^T z_k} < \alpha. \tag{13}$$

(4) *Input constraints*: for given $H_1, \ldots, H_{m_u} \in \mathcal{R}^{n_u \times n_u}$, $H_j = H_j^T \ge 0$, $h_1, \ldots, h_{m_u} \in \mathcal{R}^{n_u \times 1}$, and $\overline{u}_1, \ldots, \overline{u}_{m_u} \in \mathcal{R}$, the inputs satisfy the quadratic constraints:

$$u_k^T H_j u_k + 2h_j^T u_k \le \overline{u}_j, \quad \forall k; \text{ for } j = 1, \ldots, m_u. \tag{14}$$

(5) *State/output constraints*: for given $G_1, \ldots, G_{m_x} \in \mathcal{R}^{n \times n}$, $G_j = G_j^T \ge 0$, $g_1, \ldots, g_{m_x} \in \mathcal{R}^{n \times 1}$, and $\overline{x}_1, \ldots, \overline{x}_{m_x} \in \mathcal{R}$ the states/outputs satisfy the quadratic constraints:

$$x_{k+1}^T G_j x_{k+1} + 2g_j^T x_{k+1} \le \overline{x}_j, \quad \forall k; \text{ for } j = 1, \ldots, m_x. \tag{15}$$

An $F \in \mathcal{R}^{n_u \times n}$ satisfying these requirements will be called an admissible state feedback gain.

3. LMI Formulation of Sufficiency Conditions

The next theorem, which is the main result of this paper, derives sufficient conditions, in the form of LMIs, for the existence of an admissible F.

Theorem 3. *Let all variables, definitions, and assumptions be as above. Then there exists an admissible state feedback gain matrix F if there exists solutions $Q = Q^T \in \mathcal{R}^{n \times n}$, $Y \in \mathcal{R}^{n_u \times n}$, $\delta_j \ge 0$, $\mu_j \ge 0$, $\nu_j \ge 0$, $\overline{\Lambda} = \operatorname{diag}(\overline{\lambda}_1 I, \ldots, \overline{\lambda}_t I) > 0$, and $\overline{\Psi}_j = \operatorname{diag}(\overline{\psi}_1 I, \ldots, \overline{\psi}_t I) > 0$ to the LMIs shown in (16)–(19).*

$$\begin{bmatrix} -Q & \star & \star & \star & \star & \star & \star \\ 0 & -\alpha^2\gamma^2 I & \star & \star & \star & \star & \star \\ 0 & 0 & -\overline{\Lambda} & \star & \star & \star & \star \\ AQ + B_u Y & \alpha^2 B_w & B_p\overline{\Lambda} & -Q & \star & \star & \star \\ C_q Q + D_{qu} Y & \alpha^2 D_{qw} & 0 & 0 & -\overline{\Lambda} & \star & \star \\ C_z Q & 0 & 0 & 0 & 0 & -\alpha^2 I & \star \\ D_{zu} Y & 0 & 0 & 0 & 0 & 0 & -\alpha^2 I \end{bmatrix} < 0, \tag{16}$$

$$\begin{bmatrix} 1 & \star & \star \\ \gamma^2\overline{w}^2 & \alpha^2\gamma^2\overline{w}^2 & \star \\ x_0 & 0 & Q \end{bmatrix} \ge 0, \tag{17}$$

$$\begin{bmatrix} Q & \star & \star & \star \\ H_j^{1/2} Y & \mu_j I & \star & \star \\ -h_j^T Y & 0 & \mu_j \overline{u}_j & \star \\ 0 & 0 & \mu_j & 1 \end{bmatrix} \ge 0, \quad j = 1, \ldots, m_u, \tag{18}$$

$$\begin{bmatrix} Q & \star & \star & \star & \star & \star & \star \\ 0 & \delta_j I & \star & \star & \star & \star & \star \\ 0 & 0 & \overline{\Psi}_j & \star & \star & \star & \star \\ G_j^{1/2}(AQ + B_u Y) & \nu_j G_j^{1/2} B_w & G_j^{1/2} B_p \overline{\Psi}_j & \nu_j I & \star & \star & \star \\ C_q Q + D_{qu} Y & \nu_j D_{qw} & 0 & 0 & \overline{\Psi}_j & \star & \star \\ -g_j^T(AQ + B_u Y) & -\nu_j g_j^T B_w & -g_j^T B_p \overline{\Psi}_j & 0 & 0 & \nu_j \overline{x}_j - \delta_j \overline{w}^2 & \star \\ 0 & 0 & 0 & 0 & 0 & \nu_j & 1 \end{bmatrix} \ge 0, \quad j = 1, \ldots, m_x. \tag{19}$$

Here, $\star$ represents terms readily inferred from symmetry and the partitioning of $\overline{\Lambda}$ and $\overline{\Psi}_j$ is induced by the partitioning of Δ_k. If such solutions exist, then $F = YQ^{-1}$.

Remark 4. The variables in the LMI minimization of Theorem 3 are computed online at time k, the subscript k is omitted for convenience.

Proof. Using $u_k = Fx_k$, the dynamics in (1) become

$$x_{k+1} = \overbrace{(A + B_u F)}^{A_{cl}} x_k + B_w w_k + B_p p_k, \qquad z_k = \overbrace{\begin{bmatrix} C_z \\ D_{zu} F \end{bmatrix}}^{C_{cl}} x_k. \tag{20}$$

Consider a quadratic function $V(x) = x^T P x$, $P > 0$ of the state x_k. It follows from (20) that

$$\begin{aligned} V(x_{k+1}) - V(x_k) &= x_k^T \left[A_{cl}^T P A_{cl} - P \right] x_k + x_k^T A_{cl}^T P B_w w_k + x_k^T A_{cl}^T P B_p p_k \\ &\quad + w_k^T B_w^T P A_{cl} x_k + w_k^T B_w^T P B_w w_k + w_k^T B_w^T P B_p p_k \\ &\quad + p_k^T B_p^T P A_{cl} x_k + p_k^T B_p^T P B_w w_k + p_k^T B_p^T P B_p p_k \\ &= \begin{bmatrix} x_k^T & w_k^T & p_k^T \end{bmatrix} K \begin{bmatrix} x_k \\ w_k \\ p_k \end{bmatrix} - x_k^T C_{cl}^T C_{cl} x_k + \gamma^2 w_k^T w_k, \end{aligned} \tag{21}$$

where

$$K = \begin{bmatrix} A_{cl}^T P A_{cl} - P + C_{cl}^T C_{cl} & A_{cl}^T P B_w & A_{cl}^T P B_p \\ B_w^T P A_{cl} & B_w^T P B_w - \gamma^2 I & B_w^T P B_p \\ B_p^T P A_{cl} & B_p^T P B_w & B_p^T P B_p \end{bmatrix}. \tag{22}$$

Using $q_k = (C_q + D_{qu} F) x_k + D_{qw} w_k$,

$$\begin{aligned} q_k^T \Lambda q_k &= x_k^T \left(C_q + D_{qu} F \right)^T \Lambda \left(C_q + D_{qu} F \right) x_k \\ &\quad + x_k^T \left(C_q + D_{qu} F \right)^T \Lambda D_{qw} w_k \\ &\quad + w_k^T D_{qw}^T \Lambda \left(C_q + D_{qu} F \right) x_k \\ &\quad + w_k^T D_{qw}^T \Lambda D_{qw} w_k, \end{aligned} \tag{23}$$

where $\Lambda = \operatorname{diag}(\lambda_1 I, \ldots \lambda_t I)$.

Substituting (23) into (21), it can be verified that we can write

$$\begin{aligned} V(x_{k+1}) - V(x_k) &= \begin{bmatrix} x_k^T & w_k^T & p_k^T \end{bmatrix} \overline{K} \begin{bmatrix} x_k \\ w_k \\ p_k \end{bmatrix} + p_k^T \Lambda p_k \\ &\quad - q_k^T \Lambda q_k - x_k^T C_{cl}^T C_{cl} x_k + \gamma^2 w_k^T w_k, \end{aligned} \tag{24}$$

where $\overline{K}$ is defined in (25) and $C_{pw} := C_q + D_{qu} F$.

$$\overline{K} = \begin{bmatrix} A_{cl}^T P A_{cl} - P + C_{cl}^T C_{cl} + C_{pw}^T \Lambda C_{pw} & A_{cl}^T P B_w + C_{pw}^T \Lambda D_{qw} & A_{cl}^T P B_p \\ B_w^T P A_{cl} + D_{qw}^T \Lambda C_{pw} & B_w^T P B_w - \gamma^2 I + D_{qw}^T \Lambda D_{qw} & B_w^T P B_p \\ B_p^T P A_{cl} & B_p^T P B_w & B_p^T P B_p - \Lambda \end{bmatrix}. \tag{25}$$

Assuming that $\lim_{k \to \infty} x_k = 0$ we have

$$\sum_{k=0}^{\infty} \left[x_{k+1}^T P x_{k+1} - x_k^T P x_k \right] = -x_0^T P x_0. \tag{26}$$

We write the $\mathcal{H}_2$ cost function as

$$\| z \|_2^2 = \sum_{k=0}^{\infty} \left(x_k^T C_{cl}^T C_{cl} x_k - \gamma^2 w_k^T w_k \right) + \gamma^2 \sum_{k=0}^{\infty} w_k^T w_k. \tag{27}$$

Adding (26) and (27) and carrying out a simple manipulation gives

$$\begin{aligned} \| z \|_2^2 &= x_0^T P x_0 + \gamma^2 \| w \|_2^2 \\ &\quad + \sum_{k=0}^{\infty} \begin{bmatrix} x_k^T & w_k^T & p_k^T \end{bmatrix} \overline{K} \begin{bmatrix} x_k \\ w_k \\ p_k \end{bmatrix} + \sum_{k=0}^{\infty} \left(p_k^T \Lambda p_k - q_k^T \Lambda q_k \right), \end{aligned} \tag{28}$$

where $\overline{K}$ is defined in (25).

Setting $x_0 = 0$, it follows from (3), (12), and (28) that $\| z \|_2 < \gamma \| w \|_2$ if $\overline{K} < 0$ and $\Lambda \geq 0$. In this work, we will take $\Lambda > 0$ to simplify our solution [8]. Using (2) and Lemma 1 it can be shown that

$$\overline{K} < 0, \tag{29}$$

is also sufficient for quadratic stability of T_{zw}.

Next, we linearize the matrix inequality (29) by applying a Schur complement, to give

$$\begin{bmatrix} -P & \star & \star & \star & \star & \star & \star \\ 0 & -\gamma^2 I & \star & \star & \star & \star & \star \\ 0 & 0 & -\Lambda & \star & \star & \star & \star \\ A_{cl} & B_w & B_p & -P^{-1} & \star & \star & \star \\ C_{pw} & D_{qw} & 0 & 0 & -\Lambda^{-1} & \star & \star \\ C_z & 0 & 0 & 0 & 0 & -I & \star \\ D_{zu}F & 0 & 0 & 0 & 0 & 0 & -I \end{bmatrix} < 0. \quad (30)$$

Pre- and post-multiplying the equation above by diag$(P^{-1}, I, I, I, I, I, I)$ gives

$$\begin{bmatrix} -P^{-1} & \star & \star & \star & \star & \star & \star \\ 0 & -\gamma^2 I & \star & \star & \star & \star & \star \\ 0 & 0 & -\Lambda & \star & \star & \star & \star \\ A_{cl}P^{-1} & B_w & B_p & -P^{-1} & \star & \star & \star \\ C_{pw}P^{-1} & D_{qw} & 0 & 0 & -\Lambda^{-1} & \star & \star \\ C_zP^{-1} & 0 & 0 & 0 & 0 & -I & \star \\ D_{zu}FP^{-1} & 0 & 0 & 0 & 0 & 0 & -I \end{bmatrix} < 0, \quad (31)$$

setting $Q = \alpha^2 P^{-1}$, $F = YP\alpha^{-2} = YQ^{-1}$, $C_{pw} = C_q + D_{qu}F$ and multiplying through by α^2, the equation above becomes

$$\begin{bmatrix} -Q & \star & \star & \star & \star & \star & \star \\ 0 & -\alpha^2\gamma^2 I & \star & \star & \star & \star & \star \\ 0 & 0 & -\alpha^2\Lambda & \star & \star & \star & \star \\ AQ + B_uY & \alpha^2 B_w & \alpha^2 B_p & -Q & \star & \star & \star \\ C_qQ + D_{qu}Y & \alpha^2 D_{qw} & 0 & 0 & -\alpha^2\Lambda^{-1} & \star & \star \\ C_zQ & 0 & 0 & 0 & 0 & -\alpha^2 I & \star \\ D_{zu}Y & 0 & 0 & 0 & 0 & 0 & -\alpha^2 I \end{bmatrix} < 0. \quad (32)$$

Pre- and post-multiplying the equation above by diag$(I, I, \Lambda^{-1}, I, I, I, I)$ gives

$$\begin{bmatrix} -Q & \star & \star & \star & \star & \star & \star \\ 0 & -\alpha^2\gamma^2 I & \star & \star & \star & \star & \star \\ 0 & 0 & -\alpha^2\Lambda^{-1} & \star & \star & \star & \star \\ AQ + B_uY & \alpha^2 B_w & \alpha^2 B_p\Lambda^{-1} & -Q & \star & \star & \star \\ C_qQ + D_{qu}Y & \alpha^2 D_{qw} & 0 & 0 & -\alpha^2\Lambda^{-1} & \star & \star \\ C_zQ & 0 & 0 & 0 & 0 & -\alpha^2 I & \star \\ D_{zu}Y & 0 & 0 & 0 & 0 & 0 & -\alpha^2 I \end{bmatrix} < 0. \quad (33)$$

The equation above is a bilinear matrix inequality, thus by defining $\alpha^2\Lambda^{-1}$ as a variable $\overline{\Lambda}$, we get the LMI in (16).

Now, it follows from (3), (11), (28), and (29) that,

$$\|z\|_2^2 \le x_0^T P x_0 + \gamma^2 \|w\|_2^2 \le x_0^T P x_0 + \gamma^2 \overline{w}^2. \quad (34)$$

Thus the $\mathcal{H}_2$ constraint in (13) is satisfied if

$$x_0^T P x_0 + \gamma^2 \overline{w}^2 < \alpha^2. \quad (35)$$

Dividing by α^2, rearranging and using a Schur complement give (17) as an LMI sufficient condition for (13).

To turn (14) and (15) into LMIs, we first show that $x_k^T P x_k \le \alpha^2 \ \forall k > 0$. Since $\overline{K} < 0$, it follows from (3) and (24) that

$$x_{k+1}^T P x_{k+1} - x_k^T P x_k \le \gamma^2 w_k^T w_k. \quad (36)$$

Applying this inequality recursively, we get

$$\begin{aligned} x_k^T P x_k &\le x_0^T P x_0 + \gamma^2 \sum_{j=0}^{k-1} w_j^T w_j \\ &\le x_0^T P x_0 + \gamma^2 \overline{w}^2 < \alpha^2. \end{aligned} \quad (37)$$

It follows that

$$\left\| P^{1/2} x_k \right\|^2 < \alpha^2, \quad (38)$$

or equivalently,

$$x_k^T Q^{-1} x_k < 1, \quad \forall k > 0. \quad (39)$$

Next, we transform the constraints in (14) to a set of LMIs. Setting $F = YQ^{-1} = YP\alpha^{-2}$ and $u_k = Fx_k$,

$$\begin{aligned} e_j(u_k) &:= u_k^T H_j u_k + 2h_j^T u_k - \overline{u}_j \\ &= x_k^T Q^{-1} Y^T H_j Y Q^{-1} x_k + 2h_j^T Y Q^{-1} x_k - \overline{u}_j. \end{aligned} \quad (40)$$

Now for any $\mu_j \in \mathcal{R}$, we can write

$$\begin{aligned} e_j(u_k) = &-\mu_j\left(1 - x_k^T Q^{-1} x_k\right) - \begin{bmatrix} x_k \\ 1 \end{bmatrix}^T \\ &\times \left(\begin{bmatrix} \mu_j Q^{-1} - Q^{-1} Y^T H_j Y Q^{-1} & -Q^{-1} Y^T h_j \\ -h_j^T Y Q^{-1} & -\mu_j + \overline{u}_j \end{bmatrix} \right) \\ &\times \begin{bmatrix} x_k \\ 1 \end{bmatrix}. \end{aligned} \quad (41)$$

Therefore a sufficient condition for $e_j(u_k) \le 0$ is $\mu_j \ge 0$ and

$$\begin{bmatrix} \mu_j Q^{-1} & -Q^{-1}Y^T H_j Y Q^{-1} & -Q^{-1}Y^T h_j \\ & -h_j^T Y Q^{-1} & \overline{u}_j - \mu_j \end{bmatrix} \ge 0. \quad (42)$$

Pre- and post-multiplying by diag(Q, I) gives a bilinear matrix inequality and applying a Schur complement, we get

$$\begin{bmatrix} \mu_j Q & Y^T H_j^{1/2} & -Y^T h_j \\ H_j^{1/2} Y & I & 0 \\ -h_j^T Y & 0 & \overline{u}_j - \mu_j \end{bmatrix} \ge 0. \quad (43)$$

Pre- and post-multiplying the above bilinear matrix inequality by diag$(\mu_j^{-1/2}, \mu_j^{1/2}, \mu_j^{1/2})$ and applying a Schur complement, this is equivalent to the LMI in (18).

Finally, to obtain an LMI formulation of the state constraints (15), the following analogous steps are carried out:

$$f_j(x_{k+1}) := \begin{bmatrix} x_k \\ w_k \\ p_k \end{bmatrix}^T \begin{bmatrix} A_{cl}^T \\ B_w^T \\ B_p^T \end{bmatrix} G_j \begin{bmatrix} A_{cl} & B_w & B_p \end{bmatrix} \begin{bmatrix} x_k \\ w_k \\ p_k \end{bmatrix} + 2g_j^T \begin{bmatrix} A_{cl} & B_w & B_p \end{bmatrix} \begin{bmatrix} x_k \\ w_k \\ p_k \end{bmatrix} - \overline{x}_j. \quad (44)$$

Now for any $\nu_j, \rho_j \in \mathcal{R}$, we can write

$$f_j(x_{k+1}) = -\nu_j\left(1 - x_k^T Q^{-1} x_k\right) - \rho_j\left(\overline{w}^2 - w_k^T w_k\right) - \begin{bmatrix} x_k \\ w_k \\ 1 \end{bmatrix}^T \left(\begin{bmatrix} \nu_j Q^{-1} - \left(A_{cl} + B_p \Delta_k C_{pw}\right)^T G_j \left(A_{cl} + B_p \Delta_k C_{pw}\right) & \star & \star \\ -\left(B_w + B_p \Delta_k D_{qw}\right)^T G_j \left(A_{cl} + B_p \Delta_k C_{pw}\right) & \rho_j I & \star \\ -g_j^T \left(A_{cl} + B_p \Delta_k C_{pw}\right) & -g_j^T \left(B_w + B_p \Delta_k D_{qw}\right) & -\nu_j - \rho_j \overline{w}^2 + \overline{x}_j \end{bmatrix} \right) \begin{bmatrix} x_k \\ w_k \\ 1 \end{bmatrix}. \quad (45)$$

Therefore a sufficient condition for $f_j(x_{k+1}) \ge 0$ is $\nu_j \ge 0$, $\rho_j \ge 0$ and

$$\begin{bmatrix} \nu_j Q^{-1} & \star & \star \\ 0 & \rho_j I & \star \\ -g_j^T \left(A_{cl} + B_p \Delta_k C_{pw}\right) & -g_j^T \left(B_w + B_p \Delta_k D_{qw}\right) & \overline{x}_j - \nu_j - \rho_j \overline{w}^2 \end{bmatrix} - \begin{bmatrix} \left(A_{cl} + B_p \Delta_k C_{pw}\right)^T \\ \left(B_w + B_p \Delta_k D_{qw}\right)^T \\ 0 \end{bmatrix} G_j \left[\left(A_{cl} + B_p \Delta_k C_{pw}\right) \ \left(B_w + B_p \Delta_k D_{qw}\right) \ 0 \right] \ge 0. \quad (46)$$

Applying Schur complement to the above equation, we get

$$\begin{bmatrix} \nu_j Q^{-1} & \star & \star & \star \\ 0 & \rho_j I & \star & \star \\ -g_j^T \left(A_{cl} + B_p \Delta_k C_{pw}\right) & -g_j^T \left(B_w + B_p \Delta_k D_{qw}\right) & \overline{x}_j - \nu_j - \rho_j \overline{w}^2 & \star \\ G_j^{1/2} \left(A_{cl} + B_p \Delta_k C_{pw}\right) & G_j^{1/2} \left(B_w + B_p \Delta_k D_{qw}\right) & 0 & I \end{bmatrix} \ge 0. \quad (47)$$

When Δ is structured we proceed as follows. For norm-bounded uncertainty, we first separate the terms involving modeling uncertainties from the other terms as

$$\underbrace{\begin{bmatrix} \nu_j Q^{-1} & \star & \star & \star \\ 0 & \rho_j I & \star & \star \\ -g_j^T A_{cl} & -g_j^T B_w & \bar{x}_j - \nu_j - \rho_j \overline{w}^2 & \star \\ -G_j^{1/2} A_{cl} & -G_j^{1/2} B_w & 0 & I \end{bmatrix}}_{-T_1} + \underbrace{\begin{bmatrix} 0 \\ 0 \\ -g_j^T B_p \\ G_j^{1/2} B_p \end{bmatrix}}_{-T_2} \Delta_k \underbrace{\begin{bmatrix} C_{pw} & D_{qw} & 0 & 0 \end{bmatrix}}_{T_3} + \underbrace{\begin{bmatrix} C_{pw}^T \\ D_{qw}^T \\ 0 \\ 0 \end{bmatrix}}_{T_3^T} \Delta_k^T \underbrace{\begin{bmatrix} 0 & 0 & -B_p^T g_j & B_p^T G^{1/2} \end{bmatrix}}_{-T_2^T} \geq 0. \tag{48}$$

Equation (48) is equivalent to $-T_1 - T_2\Delta T_3 - T_3^T \Delta^T T_2^T > 0$, where $T_4 = 0$. By using (8) from Lemma 1, we have

$$\begin{bmatrix} -T_1 - T_3^T S T_3 & -T_2 \\ -T_2^T & S \end{bmatrix} > 0. \tag{49}$$

Applying Schur complement to (49), we get

$$\begin{bmatrix} -T_1 & T_3^T & -T_2 \\ T_3 & S^{-1} & 0 \\ -T_2^T & 0 & S \end{bmatrix} > 0. \tag{50}$$

Substituting the variables from (48) into (50) and swapping the third and sixth diagonal elements, we get

$$\begin{bmatrix} \nu_j Q^{-1} & \star & \star & \star & \star & \star \\ 0 & \rho_j I & \star & \star & \star & \star \\ 0 & 0 & S & \star & \star & \star \\ G_j^{1/2} A_{cl} & G_j^{1/2} B_w & G_j^{1/2} B_p & I & \star & \star \\ C_{pw} & D_{qw} & 0 & 0 & S^{-1} & \star \\ -g_j^T A_{cl} & -g_j^T B_w & -g_j^T B_p & 0 & 0 & \bar{x}_j - \nu_j - \rho_j \overline{w}^2 \end{bmatrix} \geq 0. \tag{51}$$

Pre- and post-multiplying by $\mathrm{diag}(Q, I, I, I, I, I)$ gives

$$\begin{bmatrix} \nu_j Q & \star & \star & \star & \star & \star \\ 0 & \rho_j I & \star & \star & \star & \star \\ 0 & 0 & S & \star & \star & \star \\ G_j^{1/2} A_{cl} & G_j^{1/2} B_w & G_j^{1/2} B_p & I & \star & \star \\ C_{pw} & D_{qw} & 0 & 0 & S^{-1} & \star \\ -g_j^T A_{cl} & -g_j^T B_w & -g_j^T B_p & 0 & 0 & \bar{x}_j - \nu_j - \rho_j \overline{w}^2 \end{bmatrix} \geq 0. \tag{52}$$

The above equation is bilinear and thus we pre- and post-multiply it by $\mathrm{diag}(\nu_j^{-1/2}, \nu_j^{1/2}, \nu_j^{1/2}, \nu_j^{1/2}, \nu_j^{1/2}, \nu_j^{1/2})$ to obtain

$$\begin{bmatrix} Q & \star & \star & \star & \star & \star \\ 0 & \nu_j \rho_j I & \star & \star & \star & \star \\ 0 & 0 & \nu_j S & \star & \star & \star \\ G_j^{1/2} A_{cl} & \nu_j G_j^{1/2} B_w & \nu_j G_j^{1/2} B_p & \nu_j I & \star & \star \\ C_{pw} & \nu_j D_{qw} & 0 & 0 & \nu_j S^{-1} & \star \\ -g_j^T A_{cl} & -\nu_j g_j^T B_w & -\nu_j g_j^T B_p & 0 & 0 & \nu_j \bar{x}_j - \nu_j^2 - \nu_j \rho_j \overline{w}^2 \end{bmatrix} \geq 0. \tag{53}$$

From the above equation, we can see that the variable S and its inverse appear in the matrix inequality, thus to make it uniform, we pre- and post-multiply by $\mathrm{diag}(I, I, S^{-1}, I, I, I)$ to get

$$\begin{bmatrix} Q & \star & \star & \star & \star & \star \\ 0 & \nu_j\rho_j I & \star & \star & \star & \star \\ 0 & 0 & \nu_j S^{-1} & \star & \star & \star \\ G_j^{1/2}A_{cl} & \nu_j G_j^{1/2}B_w & \nu_j G_j^{1/2}B_p S^{-1} & \nu_j I & \star & \star \\ C_{pw} & \nu_j D_{qw} & 0 & 0 & \nu_j S^{-1} & \star \\ -g_j^T A_{cl} & -\nu_j g_j^T B_w & -\nu_j g_j^T B_p S^{-1} & 0 & 0 & \nu_j \overline{x}_j - \nu_j^2 - \nu_j\rho_j\overline{w}^2 \end{bmatrix} \geq 0. \tag{54}$$

Thus the above equation is a nonlinear matrix inequality in ν_j^2 and bilinear in $\nu_j\rho_j$ and $\nu_j S^{-1}$, hence we define new variables $\overline{\Psi}_j = \nu_j S^{-1}$ and $\delta_j = \nu_j\rho_j$ and finally applying a Schur complement, we obtain the LMI of (19). □

Remark 5. The input and state constraints used in this paper are more general than those used in [9] in that we allow linear terms and so this makes it possible to include asymmetric or hyperplane constraints.

Remark 6. When there is no uncertainty, the problem reduces to disturbance rejection technique considered in [9].

Remark 7. When there is no disturbance, the results reduce to those of [7].

Remark 8. The method used in this paper guarantees recursive feasibility (see [15, Chapter 4]). Also see [16] for a different approach.

4. Numerical Examples

In this section, we present two examples that illustrate the implementation of the proposed scheme. In the first example we consider a solenoid system, and in the second example we consider the coupled spring-mass system. The solution to the linear objective minimization was computed using LMI Control Toolbox in the MATLAB® environment and α^2 was set as a variable.

4.1. Example 1. We consider a modified version of the solenoid system adapted from [17]. The system (see Figure 1) consists of a central object wrapped with coil and is attached to a rigid surface via a spring and damper, which forms a passive vibration isolator. The solenoid is one of the common actuator components. The basic principle of operation involves a moving ferrous core (a piston) that moves inside a wire coil. Normally, the piston is held outside the core by a spring and damper. When a voltage is applied to the coil and current flows, the coil builds up a magnetic field that attracts the piston and pulls it into the center of the coil. The piston can be used to supply a linear force. Application of this includes pneumatic valves and car door openers.

The system is modeled by

$$\begin{aligned} \begin{bmatrix} x_{k+1}^1 \\ x_{k+1}^2 \end{bmatrix} &= \begin{bmatrix} 0.6148 & 0.0315 \\ -0.3155 & -0.0162 \end{bmatrix} \begin{bmatrix} x_k^1 \\ x_k^2 \end{bmatrix} + \begin{bmatrix} 0.0385 \\ 0.0315 \end{bmatrix} u_k \\ &\quad + \begin{bmatrix} 0.00385 \\ 0.00315 \end{bmatrix} w_k + \begin{bmatrix} 0 \\ 10 \end{bmatrix} p_k, \\ q_k &= C_q x_k + D_{qu} u_k + D_{qw} w_k, \\ p_k &= \Delta q_k, \\ z_k &= \begin{bmatrix} C_z x_k \\ D_{zu} u_k \end{bmatrix}, \end{aligned} \tag{55}$$

where

$$C_q = \begin{bmatrix} 1 & 0 \end{bmatrix}, \qquad D_{qu} = 1, \qquad D_{qw} = 0, \tag{56}$$

where x^1 and x^2 are the position and the velocity of the plate. The cost function is specified using $C_z = \text{diag}(1, 1)$ and $D_{zu} = 10$. The magnetic force u is the control variable, and w is the external disturbance to the system, which is bounded in the range $[-1, 1]$. The initial state is given as $x_0 = [1 \;\; 0]^T$.

We choose $\gamma^2 = 0.01$ and $\gamma^2 = 1$. Figures 2 and 3 compare the closed-loop response for the high and low disturbance rejection levels, respectively, for randomly generated Δ's. The optimization is feasible, the response is stable, and the performance is good. A control constraint of $|u_k| \leq 0.5$ is imposed, which is satisfied. The computation time for 100 samples was about 10 s, making 0.1 s per sample.

4.2. Example 2. We revisit a modified version of Example 2 reported in [7]. The system consists of a two-mass-spring model whose discrete-time equivalent is obtained using Euler first-order approximation with a sampling time of 0.1 s.

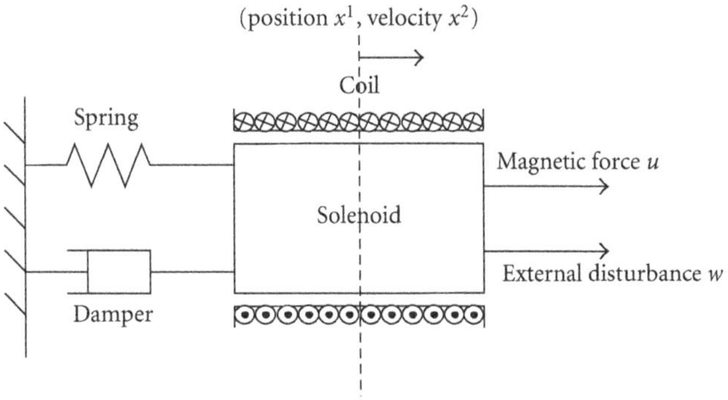

FIGURE 1: Solenoid system.

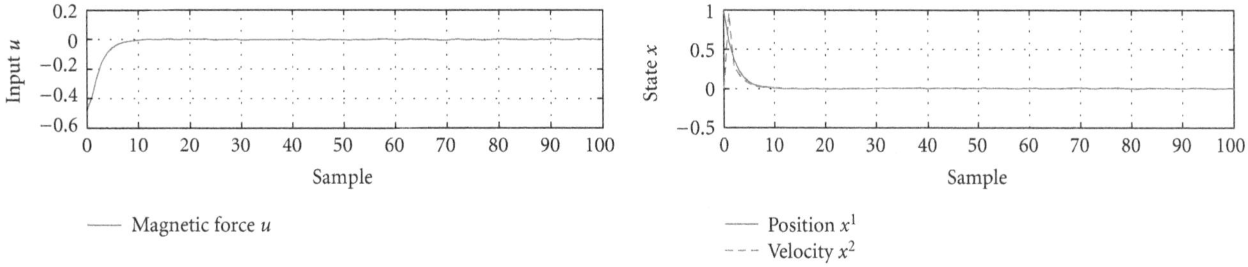

(a) (b)

FIGURE 2: Closed-loop response of the solenoid system with $\gamma^2 = 0.01$.

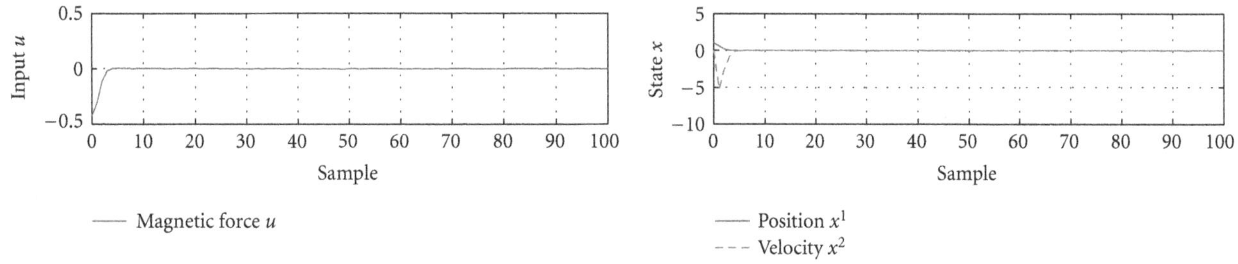

(a) (b)

FIGURE 3: Closed-loop response of the solenoid system with $\gamma^2 = 1$.

The model in terms of disturbance and perturbation variables is

$$\begin{aligned}
x_{k+1} &= \begin{bmatrix} 1 & 0 & 0.1 & 0 \\ 0 & 1 & 0 & 0.1 \\ -0.1\dfrac{K}{m_1} & 0.1\dfrac{K}{m_1} & 1 & 0 \\ 0.1\dfrac{K}{m_2} & -0.1\dfrac{K}{m_2} & 1 & 0 \end{bmatrix} x_k + \begin{bmatrix} 0 \\ 0 \\ \dfrac{0.1}{m_1} \\ 0 \end{bmatrix} u_k \\
&\quad + B_w w_k + B_p p_k, \\
q_k &= C_q x_k + D_{qu} u_k + D_{qw} w_k, \\
p_k &= \Delta q_k, \\
y_k &= \begin{bmatrix} 0 & 1 & 0 & 0 \end{bmatrix} x_k,
\end{aligned} \tag{57}$$

where

$$\begin{gathered}
B_w = \begin{bmatrix} 0 \\ 0.01 \\ 0 \\ 0 \end{bmatrix}, \qquad B_p = \begin{bmatrix} 0 \\ 0 \\ -0.1 \\ 0.1 \end{bmatrix}, \\
C_q = \begin{bmatrix} 0.475 & -0.475 & 0 & 0 \end{bmatrix}, \\
D_{qw} = 0, \qquad D_{qu=0},
\end{gathered} \tag{58}$$

where x_1 and x_2 are the positions of body 1 and 2, and x_3 and x_4 are their velocities, respectively. m_1 and m_2

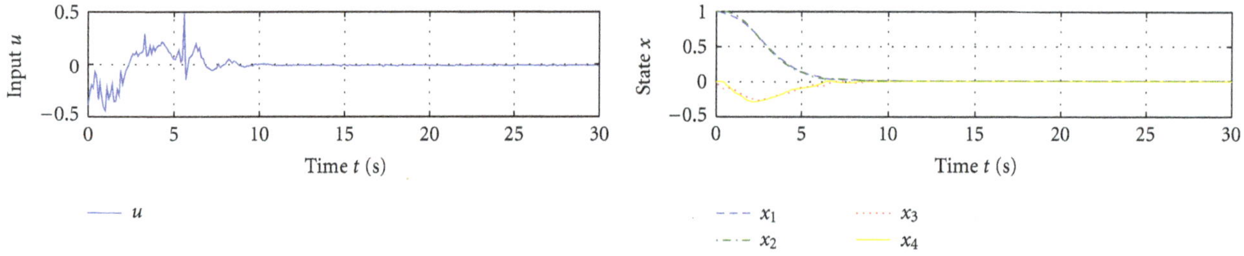

(a) (b)

FIGURE 4: Closed-loop response of the coupled spring-mass system with $\gamma^2 = 0.45$.

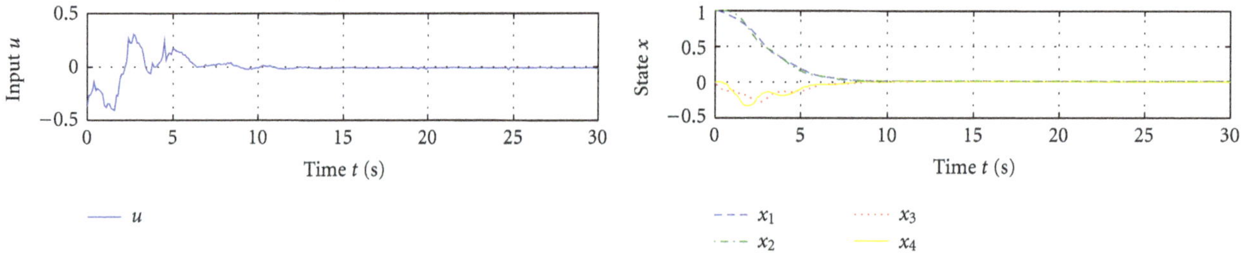

(a) (b)

FIGURE 5: Closed-loop response of the coupled spring-mass system with $\gamma^2 = 4$.

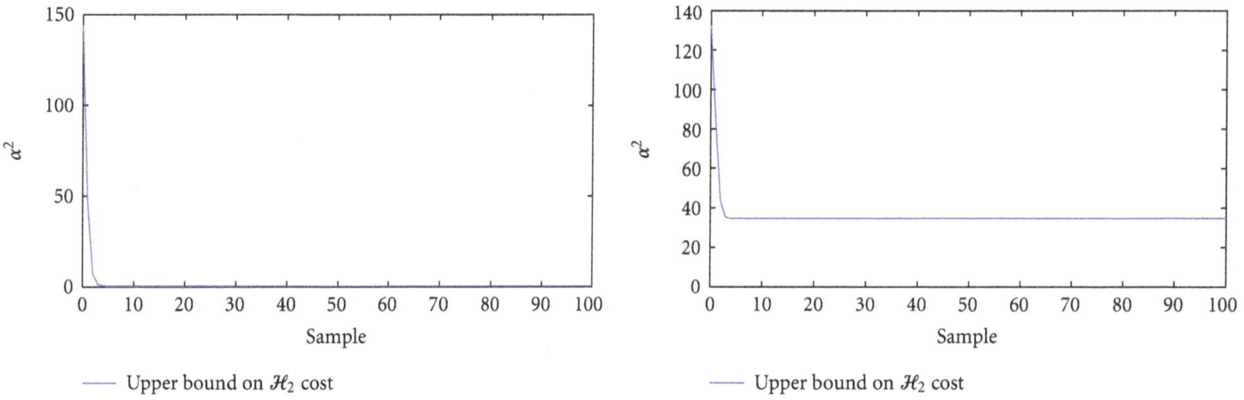

(a) System with $\gamma^2 = 0.01$

(b) System with $\gamma^2 = 1$

FIGURE 6: Upper bound on $\mathcal{H}_2$ cost function for solenoid system.

are the masses of the two bodies and K is the spring constant. The initial state is given as $x_0 = [1 \ 1 \ 0 \ 0]^T$. The cost function is specified using $C_Z = \text{diag}(1, 1, 1, 1)$, and $D_{zu} = 1$. We consider the system with $m_1 = m_2 = 1$ and $K \in [0.5, 10]$.

A persistent disturbance of the form $w_i = 0.1$ for all sample time was considered. Here we set $\gamma^2 = 0.45$ and $\gamma^2 = 4$. Figures 4 and 5 compare the closed-loop response for the high and low disturbance rejection levels, respectively, for randomly generated Δ's. The value of γ^2 for high disturbance rejection was the lowest value for which a feasible solution exists. An input constraint of $|u_k| \leq 1$ is imposed, which is satisfied. The computation time for 300 samples was about 47 s, making 0.16 s per sample.

4.3. Discussion. Note that the performance and response of the systems based on the high disturbance rejection level were better than those obtained using low disturbance rejection level, since the states and control are regulated to smaller steady state values. Constraints on the input were satisfied in both cases; however, the constraints were more conservative with respect to the control signal for the low disturbance rejection level. For example in the solenoid system, the control signal for high disturbance rejection level was 0.4826 and that for low disturbance rejection level was 0.4144. The issue of conservativeness in the mixed $\mathcal{H}_2/\mathcal{H}_\infty$ setting has been considered in [18]. For the systems considered, the upper bound on the $\mathcal{H}_2$ cost function α^2 is depicted in Figures 6 and 7 for the high and low disturbance rejection

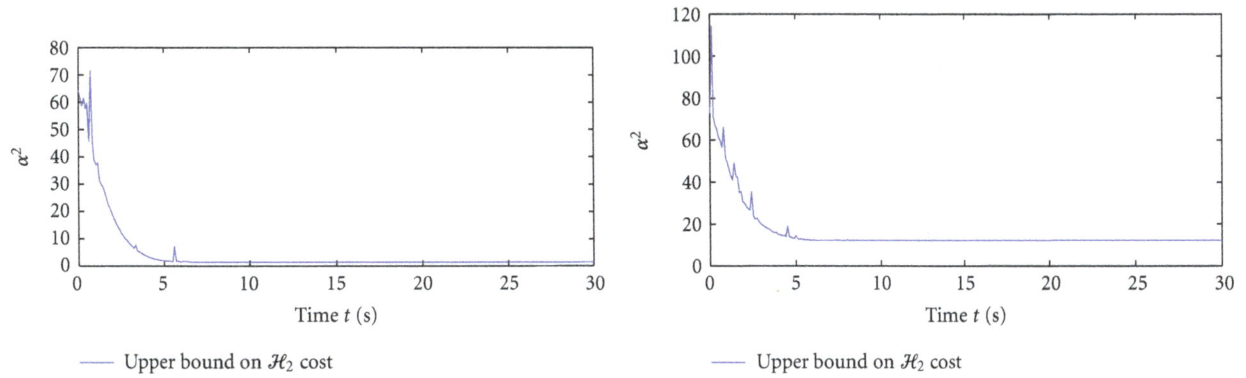

(a) System with $\gamma^2 = 0.45$

(b) System with $\gamma^2 = 4$

FIGURE 7: Upper bound on $\mathcal{H}_2$ cost function for coupled spring-mass system.

levels. It can been seen that the performance coefficient obtained for the high and low disturbance rejection levels is small. However, the offset level on the low disturbance level is higher than that for the high disturbance level. This is due to the higher value of γ^2 in $x_0^T P x_0 + \gamma^2 \overline{w}^2$. In Example 2, we have considered uncertainty in the model by using variable spring constant K.

5. Conclusion

In this paper, we proposed a robust model predictive control design technique using mixed $\mathcal{H}_2/\mathcal{H}_\infty$ for time invariant discrete-time linear systems subject to constraints on the inputs and states. This method takes account of disturbances naturally by imposing the $\mathcal{H}_\infty$-norm constraint in (14) and thus extends the work in [9] by the introduction of structured, norm-bounded uncertainty. The uncertain system was represented by LFTs. The development is based on full state feedback assumption and the on-line optimization involves the solution of an LMI-based linear objective minimization (convex optimization). Hence, the resulting state-feedback control law minimizes an upper bound on the robust objective function. The new approach reduces to that of [9] when there are no perturbations present in the system and to [7] when there are no disturbances. Thus, we have been able to show that it is possible to handle uncertainty and disturbance in the mixed $\mathcal{H}_2/\mathcal{H}_\infty$ model predictive control design approach. The two examples illustrate the application of the proposed method.

Acknowledgments

This was parts supported by Commonwealth Scholarship Commission in the United Kingdom (CSCUK) under Grant NGCS-2004-258. With regard to this paper, particular thanks go to Dr. Imad M. Jaimoukha and Dr. Eric Kerrigan.

References

[1] E. F. Camacho and C. Bordons, *Model Predictive Control*, Springer, 2nd edition, 2004.

[2] J. Richalet, A. Rault, J. L. Testud, and J. Papon, "Model predictive heuristic control. Applications to industrial processes," *Automatica*, vol. 14, no. 5, pp. 413–428, 1978.

[3] P. E. Orukpe, X. Zheng, I. M. Jaimoukha, A. C. Zolotas, and R. M. Goodall, "Model predictive control based on mixed $\mathcal{H}_2/\mathcal{H}_\infty$ control approach for active vibration control of railway vehicles," *Vehicle System Dynamics*, vol. 46, no. 1, pp. 151–160, 2008.

[4] H. S. Witsenhausen, "A minimax control problem for sampled linear systems," *IEEE Transactions on Automatic Control*, vol. 13, no. 1, pp. 5–21, 1968.

[5] R. S. Smith, "Robust model predictive control of constrained linear systems," in *Proceedings of the 2004 American Control Conference (AAC '04)*, pp. 245–250, July 2004.

[6] P. E. Orukpe and I. M. Jaimoukha, "A semidefinite relaxation for the quadratic minimax problem in $\mathcal{H}_\infty$ model predictive control," in *Proceedings of IEEE Conference on Decision and Control*, New Orleans, La, USA, 2007.

[7] M. V. Kothare, V. Balakrishnan, and M. Morari, "Robust constrained model predictive control using linear matrix inequalities," *Automatica*, vol. 32, no. 10, pp. 1361–1379, 1996.

[8] R. S. Smith, "Model predictive control of uncertain constrained linear systems: Lmi-based," Tech. Rep. CUED/F-INFENG/TR.462, University of Cambridge, Cambridge, UK, 2006.

[9] P. E. Orukpe, I. M. Jaimoukha, and H. M. H. El-Zobaidi, "Model predictive control based on mixed $\mathcal{H}_2/\mathcal{H}_\infty$ control approach," in *Proceedings of the American Control Conference (ACC '07)*, pp. 6147–6150, July 2007.

[10] P. E. Orukpe and I. M. Jaimoukha, "Robust model predictive control based on mixed $\mathcal{H}_2/\mathcal{H}_\infty$ control approach," in *Proceedings of the European Control Conference*, pp. 2223–2228, Budapest, Hungary, 2009.

[11] S. Boyd, L. El Ghaoui, E. Feron, and V. Balakrishnan, *Linear Matrix Inequalities in Systems and Control Theory*, SIAM Studies in Applied Mathematics, 1994.

[12] A. Packard and J. Doyle, "The complex structured singular value," *Automatica*, vol. 29, no. 1, pp. 71–109, 1993.

[13] K. Sun and A. Packard, "Robust $\mathcal{H}_2$ and $\mathcal{H}_\infty$ filters for uncertain LFT systems," in *Proceedings of the 41st IEEE Conference on Decision and Control*, pp. 2612–2618, Las Vegas, Nev, USA, December 2002.

[14] C. K. Li and R. Mathias, "Extremal characterizations of the Schur complement and resulting inequalities," *SIAM Review*, vol. 42, no. 2, pp. 233–246, 2000.

[15] P. E. Orukpe, *Model predictive control for linear time invariant systems using linear matrix inequality techniques*, Ph.D. thesis, Imperial College, London, UK, 2009.

[16] H. Huang, L. Dewei, and X. Yugeng, "Mixed $\mathcal{H}_2/\mathcal{H}_\infty$ robust model predictive control based on mixed $\mathcal{H}_2/\mathcal{H}_\infty$ with norm-bounded uncertainty," in *Proceedings of the 30th Chinese Control Conference*, pp. 3327–3331, Yantai, China, 2011.

[17] D. Jia, B. H. Krogh, and O. Stursberg, "LMI approach to robust model predictive control," *Journal of Optimization Theory and Applications*, vol. 127, no. 2, pp. 347–365, 2005.

[18] P. E. Orukpe, "Towards a less conservative model predictive control based on mixed $\mathcal{H}_2/\mathcal{H}_\infty$ control approach," *International Journal of Control*, vol. 84, no. 5, pp. 998–1007, 2011.

Permissions

The contributors of this book come from diverse backgrounds, making this book a truly international effort. This book will bring forth new frontiers with its revolutionizing research information and detailed analysis of the nascent developments around the world.

We would like to thank all the contributing authors for lending their expertise to make the book truly unique. They have played a crucial role in the development of this book. Without their invaluable contributions this book wouldn't have been possible. They have made vital efforts to compile up to date information on the varied aspects of this subject to make this book a valuable addition to the collection of many professionals and students.

This book was conceptualized with the vision of imparting up-to-date information and advanced data in this field. To ensure the same, a matchless editorial board was set up. Every individual on the board went through rigorous rounds of assessment to prove their worth. After which they invested a large part of their time researching and compiling the most relevant data for our readers. Conferences and sessions were held from time to time between the editorial board and the contributing authors to present the data in the most comprehensible form. The editorial team has worked tirelessly to provide valuable and valid information to help people across the globe.

Every chapter published in this book has been scrutinized by our experts. Their significance has been extensively debated. The topics covered herein carry significant findings which will fuel the growth of the discipline. They may even be implemented as practical applications or may be referred to as a beginning point for another development. Chapters in this book were first published by Hindawi Publishing Corporation; hereby published with permission under the Creative Commons Attribution License or equivalent.

The editorial board has been involved in producing this book since its inception. They have spent rigorous hours researching and exploring the diverse topics which have resulted in the successful publishing of this book. They have passed on their knowledge of decades through this book. To expedite this challenging task, the publisher supported the team at every step. A small team of assistant editors was also appointed to further simplify the editing procedure and attain best results for the readers.

Our editorial team has been hand-picked from every corner of the world. Their multi-ethnicity adds dynamic inputs to the discussions which result in innovative outcomes. These outcomes are then further discussed with the researchers and contributors who give their valuable feedback and opinion regarding the same. The feedback is then collaborated with the researches and they are edited in a comprehensive manner to aid the understanding of the subject.

Apart from the editorial board, the designing team has also invested a significant amount of their time in understanding the subject and creating the most relevant covers. They scrutinized every image to scout for the most suitable representation of the subject and create an appropriate cover for the book.

The publishing team has been involved in this book since its early stages. They were actively engaged in every process, be it collecting the data, connecting with the contributors or procuring relevant information. The team has been an ardent support to the editorial, designing and production team. Their endless efforts to recruit the best for this project, has resulted in the accomplishment of this book. They are a veteran in the field of academics and their pool of knowledge is as vast as their experience in printing. Their expertise and guidance has proved useful at every step. Their uncompromising quality standards have made this book an exceptional effort. Their encouragement from time to time has been an inspiration for everyone.

The publisher and the editorial board hope that this book will prove to be a valuable piece of knowledge for researchers, students, practitioners and scholars across the globe.

List of Contributors

Massimo Callegari, Luca Carbonari, Matteo-Claudio Palpacelli and Donatello Tina
Department of Industrial Engineering & Mathematical Sciences, Polytechnic University of Marche, 60131 Ancona, Italy

Giacomo Palmieri
e-Campus University, Faculty of Engineering, 22060 Novedrate, Italy

Long Sheng, Ya-Jun Pan and Xiang Gong
Department of Mechanical Engineering, Dalhousie University, Halifax, NS, Canada B3J 2X4

C. H. F. Silva
Cemig Gerac¸˜ao e Transmiss˜ao SA, Avenida Barbacena 1200, 16◦ Andar Ala B1 (TE/AE), 30190-131 Belo Horizonte, MG, Brazil

H. M. Henrique and L. C. Oliveira-Lopes
Faculdade de Engenharia Qu´ımica, Universidade Federal de Uberl^andia, Avenida Jo˜ao Naves de ´ Avila, 2121, Bloco 1K do Campus Santa M^onica 38408-100 Uberl^andia, MG, Brazil

P. K. Kim and S. Jung
Intelligent Systems and Emotional Engineering (I.S.E.E.) Laboratory, Department of Mechatronics Engineering, Chungnam National University, Daejeon 305-764, Republic of Korea

Xing-Ju Wang and Gui-Feng Gao
School of Traffic and Transportation, Shijiazhuang Tiedao University, Shijiazhuang, Hebei 050043, China
Traffic Safety Engineering and Emergency Management Workgroup, Traffic Safety and Control Laboratory of Hebei Province, Shijiazhuang, Hebei 050043, China

Xiao-Ming Xi
School of Traffic and Transportation, Shijiazhuang Tiedao University, Shijiazhuang, Hebei 050043, China

Xiangyu Meng and Tongwen Chen
Department of Electrical and Computer Engineering, University of Alberta, Edmonton, AB, Canada T6G 2V4

Chen-Chou Hsieh and Pau-Lo Hsu
Department of Electrical Engineering, National Chiao-Tung University, 1001 Ta Hsueh Road, Hsinchu 300, Taiwan

J.M. Perez
Cenpes, 21941-915 Rio de Janeiro, RJ, Brazil

D. Odloak
Chemical Engineering Department, University of S˜ao Paulo, CP 61548, 05424-970 S˜ao Paulo, SP, Brazil

E. L. Lima
Chemical Engineering Program/COPPE Federal University of Rio de Janeiro, CP 68502, 21941-970 Rio de Janeiro, RJ, Brazil

Chengrui Zhao, Lin Ye, Xun Yu and Junfeng Ge
Department of Control Science and Engineering, Huazhong University of Science and Technology,
1037 Luoyu Road, Hubei, Wuhan 430074, China

Alejandro Rincon
Programa de Ingenier´ıa Ambiental, Facultad de Ingenier´ıa y Arquitectura, Universidad Cat´olica de Manizales, Carrena 23 No. 60-30, Manizales 170002, Colombia

Fabiola Angulo
Departamento de Ingenier´ıa El´ectrica, Facultad de Ingenier´ıa y Arquitectura, Universidad Nacional de Colombia, Sede Manizales, Electr´onica y Computaci´on, Percepci´on y Control Inteligente, Bloque Q, Campus La Nubia, Manizales 170003, Colombia

Cong Teng
School of Mathematics and Quantitative Economics, Shandong University of Finance and Economics, Jinan, Shandong 250014, China

Wei Shanbi, Chai Yi and Li Penghua
College of Automation, Chongqing University, Chongqing 400044, China

Ahmed M. Kassem
Control Technology Department, Beni-Suef University, Beni-Suef 62511, Egypt

A. A. Hassan
Department of Electrical, Faculty of Engineering, Minia University, El Menia 61519, Egypt

Katarina Yuen, Senad Apelfröjd and Mats Leijon
Division of Electricity, Department of Engineering Science, Uppsala University, P.O. Box 534, 75121 Uppsala, Sweden

Satoshi Suzuki and Katsuhisa Furuta
Department of Robotics and Mechatronics, School of Science and Technology for Future Life, Tokyo Denki University, 5 Asahi-Chou, Senju, Adachi-Ku, Tokyo 120-8551, Japan

Eduardo Paciência Godoy
S˜ao Paulo State University, UNESP Sorocaba, SP, Brazil

Giovana Tangerino Tangerino, Rubens André Tabile and Arthur José Vieira Porto
University of S˜ao Paulo at S˜ao Carlos, S˜ao Carlos, SP, Brazil

Ricardo Yassushi Inamasu
Brazilian Agricultural Instrumentation Research Corporation, S˜ao Carlos, SP, Brazil

Yuan Ge
College of Electrical Engineering, Anhui Polytechnic University,Wuhu 241000, China
School of Automation, Southeast University, Nanjing 210096, China

Qigong Chen, Ming Jiang and Yiqing Huang
College of Electrical Engineering, Anhui Polytechnic University,Wuhu 241000, China

Reza Bohlouli, Babak Rostami and Jafar Keighobadi
Faculty of Mechanical Engineering, University of Tabriz, 29 Bahman, Tabriz, 5166614766, Iran

Chuan Wang, Michael Santone and Chengyu Cao
Department of Mechanical Engineering, University of Connecticut, Storrs, CT 06269-3139, USA

Patience E. Orukpe
Department of Electrical and Electronic Engineering, University of Benin, P.M.B 1154, Benin City, Edo State, Nigeria

www.ingramcontent.com/pod-product-compliance
Lightning Source LLC
Chambersburg PA
CBHW080255230326
41458CB00097B/5002
* 9 7 8 1 6 3 2 4 0 2 6 4 6 *